CRASH COURSE PHYSICS GRADE-12

DR RACHNA GARG

Clever Fox® PUBLISHING
Chennai • Bangalore

CLEVER FOX PUBLISHING
Chennai, India

Published by CLEVER FOX PUBLISHING 2025

ISBN: 978-93-7500-010-5

Disclaimer

Every effort has been made to ensure the accuracy of the content in this book. However, despite our best efforts, some errors or omissions may have inadvertently crept in. If you come across any mistakes or discrepancies, please let us know—we will be happy to correct them in the next edition.

For feedback: creations8@gmail.com

Preface and Acknowledgement

At Catalytic Centre, our goal is to provide the best study material that is student-friendly, easy to understand, and ideal for quick revision.

The Grade 12 Physics syllabus is vast and detailed, making it a challenge to compile all essential information within a limited space. After carefully analysing the trends of various board examinations, we have designed this book to include:

- **Memory Cards for each chapter**
- **Simplified Derivations for quick understanding**
- **Frequently Asked Past Year Board Questions**

The Memory Cards are especially helpful for rapid revision and aid in developing a *"visual" or "photogenic" memory* of the entire syllabus.

This book is intended primarily for quick revision before exams. However, students are strongly advised to first study each chapter thoroughly from the NCERT textbook and practice the NCERT questions before using this crash course as a supplement.

I extend my heartfelt gratitude to the authors of the following books, whose works have been invaluable sources of knowledge and inspiration:

- **NCERT Physics - Class 12**
- **Pradeep's Textbook of Physics - Class 12**
- **Modern's ABC of Physics - Class 12**
- **NEET Objective Physics** by **D.C. Pandey (Arihant Publications)**

These references have provided conceptual clarity and valuable guidance through the wide variety of questions and explanations presented in them.

I am deeply indebted to my students for continuously sharing the questions asked in their examinations and for offering thoughtful suggestions and feedback, which have significantly contributed to improving the quality of this book.

My sincere appreciation and thanks go to **Ms. Avelala Mounika**, our dynamic and versatile publisher **Clever Fox**, and dedicated team for their efficiency and support in bringing this project to fruition.

I sincerely hope that this book will be well received by students and teachers alike. Suggestions and feedback for further improvement will be gratefully acknowledged.

- Dr. Rachna Garg

To the Students

My Dear Students / Warriors,

Welcome to the first edition of my **Physics Crash Course for Grade 12 CBSE/ISC Board Exams – 2026.**

Preparing thoroughly for the board exams within a limited time can be a real challenge. To help you overcome it, I've compiled this concise and comprehensive guide featuring:

- **Memory cards for each chapter**
- **Simplified derivations**
- **Key definitions**
- **The most important past-year board questions (PYQs)**

The Class 12 Physics syllabus is vast, and my goal has been to present all essential information in a crisp, **easy-to-revise** format—perfect for quick and effective preparation.

✓ I would also like to share the **proper study technique** for all subjects. Many students consider simply understanding a concept as sufficient studying, but true learning involves **four essential steps**:

1. **Understanding**
2. **Analyzing**
3. **Memorizing**
4. **Writing**

To highlight the importance and interconnection of each step, I have presented the study technique in the form of a chart below.

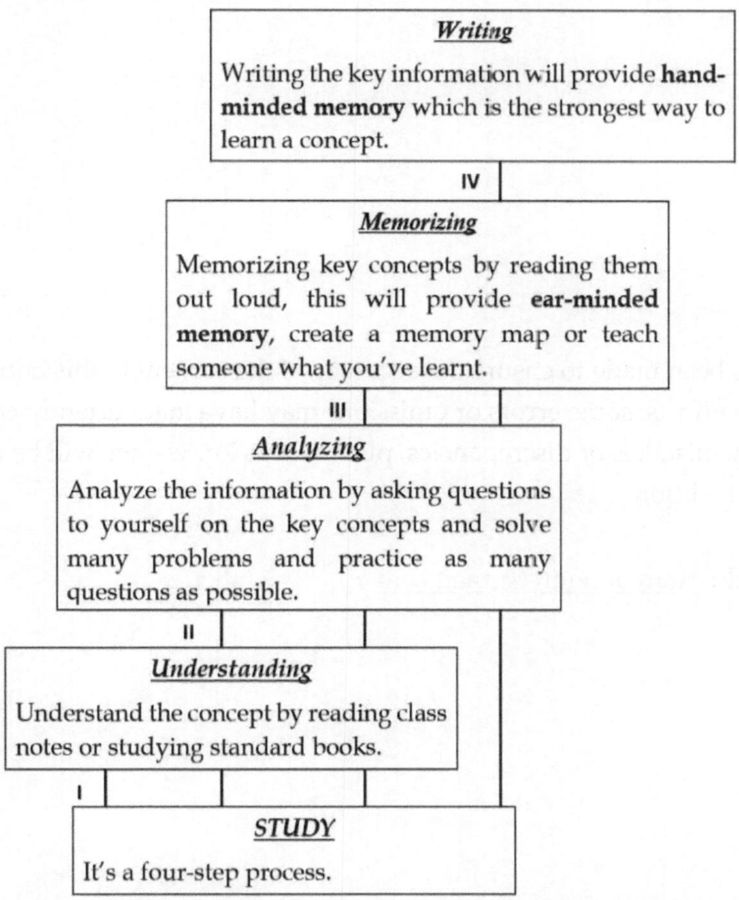

I am happy to share feedback from my students about how this study material helped them prepare for their board exams.

- **Yashasvi Kapoor (Student of computer science, The Pennsylvania State University)-**

All the resources provided to me before my board examination were extremely valuable, especially the memory cards, the consolidated list of simplified derivations and the collection of important PYQs for every topic. These greatly helped me strengthen my understanding of the concepts and made me confident about my preparation for boards!!

- **Riya Reehal (Student of Life Sciences, University of Toronto)-**

All the materials contributed to different parts of my learning journey as the examinations drew closer. The derivations helped me truly understand the concepts, the memory cards were perfect for quick reviews before tests, and the past year questions prepared me for a variety of question types. Altogether, they made studying physics more enjoyable as my knowledge kept building on strong foundations formed through constant practice and review!

(Dr. Rachna Garg)

Disclaimer

Every effort has been made to ensure the accuracy of the content in this book. However, despite our best efforts, some errors or omissions may have inadvertently crept in. If you come across any mistakes or discrepancies, please let us know – we will be happy to correct them in the next edition.

For feedback: catalyticcentre@gmail.com

Contents

Section I - Memory Cards 1-28

Section II - Derivations and Proofs

Electric Charges and Fields

1. Electric field intensity due to a point charge 29
2. Electric field intensity at a point on the axis of a dipole 29
3. Electric field intensity at a point on an equatorial line of electric dipole 30
4. Torque on a dipole in uniform electric field 30
5. Electric field due to an infinite line charge 31
6. Electric field due to a plane sheet of charge [∞ Area] 31
7. Electric field due to two infinite parallel sheets of charge 32
8. Electric field intensity due to a uniformly charged spherical shell 33

Electric Potential and Capacitance

9. Electrostatic potential due to a point charge 33
10. Electric potential due to a dipole at any point 34
11. Electric potential due to a group of point charges 35
12. Electrostatics potential energy due to a system of two-point charges 35
13. Electrostatics potential energy due to a system of 'n' point charges 36
14. Electrostatics potential energy of a dipole in external uniform electric field 36
15. Relation between E and V 37
16. Capacitance of an isolated spherical conductor 37
17. Capacitance of parallel plate capacitor 38
18. Capacitance of parallel plate capacitor with dielectric slab 38
19. Capacitor of parallel plate capacitor using conducting slab 39
20. Capacitor in series 39
21. Capacitors in parallel 40
22. Energy stored in a charged capacitor 40
23. Common potential of two capacitors when connected in parallel 41
24. Loss of energy in sharing charges by capacitors connected in parallel 41

Current Electricity

25. Drift velocity of electrons when potential difference is applied across the conductor 42
26. Relation between drift velocity and electric current 43
27. Electric mobility 43
28. Deduction of ohm's law 43
29. Relation between current density and drift velocity 44
30. Microscopic form of ohm's law 44

31. Relation between emf, internal resistance and terminal voltage **44**
32. Cells connected in series **45**
33. Cells connected in parallel **45**
34. Wheatstone bridge **47**

Moving Charges and Magnetism
35. Magnetic field at a point on the axis of circular coil carrying current **47**
36. Ampere's circuital law and its proof **48**
37. Magnetic field of long thin straight wire **49**
38. Magnetic field due to a current carrying thick circular cylinder **49**
39. Magnetic field due to a long solenoid **50**
40. Motion of a charged particle in uniform magnetic field **51**
41. Force on a current carrying conductor placed in external magnetic field **51**
42. Force between two current carrying parallel conductors **52**
43. Torque acting on a current carrying loop placed in an external magnetic field **53**
44. Moving coil galvanometer **53**
45. Conversion of galvanometer into ammeter **54**
46. Conversion of galvanometer into voltmeter **54**

Magnetism and Matter
47. Bar magnet and magnetic field intensity due to bar magnet along its axial line **55**
48. Bar magnet and magnetic field intensity due to bar magnet along its equatorial line **56**
49. Torque on a bar magnet placed in a uniform magnetic field **56**
50. Potential energy of a magnetic dipole in external magnetic field **57**
51. Magnetic dipole moment of a revolving electron **57**
52. Bar magnet as an equivalent solenoid **58**
53. Relation between magnetic permeability and susceptibility **59**
54. Magnetic flux and gauss's theorem **59**

Electromagnetic Induction
55. Motional electromotive force **59**
56. Motional induced emf in a conducting rod rotated in a uniform magnetic field **60**
57. Power dissipation in a conductor moved in a magnetic field **60**
58. Energy stored in an inductor **61**
59. Self-inductance of a long solenoid **61**
60. Mutual inductance of two long coaxial solenoids **62**

Alternating Current
61. Mean value or average value of alternating current **63**

62. Root mean square value of alternating current 64
63. AC circuit containing resistor only 64
64. AC circuit containing inductor only 65
65. AC circuit containing capacitor only 66
66. LCR series circuit 66
67. Resonance in LCR series circuit 68
68. Average power associated with an AC circuit containing resistor only 68
69. Average power associated with an AC circuit containing inductor only 69
70. Average power associated with an AC circuit containing capacitor only 69
71. Average power associated with LCR series circuit 70
72. Power factor of an AC circuit 70
73. AC generator 70
74. Transformer 71

Electromagnetic Waves

75. Displacement current 72
76. Electromagnetic waves and expressions of electric and magnetic fields 73
77. Energy density of electromagnetic waves 73
78. Intensity and momentum carried by electromagnetic waves 73
79. Radiation pressure 74
80. Prove that displacement current is ε_0 (dΦ/dt) and it is equal to conduction current 75

Ray Optics

81. Relation between f and R 75
82. Mirror formula for concave mirror real image 76
83. Mirror formula for convex mirror 77
84. Linear magnification of a spherical mirror 77
85. Refraction through a rectangular glass slab 78
86. Relation between μ_1, μ_2, u, v and R for refraction from rarer to denser medium at convex spherical surface (Real image) 78
87. Relation between μ_1, μ_2, u, v and R for refraction from denser to rarer medium at a spherical refracting surface (convex) 79
88. Lens maker's formula for convex lens 79
89. Lens formula convex lens - Real image 80
90. Linear magnification produced by a lens 81
91. Power of a lens 82
92. Combination of thin lenses in contact 82
93. Refraction through a prism - calculation of angle of deviation 83
94. Angle of minimum deviation and prism formula 83

95. Simple microscope or magnifying glass **84**
96. Compound microscope **85**
97. Astronomical telescope **86**

Wave Optics
98. Proof of laws of reflection using Huygen's principle **88**
99. Proof of laws of refraction using Huygen's principle **89**
100. Interference of light: Young's double slit experiment and conditions for constructive and destructive interference **90**
101. Proof of conservation of energy during interference **91**
102. Positions of maxima and minima on the screen and fringe width **91**
103. Diffraction of light and positions of secondary minima and secondary maxima **92**

Modern Physics
104. Photoelectric effect – stopping potential equation **93**
105. Proof for same nuclear density for all elements **94**
106. Einstein's mass energy equivalence **94**
107. De Broglie wavelength of an electron **94**
108. Distance of closest approach **94**
109. Energy of an electron in an atomic orbit **95**
110. Radius of nth orbit of an atom **95**
111. Proof of Bohr's second postulate by de – Broglie explanation **96**
112. Derivation for the frequency of radiation emitted when a hydrogen atom de-excites from level 'n' to (n-1) **96**
113. Classical frequency of revolution of an electron in n^{th} orbit **97**
114. Proof: When transition of electron takes place between higher orbitals then frequency of emitted radiation is equal to classical frequency of electron **97**

Section III – Most important past year board questions **98**

Unit - 1: ELECTROSTATICS

Coulomb's law of electrostatics:

$$F = \frac{kq_1q_2}{r^2}\ \text{N or}\ F = \frac{1}{4\pi\varepsilon_0}\cdot\frac{q_1q_2}{r^2}$$

Electric field intensity (vector quantity):

$$E = \frac{kQ}{r^2}\ \text{or}\ E = \frac{1}{4\pi\varepsilon_0}\frac{Q}{r^2}, \frac{N}{C}$$

Electric Potential (scalar quantity):

$$V = k\frac{Q}{r},\quad E = \frac{V}{d},\quad E = -\frac{dV}{dr}$$

Electric dipole:

$$\vec{P} = q \times 2\vec{\ell}$$

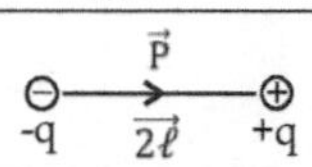

E on axial line of a dipole

$$\vec{E} = \frac{2k\vec{P}}{r^3} = \frac{2\vec{P}}{4\pi\varepsilon_0 r^3}$$

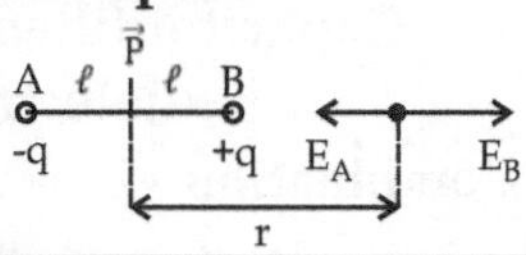

E on equatorial line-

$$E = \frac{kP}{r^3} = \frac{1}{4\pi\varepsilon_0}\cdot\frac{P}{r^3}$$

$$\vec{E} = (-)\frac{k\vec{P}}{r^3}$$

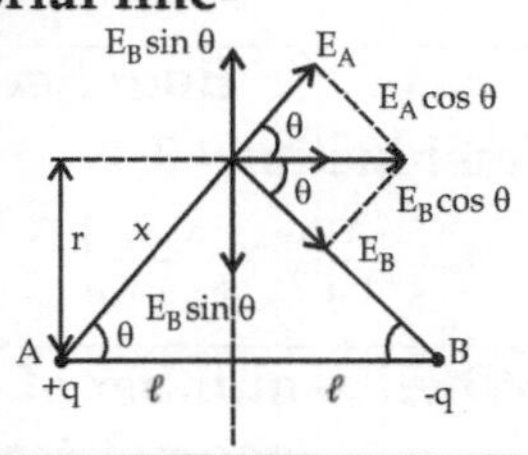

Torque on Dipole (vector quantity)

$$\tau = pE\sin\theta$$

- Work done in rotating dipole
  $W = pE(\cos\theta_1 - \cos\theta_2)$
- Potential energy of a dipole -
  $U = -pE\cos\theta$
- Stable equilibrium $\theta = 0°$, $U = -pE$
- Unstable equilibrium $\theta = 180°$, $U = +pE$

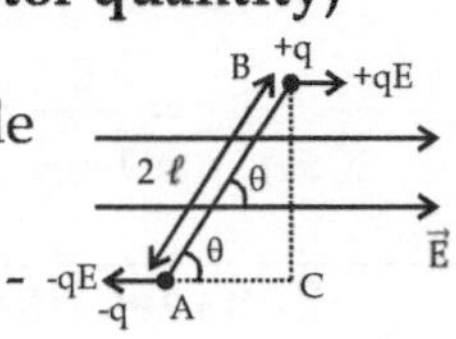

GAUSS' THEOREM -

$$\phi_E = \oint \vec{E}\cdot d\vec{S} = \frac{q}{\varepsilon_0}$$

'E' due to uniformly charged shell

$$E = \frac{kq}{R^2}\ \text{[On surface]}$$

$$E = \frac{kq}{r^2}\ \text{[Outside]}$$

$$E = 0 \text{ as } q = 0\ \text{[Inside]}$$

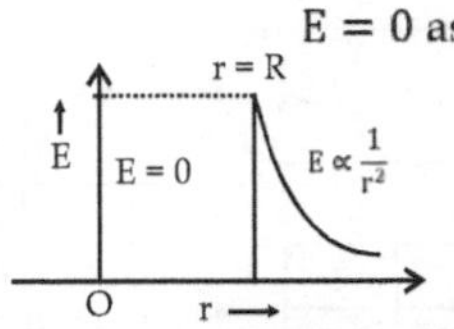

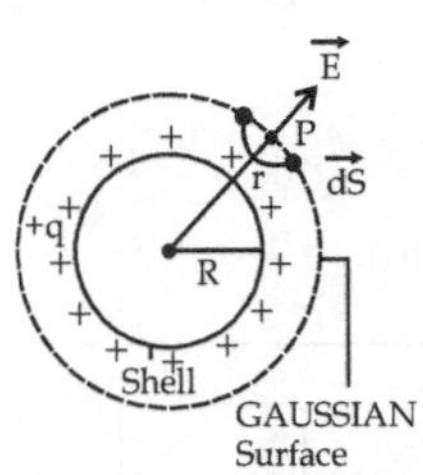

'E' due to long charged wire

$$E = \frac{2k\lambda}{r}$$

$$E = \frac{1}{4\pi\varepsilon_0}\frac{2\lambda}{r}$$

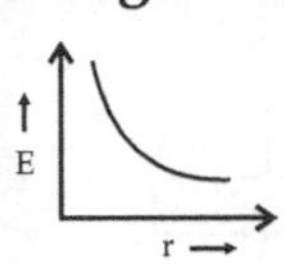

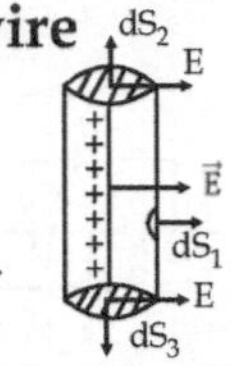

'E' due to a solid sphere of charge (Non conducting)

(i) outside the sphere

$$E = \frac{q}{4\pi\varepsilon_0 r^2}\quad \text{outside}\ E = \frac{kq}{r^2}$$

(ii) On the surface

$$E = \frac{q}{4\pi\varepsilon_0 R^2}\quad E = \frac{kq}{R^2}$$

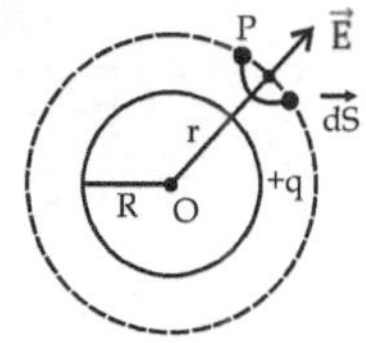

(iii) E inside the charged solid sphere

$$E = \frac{kqr}{R^3}$$

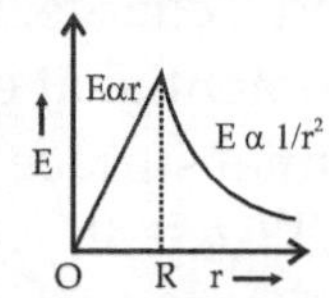

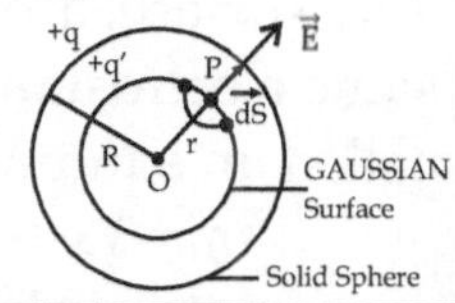

'E' due to charged plane sheet

$$E = \frac{\sigma}{2\varepsilon_0}$$

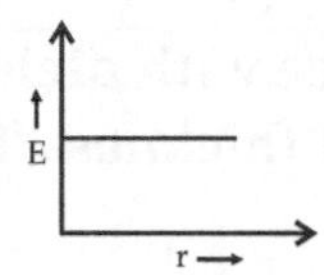

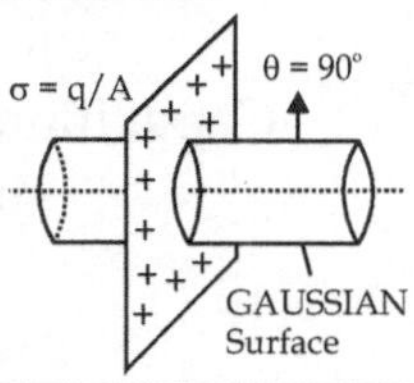

'V' due to a point charge

$$V_A = \frac{kq}{r_A}$$

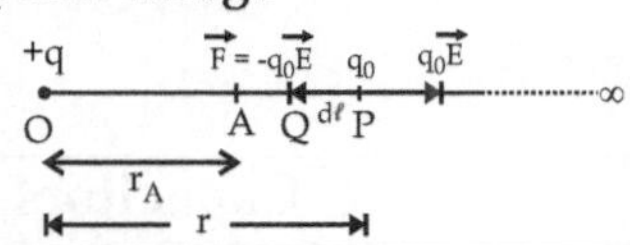

Potential Energy of a system of two charges

$$U = k\left[\frac{q_1q_2}{r_{12}}\right]$$

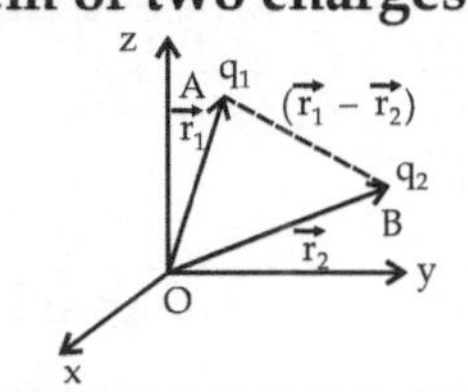

Electrical capacitance

Capacitance of a spherical conductor

$$C = \frac{r}{k} = 4\pi\varepsilon_0 r\quad \text{Units: Farad}$$

$$1\ \text{Farad (F)} = \frac{1\ \text{coulomb (c)}}{1\ \text{volt (v)}}$$

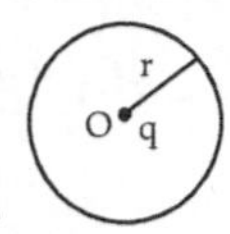

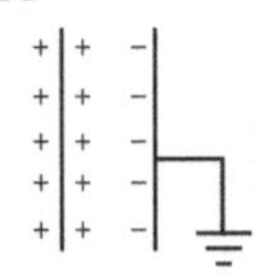

Principle - When two metallic conductors kept close to each other & one of them is earthed the other conductor can store a large amount of charge.

'E' as gradient of 'V'

$$E = -\frac{dV}{dr}$$

Potential Energy of a dipole placed in external uniform electric field

$$U = -pE\cos\theta\quad U = -\vec{p}\cdot\vec{E}$$

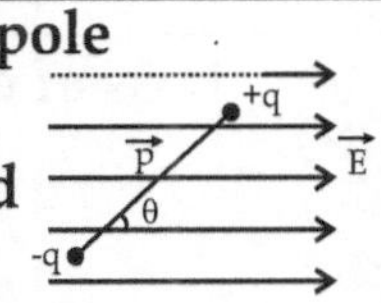

Capacitance of Parallel plate capacitor

$$C_0 = \frac{\varepsilon_0 A}{d}$$

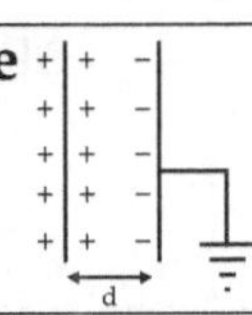

'V' at any point due to a dipole

$$\therefore V = \frac{kp\cos\theta}{r^2 - a^2\cos^2\theta}$$

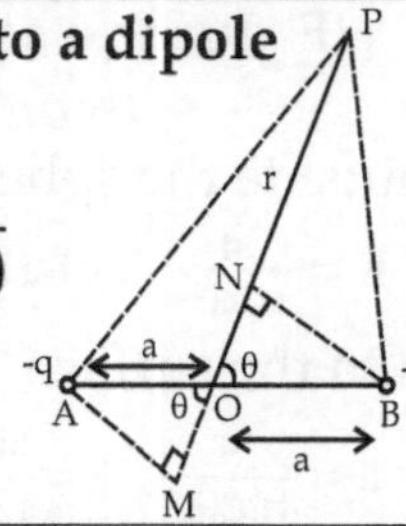

Common potential on sharing charges by two capacitors

$$V = \frac{C_1V_1 + C_2V_2}{C_1 + C_2}$$

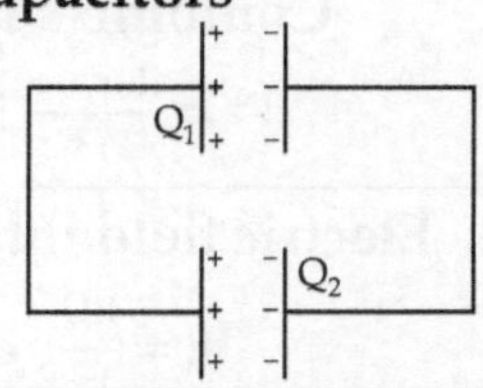

Equipotential Surface

Any surface, around a charge, which has same electrostatic potential at every point, is called an equipotential surface

$$V_B - V_A = 0, W_{AB} = 0$$

$$\therefore q_0\vec{E}\cdot d\vec{\ell} = 0$$

$$\therefore \vec{E} \perp d\vec{\ell}$$

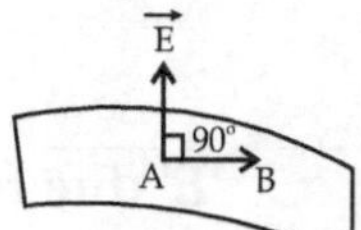

Capacitor filled with dielectric slabs in various manner

Combination - 1

$$C_s = \frac{2\varepsilon_0 AK_1K_2}{(K_1 + K_2)d}$$

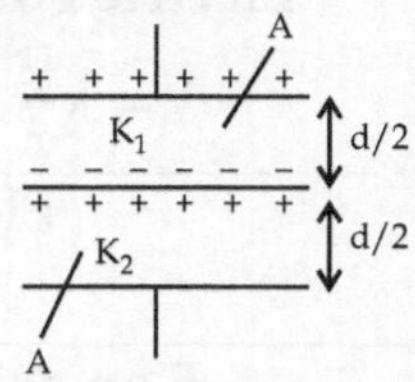

Capacitance with dielectric slab of thickness 't'

$$C = \frac{\varepsilon_0 A}{\left[d - t + \frac{t}{K}\right]}$$

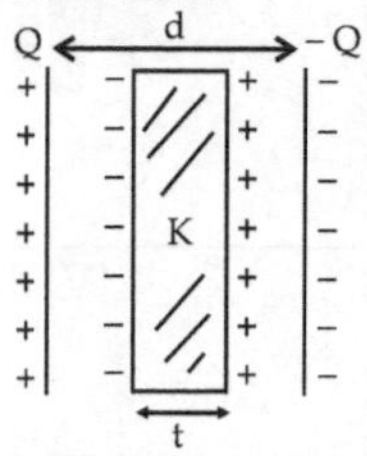

Capacitor filled with different dielectric slabs

Combination - 2

$$C = C_1 + C_2 = \frac{\varepsilon_0 A(K_1 + K_2)}{2d}$$

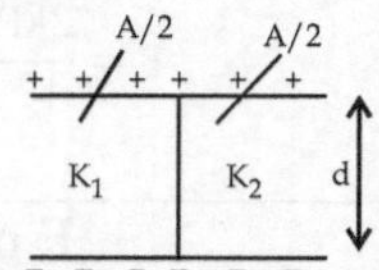

Both Series & Parallel

Combination - 3

$$C = \frac{\varepsilon_0 A}{d}\left[\frac{K_1}{2} + \frac{K_2K_3}{K_2 + K_3}\right]$$

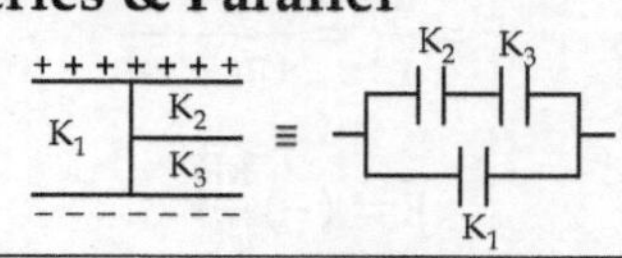

Capacitors in series

$$\frac{1}{C_S} = \frac{1}{C_1} + \frac{1}{C_2} + \frac{1}{C_3}$$

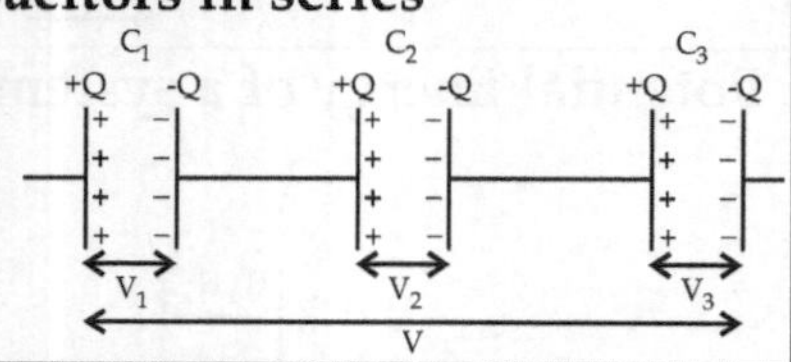

When ∞ number of capacitors arranged in geometrical progression

$$C_{equi} = \frac{C}{1 - \frac{1}{2}} = 2C$$

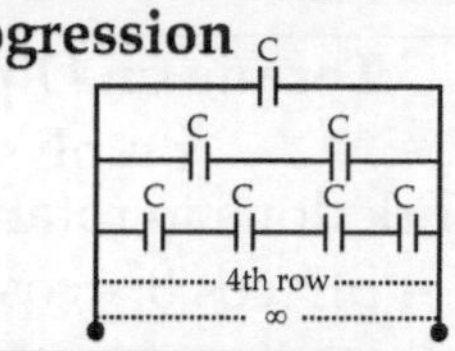

Capacitors in Parallel

$$C_p = C_1 + C_2 + C_3$$

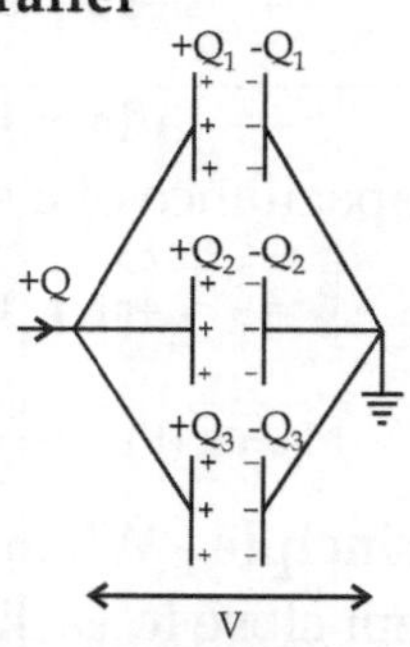

Behaves like parallel combination

$$C = \frac{2\varepsilon_0 A}{d}$$

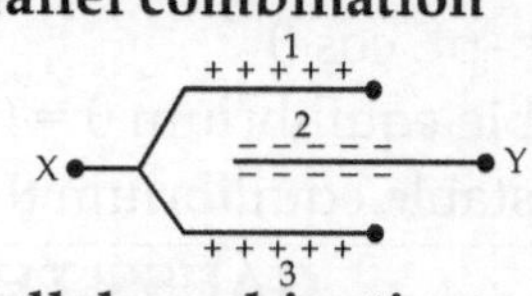

Behaves like parallel combination

$$C = \frac{2\varepsilon_0 A}{d}$$

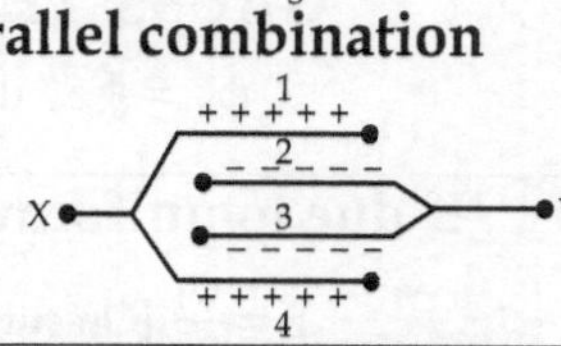

Energy stored in charged capacitor

$$U = \frac{1}{2}\frac{Q^2}{C} = \frac{1}{2}CV^2 = \frac{1}{2}QV$$

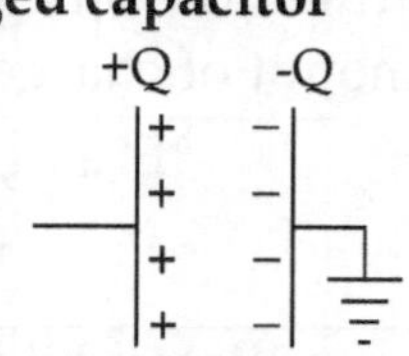

Behaves like 3 capacitors in parallel combination

$$C = \frac{3\varepsilon_0 A}{d}$$

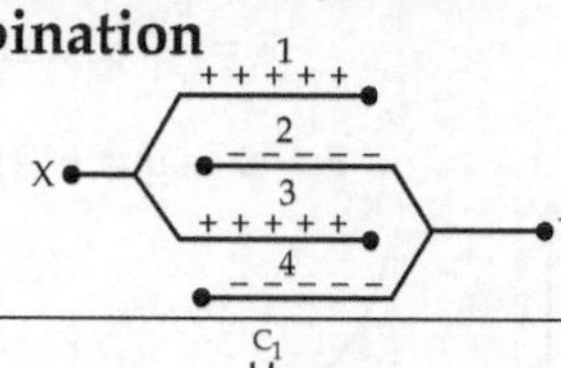

Loss of energy on sharing charges by two capacitors

$$\Delta U = \frac{C_1C_2(V_1 - V_2)^2}{2(C_1 + C_2)}$$

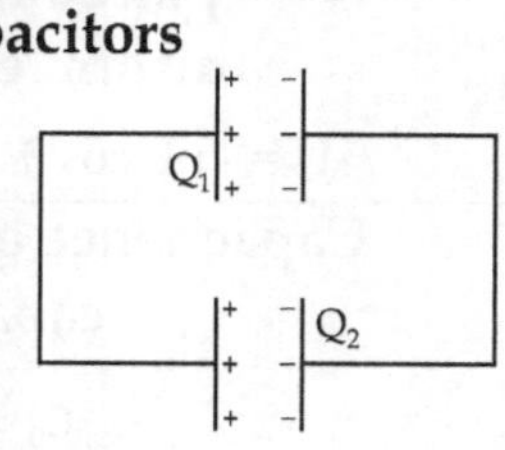

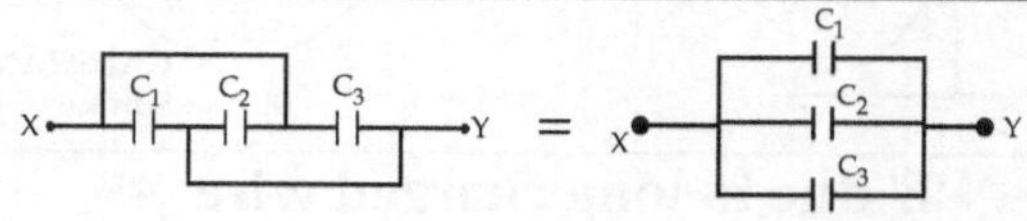

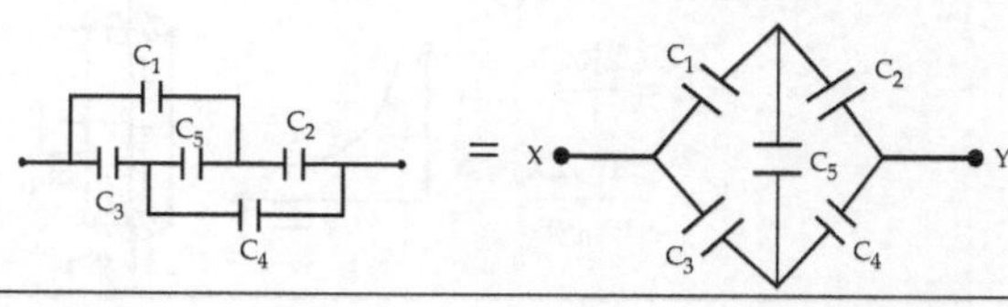

Unit - 2: Current Electricity

Electric Current:

$$I = \frac{ne}{t}$$

Relaxation time (τ) - The short time for which a free electron accelerates before it undergoes a collision with the positive ion in the conductor, is called relaxation time.

Drift velocity (v_d) - The average velocity with which an electron is drifted in a conductor under the influence of external electric field.

$$\therefore \vec{v}_d = (-)\frac{e\vec{E}\tau}{m}$$

Electrical Energy and Power -

$$P = \frac{W}{t} = \frac{VIt}{t} = VI = I^2R = \frac{V^2}{R}$$

Relation between v_d and I

$$I = neAv_d$$

$$\therefore I = \frac{nAe^2E\tau}{m}$$

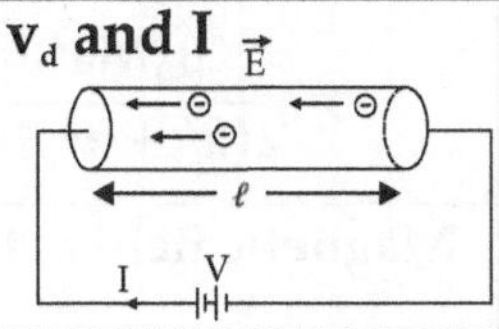

Resistance:

$$\therefore R = \frac{m\ell}{nAe^2\tau} = \frac{m}{ne^2\tau}\frac{\ell}{A}$$

Here, resistivity ρ:

$$\rho = \frac{m}{ne^2\tau} \qquad \therefore R = \rho\frac{\ell}{A}$$

Electron mobility (μ) -

It is the drift velocity acquired per unit strength of electric field which is applied across the conductor.

$$\mu = \frac{v_d}{E} \quad \text{or} \quad \mu = \frac{e\tau}{m}$$

Current density (J)

$$\vec{J} = \frac{I}{\vec{A}} \quad \text{or} \quad \vec{J} = -ne\vec{v_d}$$

Temperature dependence of Resistivity (ρ)

$$\alpha = \frac{R_t - R_0}{R_0 \times \Delta T} \text{ per } ^\circ c$$

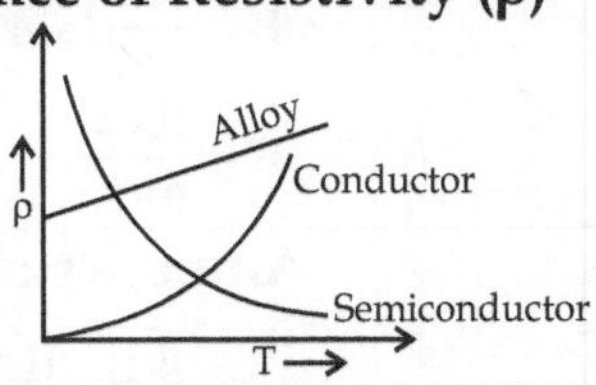

Combination of resistance in Series-

$$R_{eq} = R_1 + R_2 + R_3$$

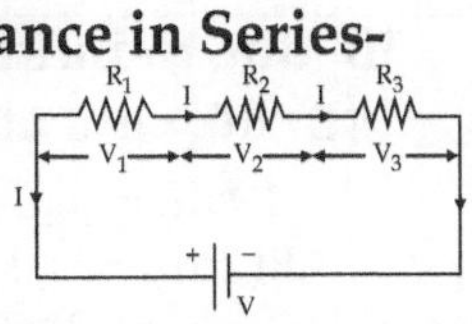

Parallel combination

$$\frac{1}{R_{eq}} = \frac{1}{R_1} + \frac{1}{R_2} + \frac{1}{R_3}$$

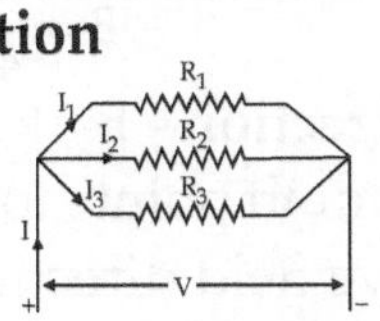

Conductance (G):

$$G = \frac{1}{R}$$

Conductivity (σ):

$$\sigma = \frac{1}{\rho}$$

Vector form of Ohm's Law: or Microscopic form of Ohm's Law:

$$\vec{J} = \sigma\vec{E}$$

Ohm's Law:

$$V = IR$$

V - I characteristics:

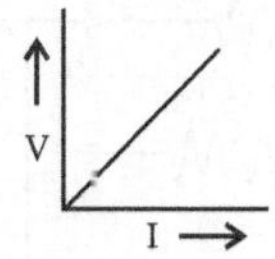

(i) Ideal conductor

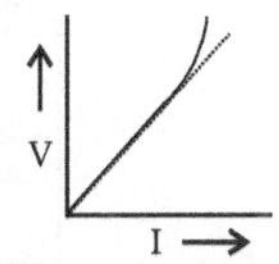

(ii) Good conductor

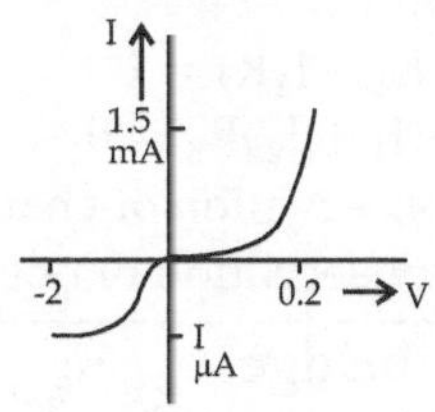

(iii) For diode

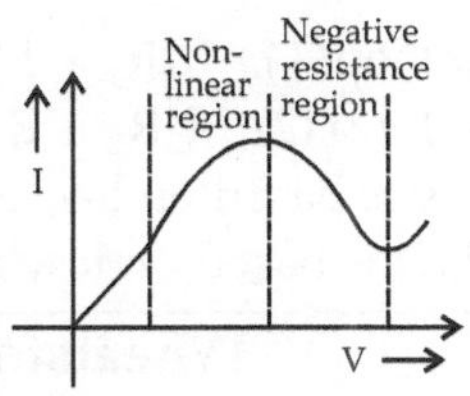

(iv) For GaAs Semiconductor

Electromotive force of a cell (emf)

$$E = \frac{dW}{dq}, \quad \text{unit: Volt} = J/C$$

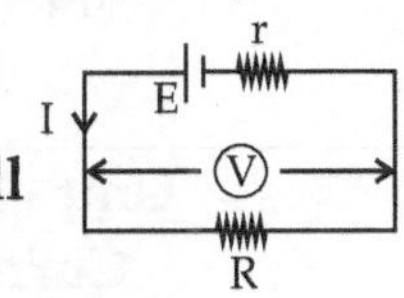

Terminal Voltage of the cell

$$V = E - Ir, \; V = IR$$

Here, r is internal resistance of the cell

$$r = \left(\frac{E}{V} - 1\right)R$$

Combinations of cells in Series:

$$E_{eq} = E_1 + E_2$$

$$r_{eq} = r_1 + r_2$$

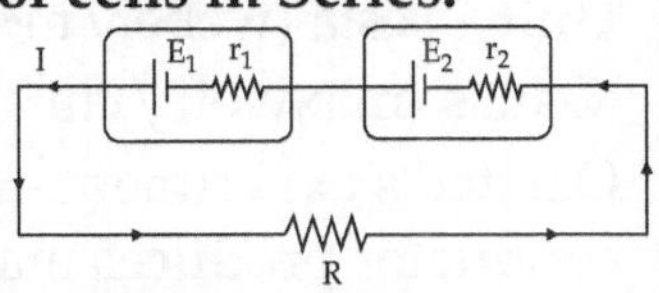

If there are 'n' cells of emf E and internal resistance r then

$E_{eq} = nE$

$r_{eq} = nr$

Total resistance = (nr + R)

$$I = \frac{nE}{nr+R}$$

Combinations of cells in Parallel:

$$\text{Parallel - } \frac{1}{r_p} = \frac{n}{r}$$

$$\text{Total Resistance } = \frac{r}{n} + R$$

$$\therefore I = \frac{E}{\left(\frac{r}{n} + R\right)} = \frac{nE}{(r + nR)}$$

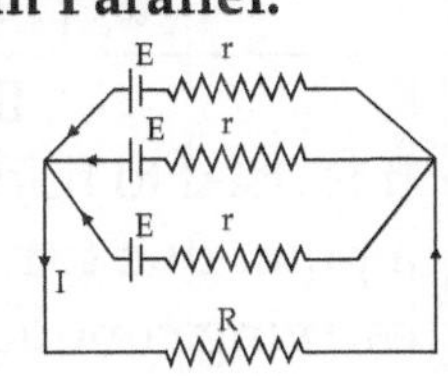

Parallel combination of two cells:

$$V = E_{eq} - Ir_{eq}$$

$$E_{eq} = \frac{E_1r_1 + E_2r_2}{r_1 + r_2}$$

$$r_p = \frac{r_1r_2}{r_1 + r_2}$$

Kirchhoff's Laws

(I) Junction rule

$$I_1 + I_2 = I_3 + I_4 + I_5$$

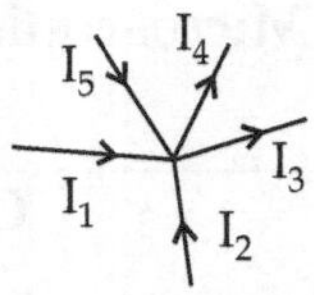

(II) Loop rule

$\sum \Delta V = \sum IR = 0$ around any closed loop.

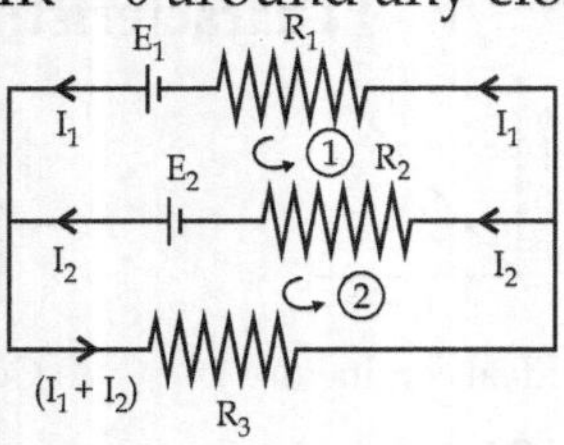

For loop(1) $-I_1R_1 + E_1 - E_2 + I_2R_2 = 0$

For loop (2) $-I_2R_2 + E_2 - (I_1 + I_2)R_3 = 0$

I law is based on law of conservation of charge.

II law is based on law of conservation of energy.

Wheatstone bridge

$$\frac{I_1P}{I_1Q} = \frac{I_2R}{I_2S} \text{ or } \frac{P}{Q} = \frac{R}{S}$$

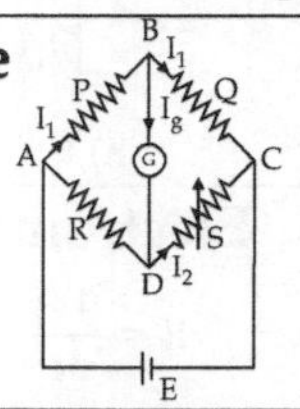

Unit - 3: Magnetic effects of Current and Magnetism

Magnetic field (B)

Produced by a magnet, moving charges or current carrying conductor.

Unit - Tesla (weber/m²).

Gauss (maxwell / cm²) $1T = 10^4G$

Oersted's experiment - A current carrying conductor produces magnetic field around it.

The direction of magnetic field can be determined by right hand thumb rule.

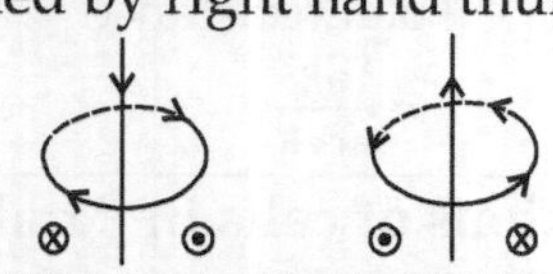

Biot Savart's law

It is used to find magnetic field at a point due to a current carrying conductor.

$$dB = \frac{\mu_0}{4\pi}\frac{Id\ell \sin\theta}{r^2}$$

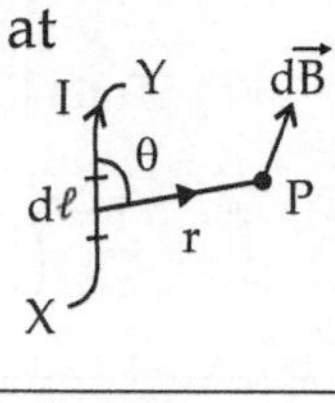

Vector Form -

$$\overrightarrow{dB} = \frac{\mu_0}{4\pi}\frac{I\overrightarrow{dl} \times \vec{r}}{|\vec{r}|^3}$$

Magnetic field at a point on the axis of a current carrying circular loop

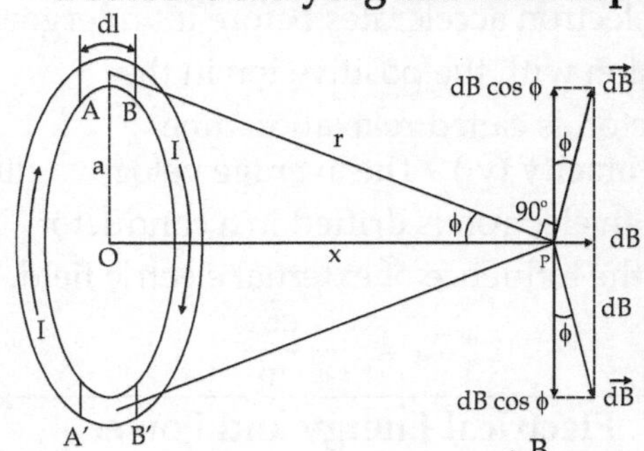

$$\therefore B = \frac{\mu_0}{4\pi}\frac{2\pi NIa^2}{(a^2 + x^2)^{3/2}}$$

$$= \frac{\mu_0 NIa^2}{2(a^2 + x^2)^{3/2}}$$

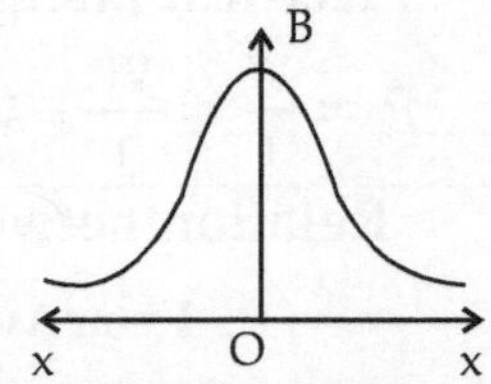

Magnetic field at the centre of a circular coil -

$$B = \frac{\mu_0}{2r}NI$$

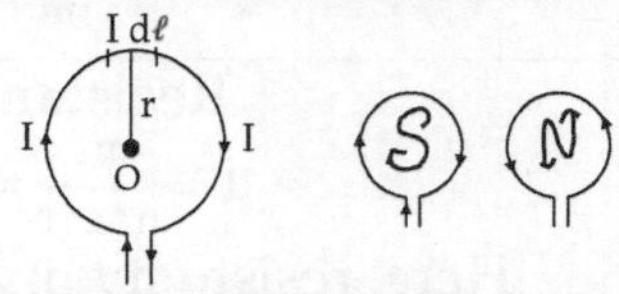

'B' due to finitely long wire From Biot Savart's law

$$B = \frac{\mu_0}{4\pi}\cdot\frac{I}{r}(\sin\phi_1 + \sin\phi_2)$$

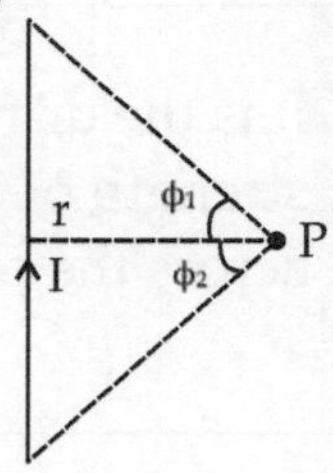

If the conductor is infinitely long

$$\phi_1 = 0°, \phi_2 = \frac{\pi}{2}$$

$$B = \frac{\mu_0 I}{4\pi r}\left(\sin 0° + \sin\frac{\pi}{2}\right)$$

$$\therefore B = \frac{\mu_0 I}{4\pi r}$$

If the point ' P ' is lying at its one end then

$$\phi_1 = \phi_2 = \pi/2$$

$$\therefore B = \frac{\mu_0 I}{4\pi r}\left(\sin\frac{\pi}{2} + \sin\frac{\pi}{2}\right)$$

$$B = \frac{\mu_0}{4\pi}\frac{2I}{r}$$

Ampere's circuital law

$$\oint B \cdot d\ell = \mu_0 I$$

'B' due to infinitely long wire

$$\oint B \cdot d\ell = \mu_0 I(\text{ at P})$$

$$B\oint d\ell = \mu_0 I$$

$$B(2\pi r) = \mu_0 I$$

$$B = \frac{\mu_0}{4\pi}\frac{2I}{r}$$

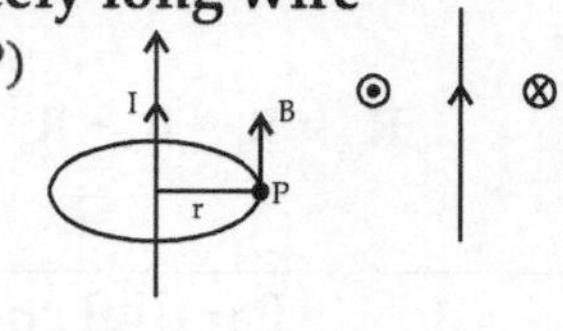

Direction - By Right hand thumb rule. Thumb points towards current, fingers give field direction.

'B' due to current through a very long circular cylinder

Case -I - Point ' P ' lying outside, i.e., r > R

Acc. to ampere's circuital law

$$\oint \vec{B} \cdot \vec{d\ell} = \mu_0 I$$

$$\text{or } B \times 2\pi r = \mu_0 I$$

$$B = \frac{\mu_0 I}{2\pi r};\ B \alpha \frac{1}{r}$$

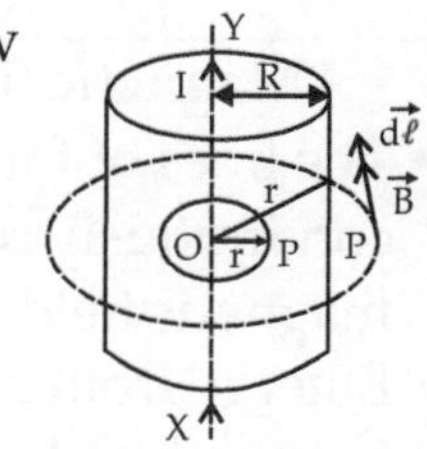

Case - II: Point ' P ' lying inside i.e., r < R. In this case, we have to calculate current passing through this cross section.

$$\therefore I' = \frac{I}{\pi R^2} \times \pi r^2 = \frac{I r^2}{R^2}$$

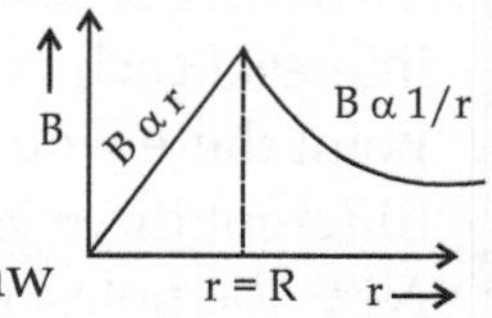

Acc. to ampere's circuital law

$$\oint \vec{B} \cdot d\vec{\ell} = \mu_m I';\ B \times 2\pi r = \mu_m \frac{I r^2}{R^2}$$

$$\text{or } B = \frac{\mu_m I r}{2\pi R^2};\quad B \alpha r$$

Here $\mu_m = \mu_0 \mu_r$

'B' due to a solenoid

$$B = \mu_0 n I$$

n = number density $\left(\frac{N}{\ell}\right)$

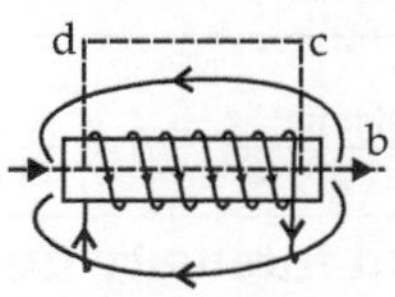

Relation between permeability of free space (μ_0) and permittivity of free space (ε_0)

$$\mu_0 \varepsilon_0 = \frac{1}{c^2} \quad \text{or} \quad \mu_0 \varepsilon_0 = \frac{1}{9 \times 10^{16}}\ s^2\ m^{-2}$$

Force on a moving charge in a uniform magnetic field

$$\vec{F} = q(\vec{v} \times \vec{B})$$

Or $F = qvB\sin\theta$

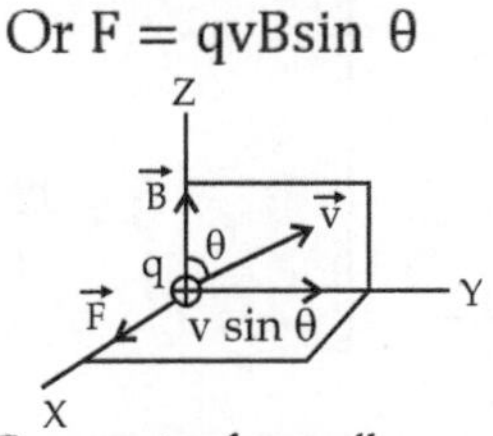

"Magnetic Lorentz force"

$$B = \frac{F}{qv\sin\theta};\ 1\text{ Tesla} = \frac{1N}{1C \times 1\ m/s \times 1}$$

$\therefore$ 1 Tesla $= 1NA^{-1}\ m^{-1}$

The direction of the force on the charged particle is determined by Fleming's left hand rule.

Motion of a charged particle in combined electric and magnetic field

$$v = E/B$$

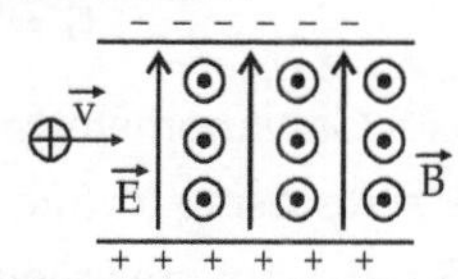

Motion of charged particle in uniform magnetic field

$$\therefore r = \frac{mv_\perp}{qB} = \frac{mv\sin\theta}{qB}$$

$$T = \frac{2\pi m}{qB}$$

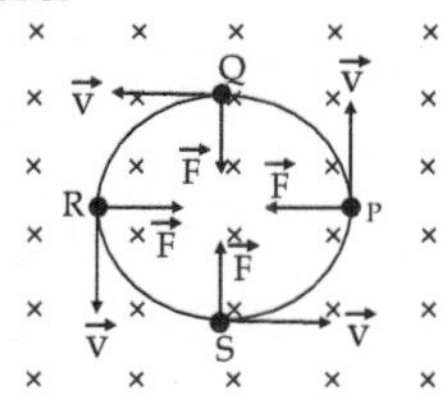

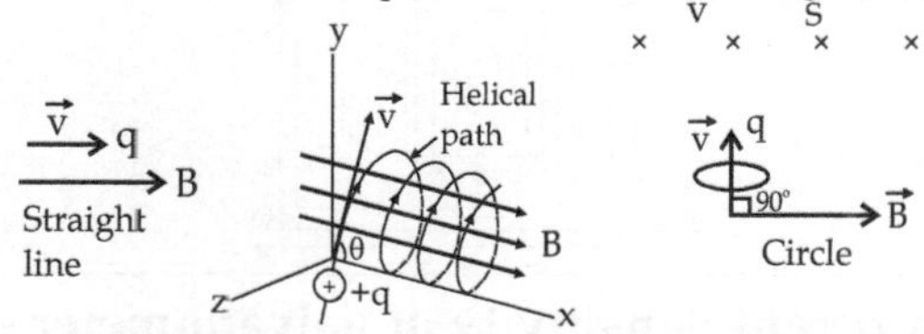

Force on a current carrying conductor placed in a uniform magnetic field

$$F = I\ell B\sin\theta$$

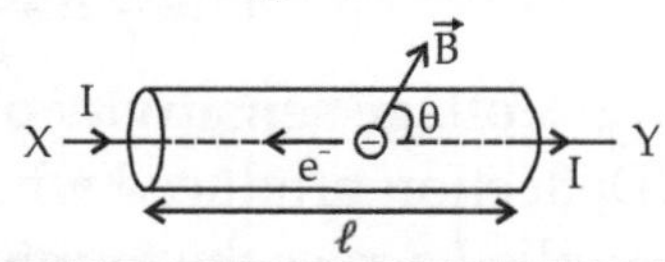

Force per unit length between two parallel current carrying wires

$$\frac{F_1}{\ell} = \frac{F_2}{\ell} = \left(\frac{\mu_0}{4\pi}\right)\frac{2I_1I_2}{r} \qquad \text{Units: } \frac{N}{m}$$

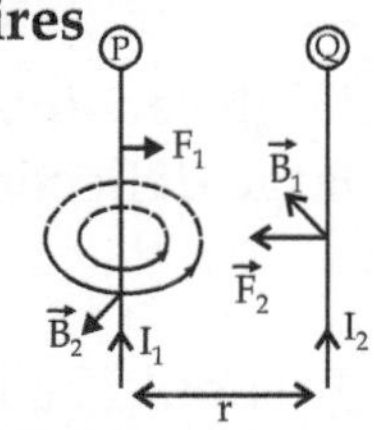

'τ' on a current carrying loop placed in a magnetic field

$$\tau = nIAB\sin\theta = MB\sin\theta$$

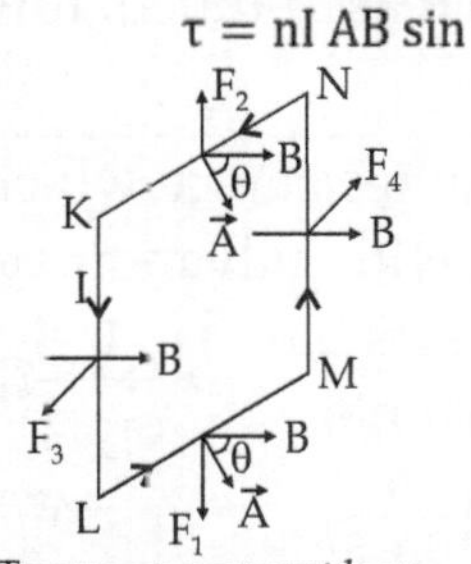

Torque on a current loop placed in a magnetic field

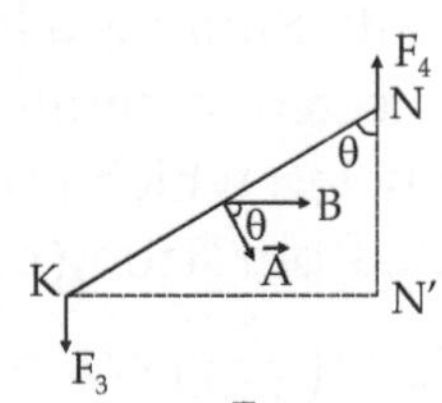

Torque

Current Carrying Loop as a Magnetic Dipole

Magnetic dipole moment (M): Magnetic dipole moment

$$M = nIA$$

n = number of turns in the loop

I = Current in the loop

A = Area of the loop

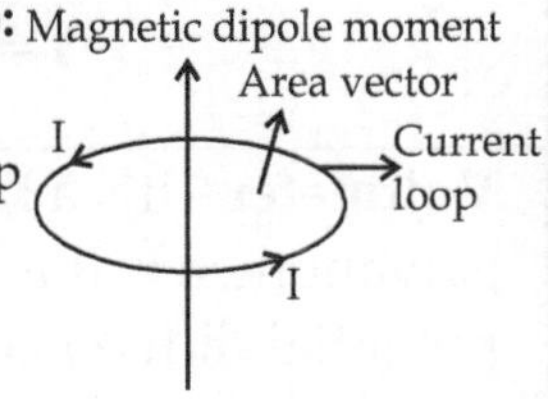

Moving Coil Galvanometer

$$G = \frac{k}{nAB}$$

G = Galvanometer constant

k = restoring torque for unit twist

B = External magnetic field

A = Area of each turn of the coil

$$\tau = nI\,AB\sin\theta$$

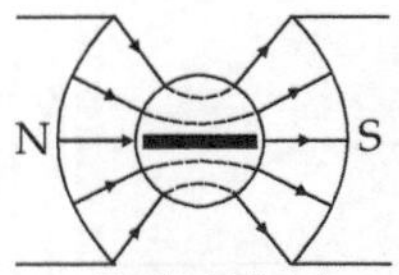

Radial Field (Top view of the galvanometer)

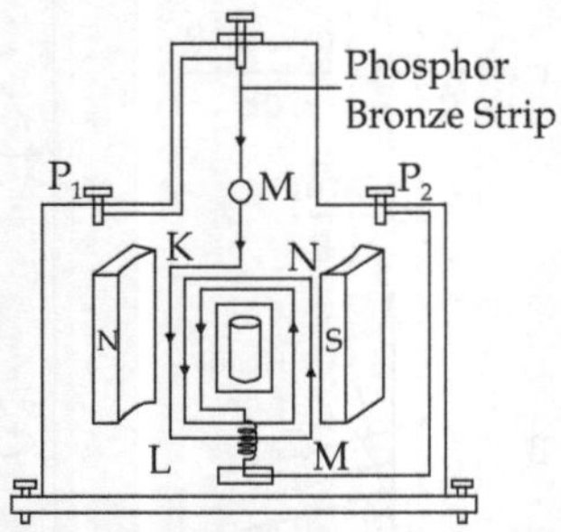

Current Sensitivity of galvanometer - Deflection produced when unit current flows.

$$I_s = \frac{\theta}{I} = \frac{nAB}{k}$$

Voltage sensitivity of galvanometer - Deflection produced when unit voltage is applied across the terminals.

$$V_s = \frac{\theta}{V} = \frac{\theta}{IR} = \frac{nBA}{kR} = \frac{I_s}{R}$$

High sensitivity - When for small current, large deflection is produced.

$$\because \theta = \frac{nBA}{k} I$$

'B' should be high

'k' can be decreased, by using quartz or phosphor bronze strip.

'A' & 'n' cannot be increased much. It will make it bulky.

Shunt - Shunt is a low resistance which is connected in parallel with galvanometer/ ammeter, which protects the same from strong currents.

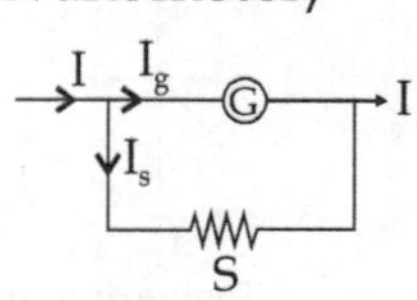

$$I_g = I\left(\frac{S}{G+S}\right) \text{ or } I_S = I\left(\frac{G}{G+S}\right)$$

Ammeter - It's a low resistance galvanometer. It is used to measure the current in a circuit.

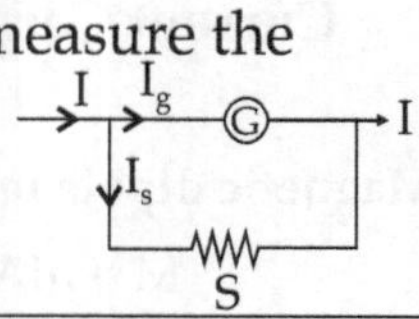

$$S = \left[\frac{I_g}{I - I_g}\right] G$$

Voltmeter - It's a high resistance galvanometer. It is used to measure the potential difference between any two points in the circuit.

$$R = \frac{V}{I_g} - G$$

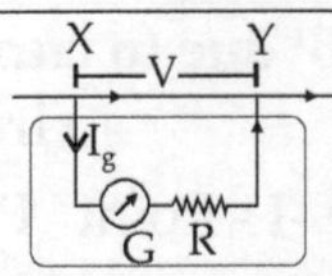

Magnetic field lines of a bar magnet:

- The lines of force in a magnetic field, which continuously represent the direction of magnetic field.
- Emerge from north, follow a curved path, enter the south pole. Closed loops.
- Two lines of force never intersect each other, if they do then at that point there would be two magnetic fields in different directions which is impossible.
- Near the poles, magnetic field is stronger.
- In uniform magnetic field, the lines of force are parallel and equidistant.

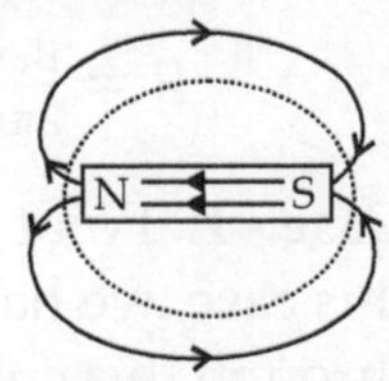

Bar magnet is equivalent to current carrying solenoid:

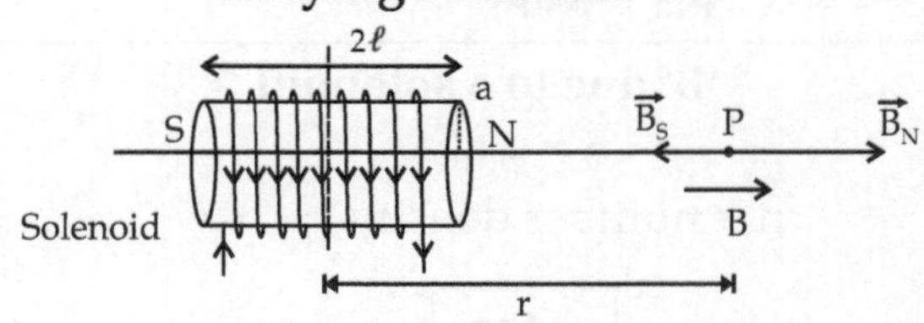

$$dB = \frac{\mu_0}{4\pi}\frac{2\pi(ndx)ia^2}{[a^2+(r-x)^2]^{3/2}} \qquad B = \frac{\mu_0}{4\pi}\frac{2M}{r^3}$$ [Same as bar magnet]

Magnetic field intensity due to magnetic dipole (Bar magnet):

(i) Axial Line:

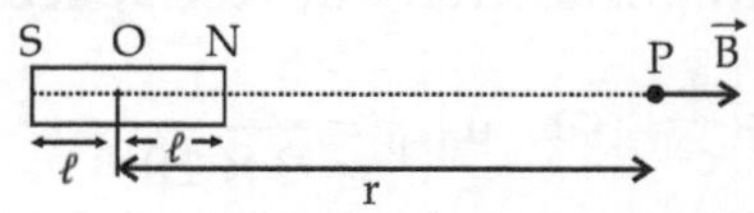

$$B = \frac{\mu_0}{4\pi}\frac{2M}{r^3}$$

(II) On equatorial Line

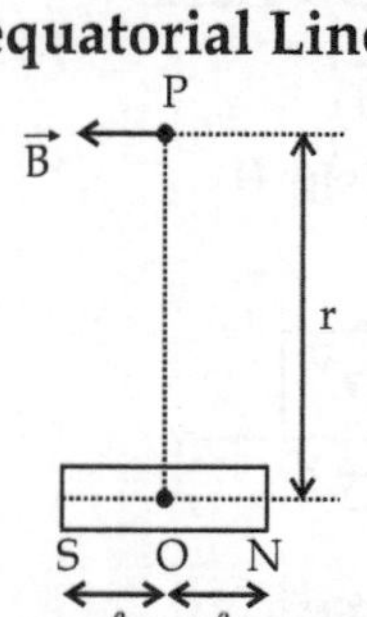

$$B = \frac{\mu_0}{4\pi}\frac{M}{r^3}$$

Torque on a bar magnet in a uniform magnetic field-

$$\therefore \tau = MB\sin\theta$$

$$\text{or } \vec{\tau} = \vec{M} \times \vec{B}$$

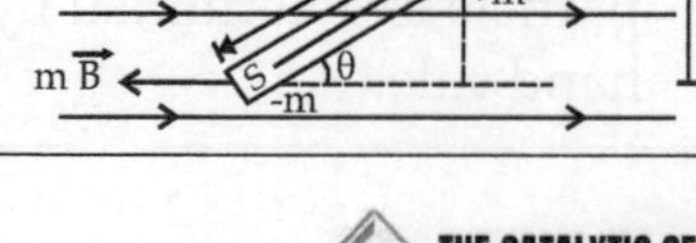

Magnetic susceptibility (χ_m)

It is a measure of how easily a substance is magnetised in a magnetising field. It is the ratio of the intensity of magnetisation to the magnetic intensity of the magnetising field.

$$\mu_r = 1 + \chi_m$$

Diamagnetic substances

Feebly repelled by a magnet, when placed in external field, they get magnetised in a direction opposite to that of field. χ_m small and negative. It is independent of temperature. Ex - Bi, Cu, Pb, S_i, N_2, H_2O, NaCl

Paramagnetic substances

Feebly attracted by a magnet. When placed in external field they get feebly attracted and get magnetised in the direction of magnetic field. χ_m small and positive. $\mu_r > 1$.

Ex - Al, Na, Ca, O_2, $CuCl_2$ $\quad \chi_m \propto \frac{1}{T}$ curie's law

Ferromagnetic substances

Strongly attracted by a magnet & strongly magnetised in the direction of magnetic field. χ_m large and positive. Ex - Fe, Co, Ni, Gd

Meissner Effect

Superconductors act like perfect diamagnetic substances at very low temperature it is called **Meissner Effect.**

$$\chi = -1 \text{ and } \mu_r = 0$$

Unit - 4: Electromagnetic induction and Alternating Current

Electromagnetic induction

Electromagnetic induction is the phenomenon of production of electric current (or emf) in a coil, when the magnetic flux linked with the coil is changed.

Faraday's Laws of electromagnetic induction

1) Whenever magnetic flux changes with a circuit, emf is induced.

2) Induced e.m.f. lasts as long as the change in magnetic flux continues.

3) Induced e.m.f. magnitude is directly proportional to the rate of change of magnetic flux linked with the circuit.

$$\varepsilon \propto \frac{\phi_2 - \phi_1}{t} \text{ or } \varepsilon = -k\frac{\phi_2 - \phi}{t}$$

k = 1 in S.I. system.

$$\therefore \varepsilon = -\frac{d\phi}{dt} \text{ or } \varepsilon = -N\frac{d\phi}{dt}$$

N = number of turns in the coil

Lenz's Law:

The induced emf (or current) produced in a circuit is always in such a direction that it opposes the change or the cause that produces it. Lenz's law follows law of conservation of energy. The mechanical work done in moving the magnet with respect to coil gets converted into electrical energy.

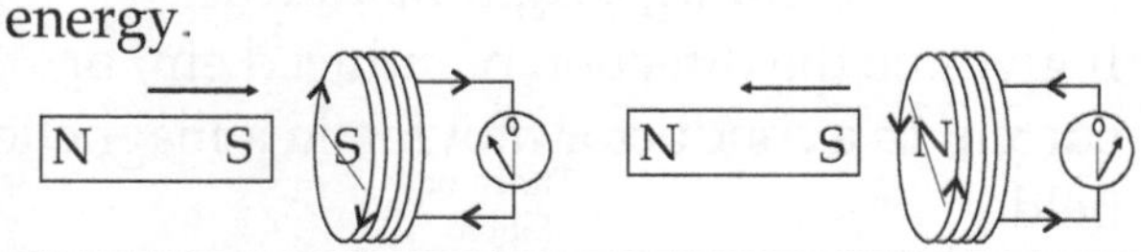

Magnetic flux (Ø)

The magnetic flux linked with a surface, when held in a magnetic field is the number of magnetic field lines crossing the surface normally and is measured as the product of component of the magnetic field normal to the surface and surface area vector. It is a scalar quantity.

$$\phi = \vec{B} \cdot \vec{A}$$

$$\phi = BA\cos\theta$$

$$\phi = \frac{F}{qV} \times A[M^1L^2T^{-2}A^{-1}]$$

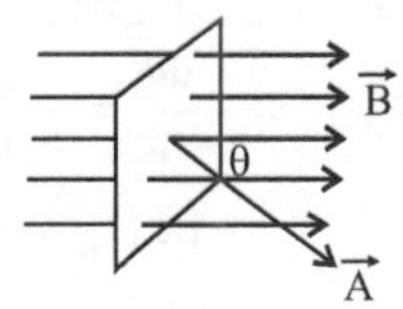

Various methods of producing emf

1. By changing the magnitude of magnetic field 'B'.
2. By changing the area
3. By changing 'θ' between $\vec{B}$ and $\vec{A}$.

$$\phi = BA\cos\theta$$

Motional induced e.m.f. in a conducting rod rotated in magnetic field –

Rod of length 'ℓ' moving with uniform angular velocity ω in the plane of paper.

$$\varepsilon = \frac{1}{2}B\omega\ell^2$$

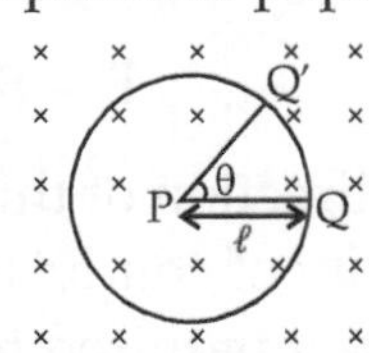

Motion of a rod in magnetic field

PQ is moving with constant velocity from x = 0 to x = b then to x = 2b. How force, emf, flux & power varies with distance shown here.

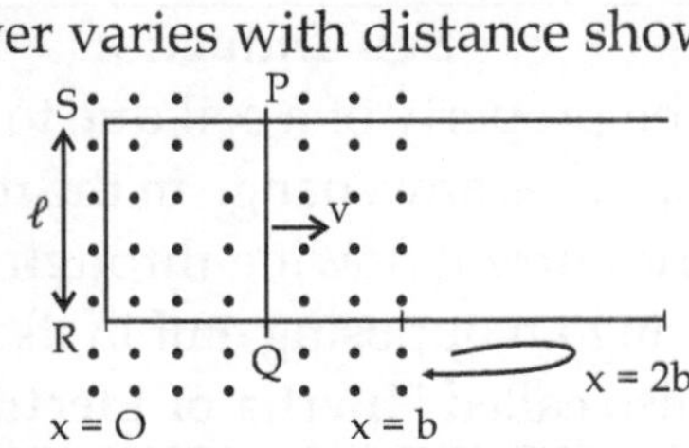

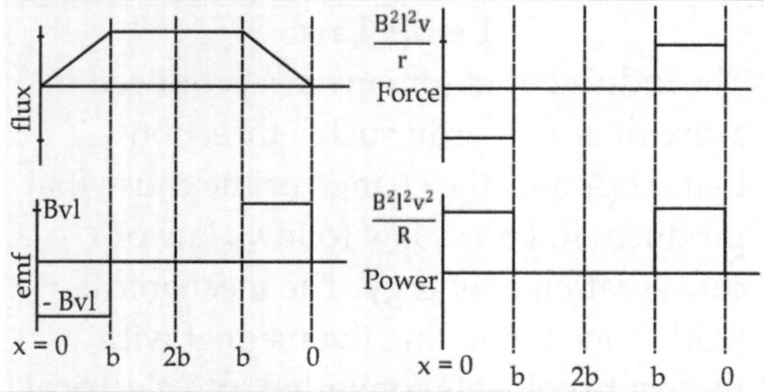

Fleming's right hand rule

It gives us the direction of induced emf or current in a conductor moving in a magnetic field.

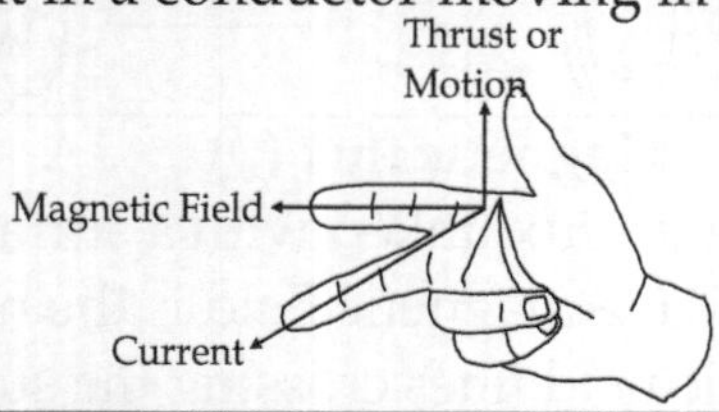

Motional Electromotive Force

The flux associated with KLMN will be

$$\phi_B = B \cdot A \cdot = B x \ell$$

'x' is changing with time

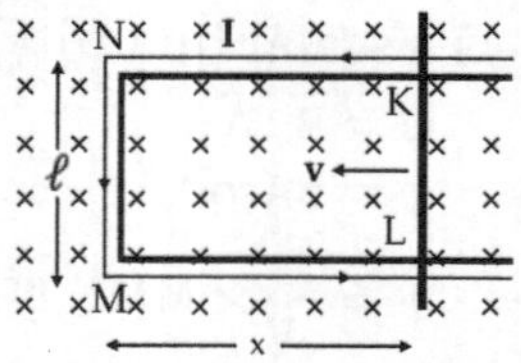

$$\varepsilon = -\frac{d\phi}{dt} = -\frac{d}{dt}(Bx\ell)$$

$$\varepsilon = -B\ell\left(\frac{dx}{dt}\right) = -B\ell \times -v$$

$$\varepsilon = Bv\ell$$

Power dissipated in a conductor moving in magnetic field

When a conductor is moving in a uniform magnetic field the emf induced is

$$\varepsilon = Bv\ell$$

If the resistance of moving arm is 'R' then current

$$I = \frac{\varepsilon}{R} = \frac{Bv\ell}{R}$$

Now, the magnetic force on the conductor

$$F = BI\ell$$

$$\therefore F = B\ell\left[\frac{Bv\ell}{R}\right] = \frac{B^2\ell^2 v}{R}$$

The direction of this force is opposite to the velocity of conductor.

$\therefore$ Power required to push the conductor

$$P = F \times v$$

$$P = \frac{B^2\ell^2 v}{R} \times v = \frac{B^2\ell^2 v^2}{R}$$

Self-Induction

It is the property of a coil due to which, the coil opposes any change in the magnitude of electric current flowing through it by inducing an opposing emf in itself.
It is also called **"Inertia of Electricity"**.

Magnetic flux linked with the coil is directly proportional to the current passing through it.

$$\phi \propto I \text{ or } \phi = LI$$

L = coefficient of Self induction

$$\because \varepsilon = -\frac{d\phi}{dt} = -\frac{d}{dt}(LI)$$

$$\text{or } \varepsilon = -L\frac{dI}{dt}$$

(-) sign shows that induced emf is opposing

If $\frac{dI}{dt} = 1$ then $\varepsilon = L$

Self Induction

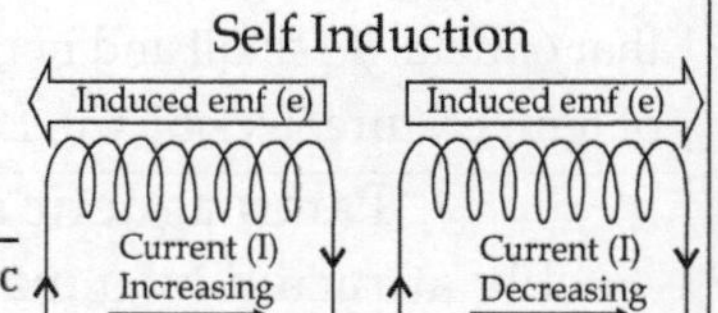

Unit-Henry

$$1 \text{ Henry } = \frac{1 \text{ Volt}}{1 \text{ amp/sec}}$$

$$L = [M^1L^2T^{-2}A^{-2}]$$

Inductors in Series

Suppose at any instant dI/dt is the rate of change of current through them.

$$\varepsilon_1 = -L\frac{dI}{dt} \text{ and } \varepsilon_2 = -L_2\frac{dI}{dt}$$

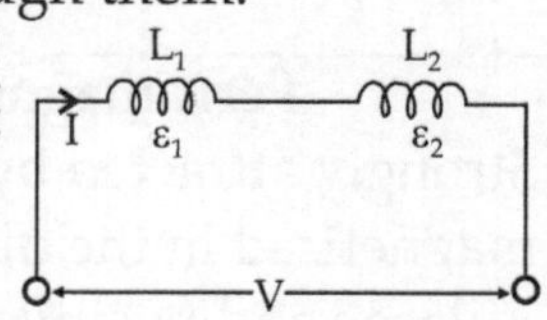

$$\text{Net emf } = \varepsilon_1 + \varepsilon_2$$

$$\therefore -L_s\frac{dI}{dt} = -L_1\frac{dI}{dt} - L_2\frac{dI}{dt}$$

$$\therefore L_s = L_1 + L_2$$

Inductors in parallel

$$\varepsilon = -L_1\frac{dI_1}{dt} \text{ and}$$

$$\varepsilon = -L_2\frac{dI_2}{dt}$$

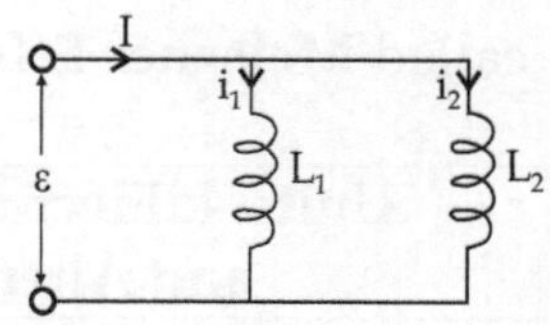

$$\frac{dI_1}{dt} = -\frac{\varepsilon}{L_1} \text{ and } \frac{dI_2}{dt} = -\frac{\varepsilon}{L_2}$$

$$\because I = I_1 + I_2$$

$$\therefore \frac{dI}{dt} = \frac{dI_1}{dt} + \frac{dI_2}{dt}$$

$$\text{or } \frac{dI}{dt} = -\frac{\varepsilon}{L_1} - \frac{\varepsilon}{L_2} \text{ ..... (1)}$$

$$\text{Also } \varepsilon = -L_p\frac{dI}{dt} \text{ or } \frac{dI}{dt} = -\frac{\varepsilon}{L_p} \text{ .... (2)}$$

$$\therefore -\frac{\varepsilon}{L_p} = -\frac{\varepsilon}{L_1} - \frac{\varepsilon}{L_2}$$

$$\text{or } \frac{1}{L_p} = \frac{1}{L_1} + \frac{1}{L_2}$$

Energy stored in inductor

Source emf has to spend energy in sending current through the circuit against induced emf. This energy is stored in the inductor as magnetic field.

$$dW = EIdt = \left(L\frac{dI}{dt}\right)Idt = LIdI$$

$$W = \int_0^I LIdt = L\int_0^I IdI = \frac{1}{2}LI^2$$

$$W = U = \frac{1}{2}LI^2$$

Self-inductance of a long solenoid

For a solenoid $B = \mu_0 nI$

n = number of turns per unit length

Now magnetic flux passing through each turn $\phi = \mu_0 nI \times A$

Total turns = $n \times \ell$

∴ Total flux linked with the coil

$\phi = \mu_0 nIA \times n\ell = \mu_0 n^2 \ell AI$

$\phi = LI$; $\mathbf{L = \mu_0 n^2 \ell A}$

If the core is having any other magnetic material then $L = \frac{\mu_0 \mu_r N^2 A}{\ell}$

$[\text{Here, } n = \frac{N}{\ell}]$

$\because \mu = \mu_0 \mu_r$

Mutual Induction

Mutual induction is that property of a coil due to which it opposes any change in the strength of current flowing through another coil placed nearby by developing an induced emf.

$$\varepsilon = -M\frac{dI}{dt}$$

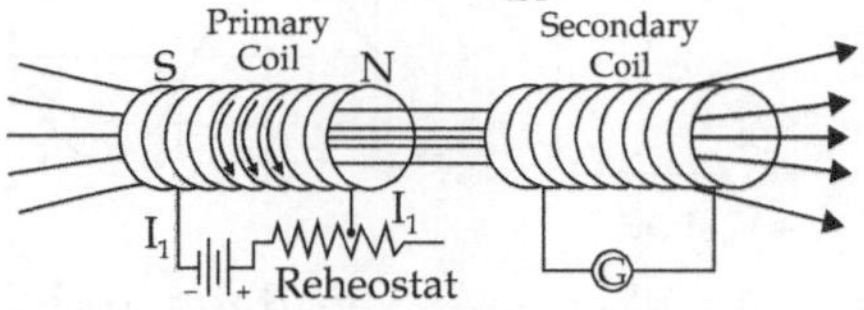

Mutual Induction

Mutual inductance of two long coaxial solenoids

$\phi_1 = M_{12} I_2$

$\phi_1 = \mu_0 n_1 n_2 \pi r_1^2 \ell I_2$

$\therefore M_{12} = \mu_0 n_1 n_2 \pi r_1^2 \ell$

$\phi_2 = M_{21} I_1$

$\phi_2 = \mu_0 n_1 n_2 \pi r_1^2 \ell I_1$

$M = \frac{\mu_0 N_1 N_2 A}{\ell}$

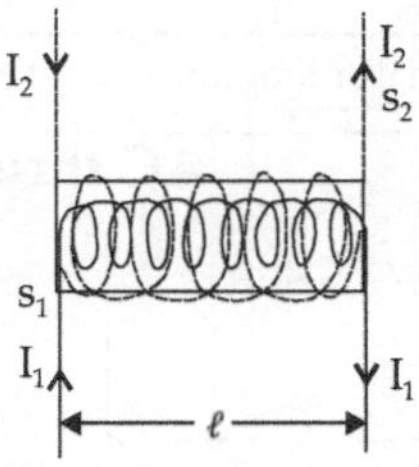

A.C. generator/A.C. Dynamo

This device convert mechanical energy into electrical energy.

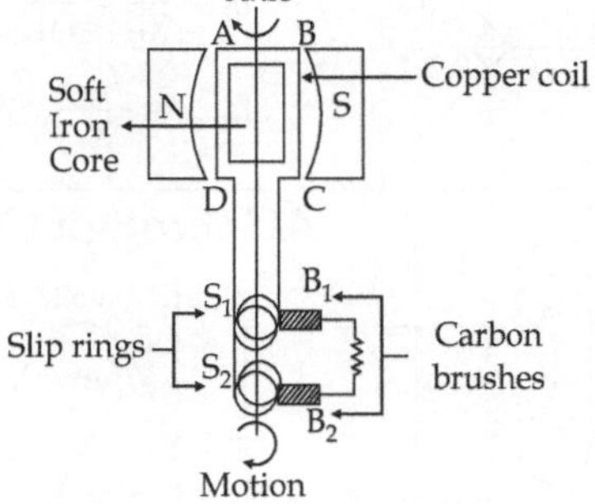

Direction of induced current is determined by Fleming's right hand rule.

$e = NBA\omega \sin \omega t$

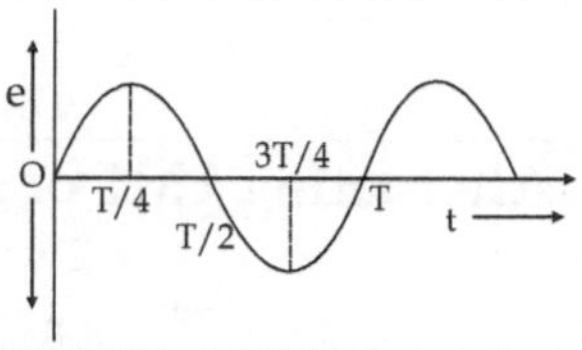

Graphs

If the current flowing through an inductor is increasing then,

$$U = \frac{1}{2}LI^2$$

(i) $\phi = LI$

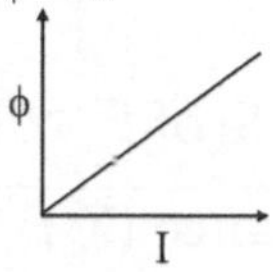

(ii) $e = \left|\frac{LdI}{dt}\right| \rightarrow$

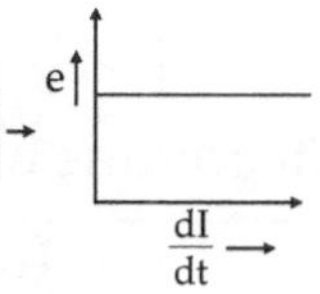

(iii) Energy = 1/2 Li^2 →

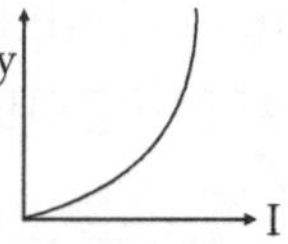

Notes - If the current flowing through straight wire is increasing with time, then the induced current in loop (1) and (2) will be clockwise and anticlockwise respectively.

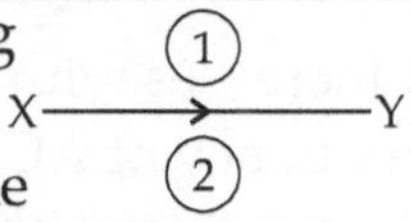

→ If the current flowing through the straight wire is decreasing with time then according to Lens's law, the induced current in loop (1) and (2) will be clockwise and anticlockwise respectively.

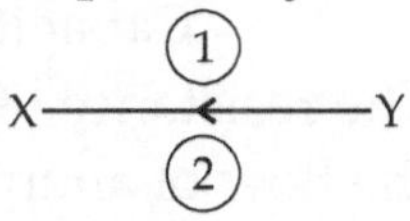

Alternating current (AC)

The electric current which oscillates between positive and negative values sinusoidally is called alternating current.

$$I = I_0 \sin \omega t$$

$$I = I_0 \sin \frac{2\pi}{T} t$$

Average or mean value of alternating current

It is that value of AC which sends same amount of charge through a circuit in half cycle (T/2) as sent by steady current for the same time.

$$I_{av} = \frac{2I_0}{\pi} = 0.637 I_0$$

Effective value or RMS value of AC

It is that value of AC in one complete cycle which generates same amount of heat through a circuit as produced by steady current in the same time and same circuit.

$$I_{rms} = \frac{I_0}{\sqrt{2}} = 0.707 I_0$$

Alternating EMF or Alternating Voltage

$$V_{av} = \frac{2V_0}{\pi} = 0.637 V_0$$

The sine curve for AC & AV

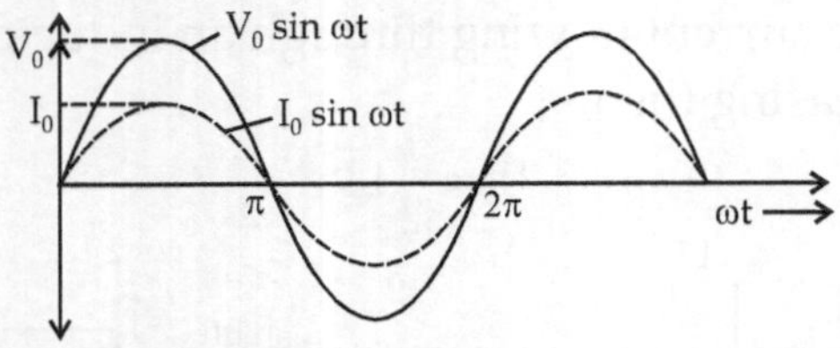

Magnitude of V > magnitude of I $\because I = \frac{V}{R}$

Inductive Reactance (X_L)

The resistance offered by an inductor to the flow of alternating current is called Inductive reactance (X_L).

$$X_L = \omega L = 2\pi\upsilon L$$

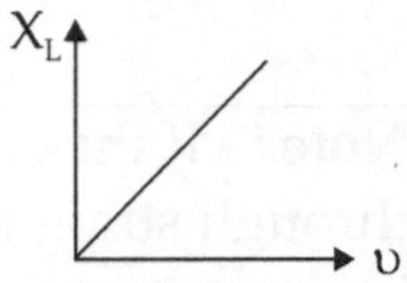

It increases with increase in frequency of AC.

If $\upsilon = 0$, $X_L = 0$, that means an inductor offers an easy path to direct current.

Capacitative Reactance (X_C)

The resistance offered by a capacitor to the flow of alternating current is called capacitative reactance (X_C).

$$X_C = \frac{1}{\omega C} = \frac{1}{2\pi\upsilon C}$$

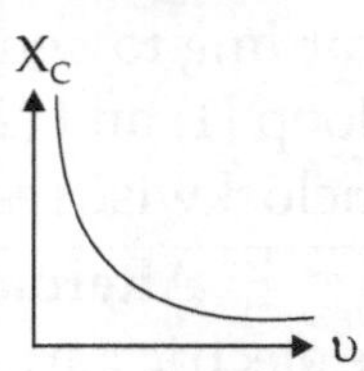

If $\upsilon = 0$, $X_C = \infty$

A capacitor offers infinite resistance to direct current.

Phasors

In AC circuits, when two or more elements are present, and we need to analyze the circuit we use rotating vectors called phasors. The diagram representing these phasors are called phasor diagrams.

The length of a phasor represents the peak value. Their projections on fixed axis give instantaneous value of alternating current.

Power in an A.C. Circuit

The average power in one complete cycle is given by-

$$P_{av} = V_{rms} I_{rms} \cos\phi$$

'ϕ' is the angle between voltage phasor and current phasor.

Power factor

In the expression of average power consumption in A.C. circuit over complete cycle, term '$\cos\phi$' is called power factor.

$$\cos\phi = \frac{R}{Z}$$

Wattless Current

In pure resistor circuit the average power consumption is maximum because the voltage phasor and current phasor are in same phase $\phi = 0°$.

In pure inductor or pure capacitative circuit, the voltage phasor and current phasor are perpendicular to each other and average power of the circuit is zero.

$$P_{av} = V_{rms} I_{rms} \cos\phi = 0$$

In such circuit, the current is called Wattless Current.

AC through Resistor

$$V = V_0 \sin \omega t, I = I_0 \sin \omega t$$

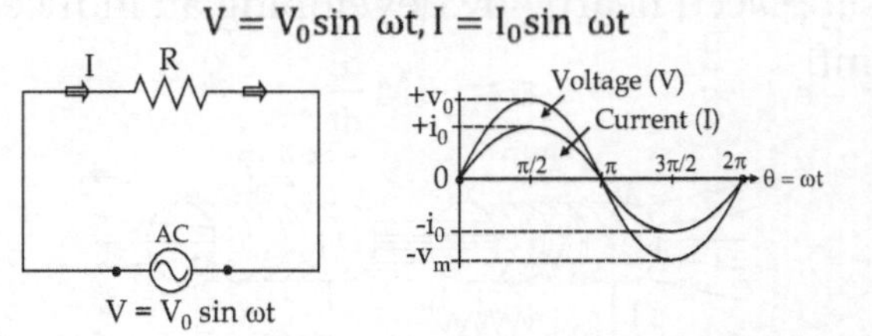

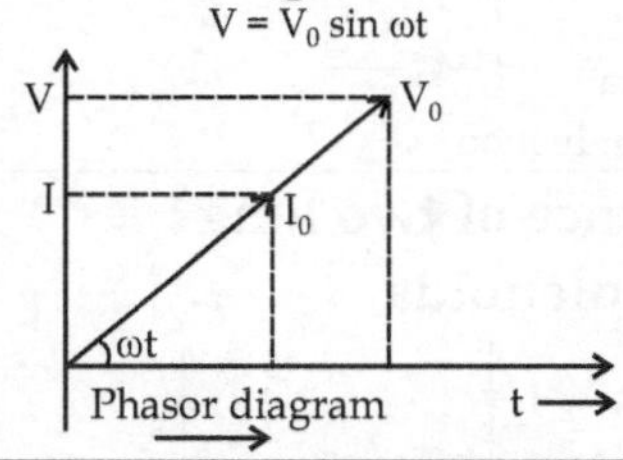

Voltage and Current are in the same phase.

$$I_0 = \frac{V_0}{R}, I_{rms} = \frac{V_{rms}}{R}$$

AC through Inductor

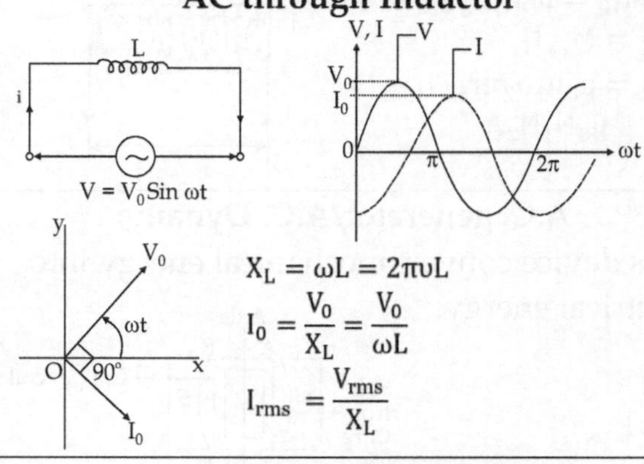

$$X_L = \omega L = 2\pi\upsilon L$$

$$I_0 = \frac{V_0}{X_L} = \frac{V_0}{\omega L}$$

$$I_{rms} = \frac{V_{rms}}{X_L}$$

AC through Capacitor

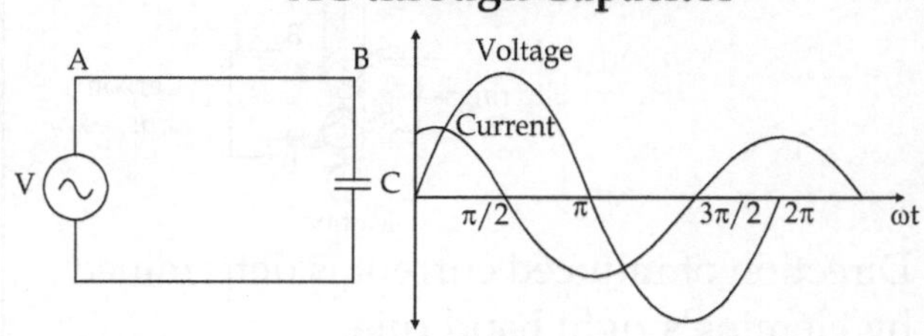

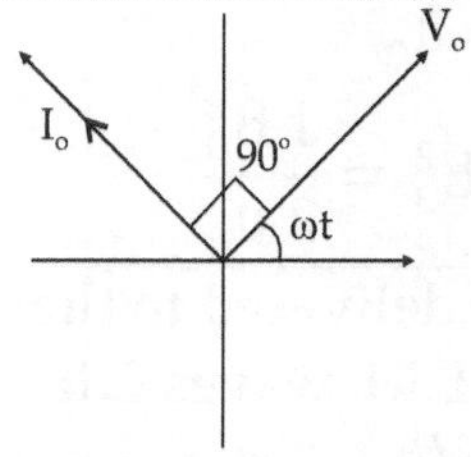

$$X_C = \frac{1}{\omega c} = \frac{1}{2\pi\upsilon C}$$

$$I_0 = \frac{V_0}{X_C} = V_0\omega C$$

Current leads voltage by $\pi/2$
Average power consumption = 0
$\cos\phi = \cos 90° = 0$

L-C-R series circuit

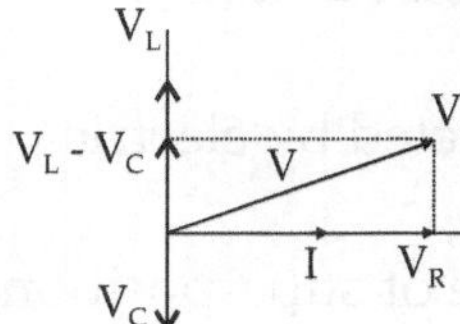

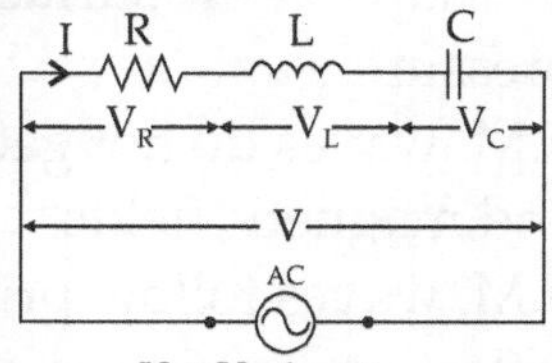

$V = V_0 \sin \omega t$

Impedance -

$$Z = \sqrt{R^2 + (X_L - X_C)^2}$$

$$Z = \frac{V_{rms}}{I_{rms}}$$

$$\tan\phi = \frac{X_L - X_C}{R}$$

Assuming $V_L > V_C$

$$V = \sqrt{V_R^2 + (V_L - V_C)^2}$$

- If $X_L > X_c$, $\tan\phi$ is positive
  $I = I_0 \sin \omega t$
  $V = V_0 \sin(\omega t + \phi)$ [Voltage leads]
- If $X_L < X_C$,
  ϕ = negative [Current leads]
  $I = I_0 \sin \omega t$
  $V = V_0 \sin(\omega t - \phi)$
- If $X_L = X_C$, $\phi = 0$
  Current and Voltage are in phase.
  Impedance (Z) = R (minimum)
  It is called resonant LCR series circuit.
- Resonant angular frequency

$$\omega_r = \frac{1}{\sqrt{LC}}$$

$$\text{or } \upsilon_r = \frac{1}{2\pi\sqrt{LC}}$$

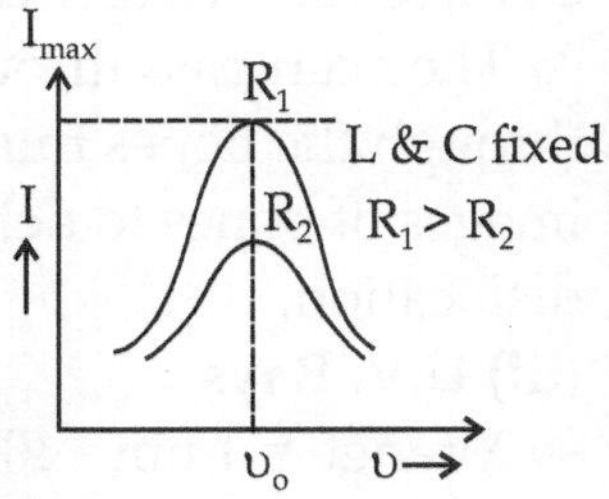

Average power consumption

$P_{av} = V_{rms} \times I_{rms} \cos\phi$
$\because \phi = 0 \therefore \cos\phi = 1$
$P_{av} = V_{rms} \times I_{rms}$ (maximum)
Current is maximum in the Circuit

$$I_{max} = \frac{V_{rms}}{R}$$

Quality Factor (Q - Factor)

It is the ratio of voltage drop across inductor (or capacitor) to the applied voltage.

$$Q = \frac{V_L \text{ or } V_C}{V_R}$$

$$Q = \frac{IX_L}{IR} = \frac{IX_C}{IR}$$

$$Q = \frac{\omega L}{R} = \frac{1}{\omega CR}$$

Energy stored in capacitor

$$U = \frac{1}{2}\frac{q^2}{C} = \frac{1}{2}CV^2 = \frac{1}{2}qV$$

Energy stored in Inductor

$$U = \frac{1}{2}LI^2$$

Transformer

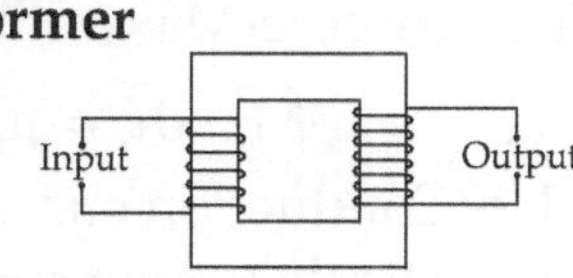

- Works on the principle of electromagnetic induction.
- It changes low alternating voltage at high current into high voltage at low current or vice-versa.

$$\frac{E_s}{E_p} = \frac{I_p}{I_s} = \frac{N_s}{N_p} = K \left[\text{Transformation ratio}\right]$$

- Step-up transformer

$$E_s > E_P \quad \therefore N_S > N_P$$

- Step-down transformer

$$E_S < E_P \therefore N_S < N_P$$

$$\text{Efficiency } (\eta) = \frac{E_S I_S}{E_P I_P} \times 100$$

- Core should be laminated and made of soft iron to avoid hysteresis loss.
- The transmission of A.C. is done always at high voltage & low current to avoid heat loss during transmission.

Unit - V
Electromagnetic Waves

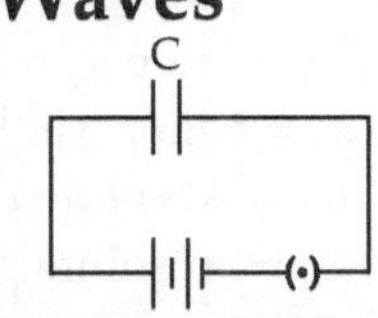

Electromagnetic Waves

1. Displacement current

- When a capacitor is being charged, the electric field between the plates change with time that means electric flux changes with time. It produces displacement current between the plates.

$$I_D = \varepsilon_0 \frac{d\phi_E}{dt}$$

- Displacement current was introduced by Maxwell because Ampere's circuital law for conduction current during the charging of a capacitor was found inconsistent.

2. Maxwell's equations -

These are the basic laws of electricity and magnetism.

(a) Gauss' Law in electrostatics

$$\oint \vec{E} \cdot d\vec{S} = \frac{q}{\varepsilon_0}$$

(b) Gauss' theorem in magnetostatics-

$$\oint \vec{B} \cdot d\vec{S} = 0$$

(c) Faraday's law of electromagnetic induction

$$\oint \vec{E} \cdot d\vec{\ell} = -\frac{d\phi_B}{dt}$$

(d) Ampere-Maxwell's circuital law

$$\oint \vec{B} \cdot d\vec{\ell} = \mu_0(I_c + I_D)$$

I_C = Conduction current

I_D = Displacement current

3 - Electromagnetic waves

The waves radiated by an accelerated or oscillating charged particles in which a variable magnetic field produces electric field and variable electric field Produces magnetic field and these two propagate perpendicular to each other and perpendicular to the direction of propagation of wave. These are transverse waves

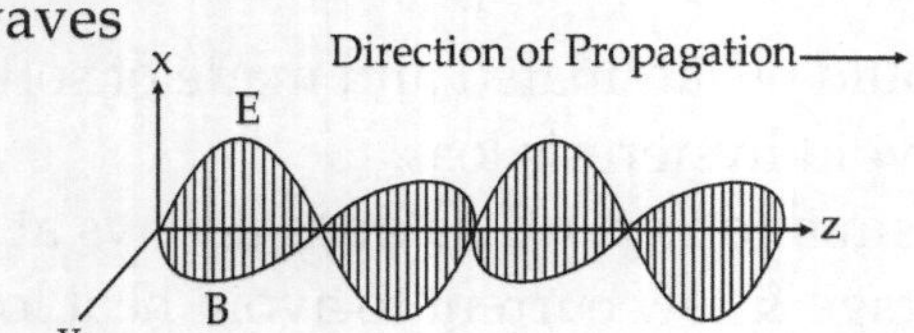

$$c = \frac{E_0}{B_0} = 3 \times 10^8 \text{ m/s}$$

$$c = \frac{1}{\sqrt{\mu_0 \varepsilon_0}}$$

Here $\mu_0 = 4\pi \times 10^{-7}$ T A^{-1} m

$\varepsilon_0 = 8.85 \times 10^{-12} C^2 N^{-1} m^{-2}$

μ_0 = Absolute permeability of free space

ε_0 = Absolute permittivity of free space

$E = E_0 \sin(kx - \omega t)\hat{j}$

$B = B_0 \sin(kx - \omega t)\hat{k}$

Energy associated with electromagnetic wave (U)

$$U = \frac{1}{2}\varepsilon_0 E^2 + \frac{1}{2}\frac{B^2}{\mu_0}$$

Average energy density (u)

$$u_{av} = \frac{1}{4}\varepsilon_0 E_0^2 + \frac{1}{4}\frac{B_0^2}{\mu_0}$$

Energy in electromagnetic waves is divided equally into electric and magnetic fields

$$u_e = u_m$$

$$u_{av} = \frac{1}{2}\varepsilon_0 E_0^2 = \frac{1}{2}\frac{B_0^2}{\mu_0}$$

Linear momentum delivered to the surface on which E.M. waves fall

$$p = U/c$$

U = Total energy transmitted by E.M. waves

c = speed of Light

- Since they have momentum, they impart pressure on the surface called Radiation pressure.
- E.M. waves do not get deflected by electric and magnetic fields.
- E.M. waves follow principle of superposition, reflection, refraction, interference, diffraction and polarisation.

Electromagnetic Spectrum

γ-Ray	X-Ray	U.V.	Visible	I.R.	Micro	Radio

(i) γ-Rays

→ $\lambda < 10^{-3}$ nm

→ Produced by decay of the nucleus (Radioactive decay)

→ Can be detected by photographic film and ionisation chamber.

→ Used in radiation therapy to kill tumour and cancer.

→ For food preservation and killing of germs.

(ii) X – Rays

→ λ range 10^{-3} nm - 1 nm

→ Produced in X-ray tubes & inner shell electrons.

→ Detected by photographic film, ionisation chamber and Geiger tubes.

→ They can pass through muscles and not through the bones thus it is used to take images of bones to detect any crack or dislocation.

(iii) U.V. Rays

→ λ-range = 1 nm - 400 nm

→ Produced when electrons move from one energy level to another energy level.

→ Detected by Photocells & photo graphic film.

→ To kill germs.

→ Used in Burglar alarm.

→ To study the molecular structure

(iv) Visible Light
→ λ-range = 400 nm - 800 nm
→ Emitted during transition of electrons from one energy level to another energy level.
→ Detected by eye, photo-cells and photographic film.
(v) Infrared radiations
→ λ-range = 700 nm - 1 mm
→ Produced by hot bodies and molecules
→ Detected by thermopile, Infrared photographic film, and Bolometers.
→ Used in treatment of muscular complaints, in remote control & in green houses to maintain the temperature
(vi) Microwaves
→ λ range = 1 mm - 0.1 m
→ Produced by magnetron valve klystron value, and Gunn diodes
→ Detected by diodes.
→ Used in RADAR communication, analysis of molecular structure, cooking and in geo-stationary satellites.
(vii) Radio waves
→ λ range > 0.1 m
→ Produced by acceleration of charged particles.
→ Detected by receiver antenna.
→ Used in radio/TV communication and navigation systems. Also, in bluetooth and Wifi connectivity.

Unit - VI
Optics

Ray Optics and Optical instruments

Reflection of Light - The bouncing back of light in the same medium when it strikes any surface (smooth or rough).
Laws of Reflection -
I - The incident ray reflected ray and normal at the point of incidence all lie in the same plane.

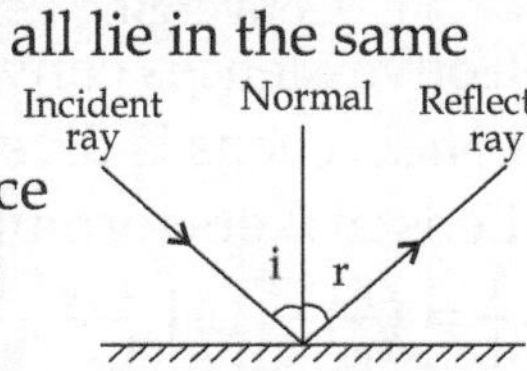

II - The angle of incidence (i) is always equal to the angle of reflection (r).
- This is true for both regular and irregular reflection.

Types of Mirrors -
(1) Plane mirror- Always forms virtual, erect and same size image. The image is as far behind the mirror as the object is in front of it.

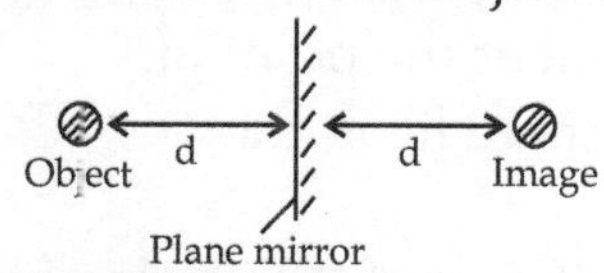

(2) Spherical mirrors -

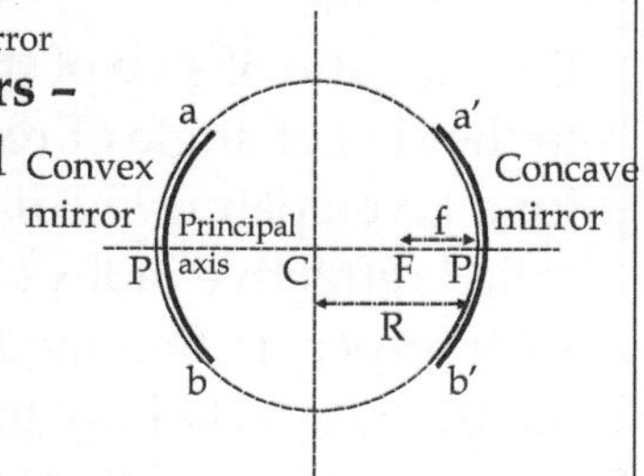

They have spherical reflecting surfaces.
Two types-
(a) convex mirror
(b) concave mirror
P = Pole, C = centre of curvature or centre of sphere whose spherical mirror is a part.
- The line joining pole and centre of curvature is called principal axis.

Focal length f = R/2
Object distance = u (Always negative)
Image distance = v (May be positive or negative)
Focal length for concave = negative
Focal length for convex = positive
- New cartesian sign convention

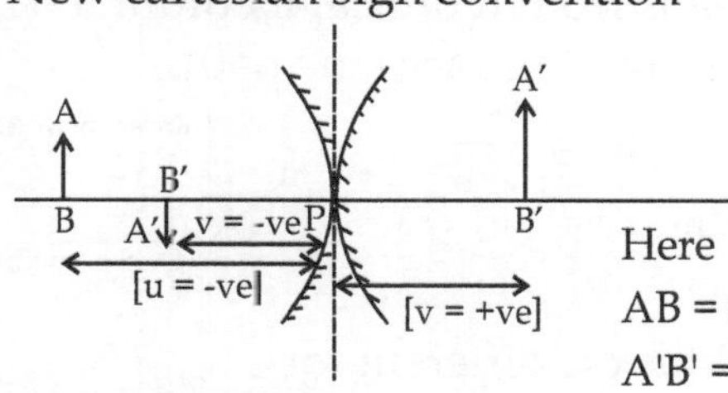

Here
AB = Object
A'B' = Image

Object height h_0 = +ve
Image height h_i = -ve for real image)
h_i = +ve (for virtual image)
Mirror formula → $\frac{1}{v}+\frac{1}{u}=\frac{1}{f}$
- In this formula place numerical values with sign.
- Linear magnification (m) -

$$m=-\frac{v}{u}=\frac{h_i}{h_0}$$

- Other formulas of linear magnification -

$$m=\frac{f}{f-u}=\frac{f-v}{f}$$

Refraction of Light

The phenomenon of deviation of path of light when it goes from one optical medium to other optical medium is called refraction of light.
Cause of refraction - Due to different optical density, the speed of light is different in different medium.

Laws of refraction

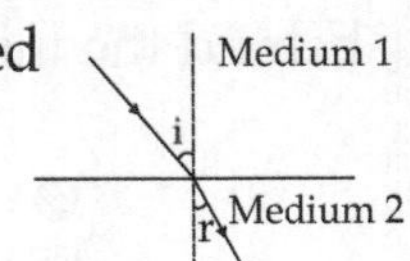

I- The incident ray, the refracted ray and normal at the point of incidence all three lie in the same plane.

II - The ratio of sine of the angle of incidence to the sine of angle of refraction is constant for a given pair of media. This constant is called refractive index of second medium with respect to first medium. In second medium refracted ray travels. $_1\mu_2 = \frac{\sin i}{\sin r}$

Absolute refractive index - The refractive index of a medium with respect to air or vacuum is called absolute refractive index.

$$\mu = \frac{c}{v}$$

c → speed of light in vacuum

v → speed of light in medium

Principle of reversibility- When light goes from one optical medium to another optical medium then if we reverse the path of light ray that means from II optical medium to I optical medium, it retraces the path.

$$_1\mu_2 = \frac{1}{_2\mu_1}$$

$$_1\mu_2 = \frac{\sin i}{\sin r} = \frac{v_1}{v_2} = \frac{\mu_2}{\mu_1} = \frac{\lambda_1}{\lambda_2}$$

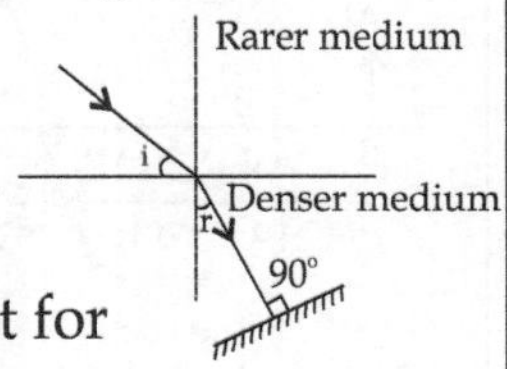

- Refractive index is different for different wavelengths.

$$\mu \propto \frac{1}{\lambda}$$

$$\because \lambda_V < \lambda_R$$

$$\therefore \mu_V > \mu_R$$

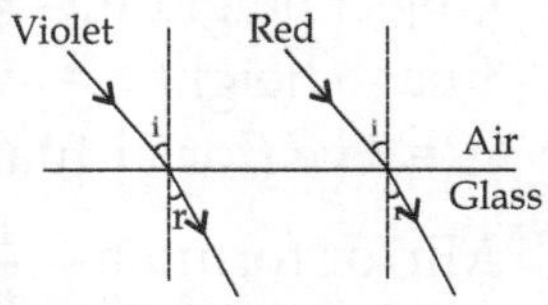

Red deviates less

- Violet colour deviates more than the red colour. Due to this splitting of white light takes place which is called dispersion of light.

Critical angle (i_c)

- When light goes from denser to rarer medium, it bends away from the normal.
- **It is that angle of incidence in the denser medium for which angle of refraction is 90°.**

$$\mu_1 \sin i_c = \mu_2 \sin 90°$$

$$\text{or } \sin i_c = \frac{\mu_2}{\mu_1}$$

$$_1\mu_2 = \sin i_c$$

$$\text{and } _2\mu_1 = \frac{1}{\sin i_c}$$

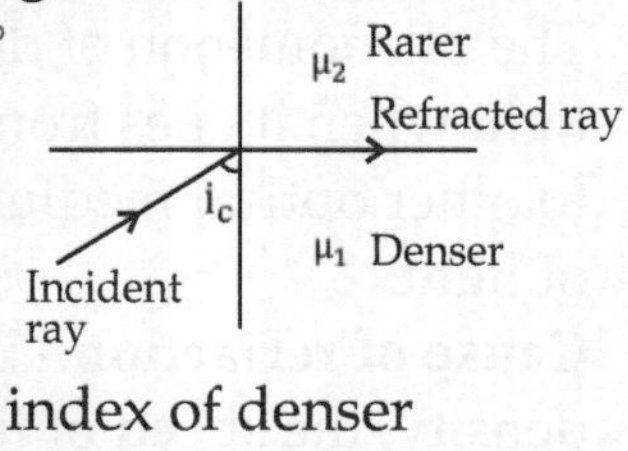

Here $_2\mu_1$ is Refractive index of denser medium with respect to rarer medium.

Total internal reflection (TIR)

- When light goes from denser medium to rarer medium it bends away and if it falls at critical angle in denser medium, it goes along the interface of two media, which means ∠r = 90°.
- **"When light goes from denser to rarer medium and angle of incidence is greater than critical angle, it gets reflected in the same medium making angle 'r' which is equal to angle of incidence. This phenomenon is called total internal reflection".**

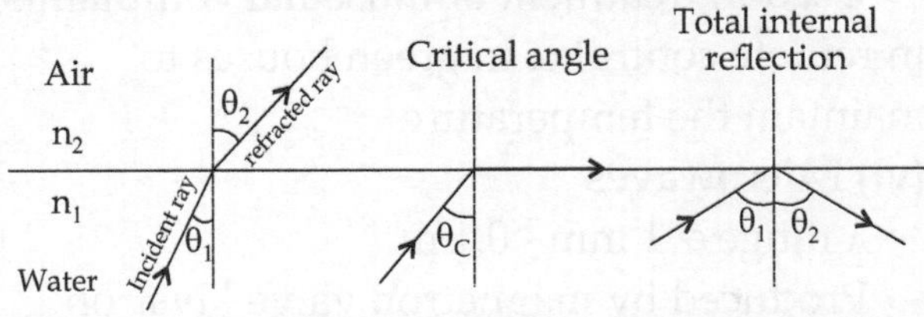

- Observed in the formation of mirage, and in diamonds.

Refraction at spherical surfaces -

→ When object is in rarer medium -

That means refraction is taking place from rarer to denser medium.

$$\frac{\mu_2}{v} - \frac{\mu_1}{u} = \frac{\mu_2 - \mu_1}{R}$$

→ Refraction from denser to rarer medium -

When object is in denser medium and image is formed in rarer medium -

$$\frac{\mu_1}{v} - \frac{\mu_2}{u} = \frac{\mu_1 - \mu_2}{R}$$

Lens

It is a transparent refracting medium which is bounded by two surfaces in which either both the surfaces are spherical or one plane & one spherical.

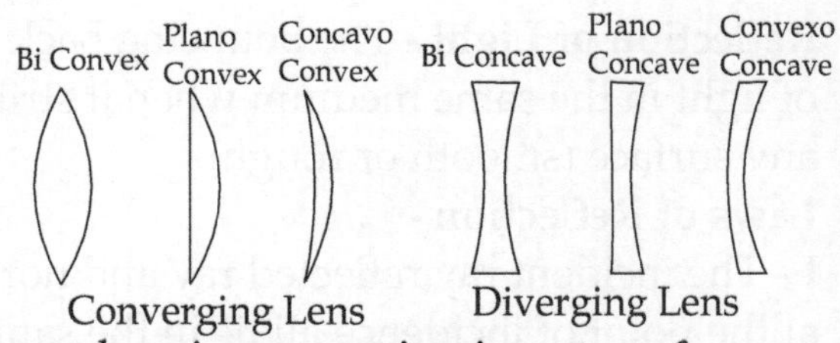

- Convex lens is converging in nature & concave lens is diverging in nature.
- Lens makers formula -

$$\frac{1}{f} = \left(\frac{\mu_2}{\mu_1} - 1\right)\left(\frac{1}{R_1} - \frac{1}{R_2}\right)$$

- Lens formula

$$\frac{1}{f} = \frac{1}{v} - \frac{1}{u}$$

Sign convention-

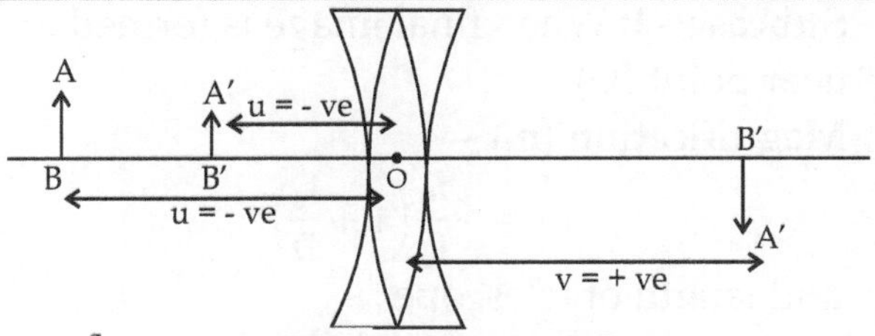

$f_{convex} = +ve$ AB = Object

$f_{concave} = -ve$ A'B' = Image

Object distance 'u' = -ve (always)

Image distance 'v' = +ve (For real images)

v = -ve (For virtual images)

Linear magnification (m)

$$m = \frac{h_i}{h_0} = \frac{v}{u}$$

h_i = Image height

(Negative for real images & positive for virtual images)

h_0 = object height (always positive)

Other formulas for magnification -

$$m = \frac{v}{u} = \frac{h_i}{h_0} = \frac{f-v}{f} = \frac{f}{f+u}$$

Note - In all the formulas, we need to put numerical values with sign.

- Change in focal length when a lens is immersed in other medium

$$\frac{1}{f_m} = \left[\frac{\mu}{\mu'} - 1\right]\left[\frac{1}{R_1} - \frac{1}{R_2}\right]$$

μ' is the refractive index of the medium in which lens is immersed.

- Please note if $\mu = \mu'$, $f_m = \infty$, the lens would behave like a glass slab.
- If $\mu < \mu'$ then $f_m > f$ and the nature of lens will be opposite i.e., a concave lens will behave like convex lens and vice versa.
- If $\mu > \mu'$ then $f_m > f$ and the nature remain the same that means convex lens behaves like convex only & concave behaves like concave lens only.

Power of lens

→ Power of a lens is its ability to converge or diverge the rays falling on it.

$P = \frac{1}{f}$ when ' f ' in meter

$P = \frac{100}{f}$ when f is in centimeter

→ Combination of lenses.

$P = P_1 + P_2 + P_3 \cdots\cdots$

or $\frac{1}{f} = \frac{1}{f_1} + \frac{1}{f_2} + \frac{1}{f_3}$

→ Magnification of combination of lenses -

$$m = m_1 \times m_2 \times m_3$$

Prism

It is a transparent refracting medium in which two plane surfaces are meeting at an angle with respect to each other as shown in the figure.

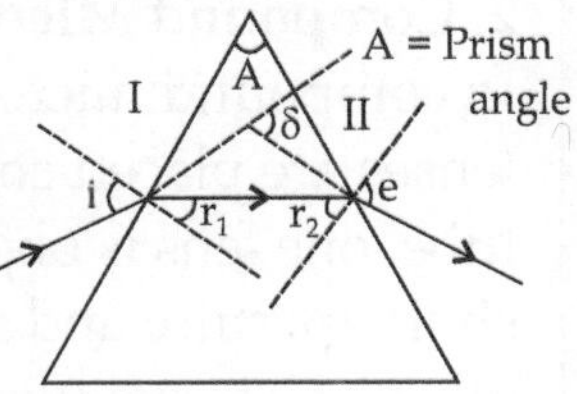

δ = Angle of deviation

e = Angle of emergence

r_1 and r_2 are angles of refraction at I & II face.

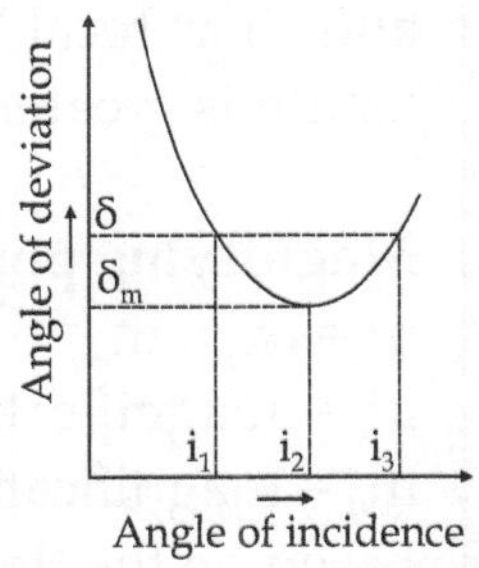

$r_1 + r_2 = A$

$\delta = i + e - A$

or $\delta = i + e - A$

At minimum deviation (δ_m)

$i_1 = i_2 = i$ & $r_1 = r_2$

Then

$$\mu = \frac{\sin{(A+\delta_m)/2}}{\sin A/2}$$

$$\text{and } \delta_m = (\mu - 1)A$$

(For very thin prism)

Angle of deviation depends on

(a) Angle of prism (A)

(b) material of prism (μ)

(c) angle of incidence (i)

Optical instruments

1. Simple Microscope -

It is made of convex lens which acts like magnifying glass when object is placed between focal point and optical centre of lens. We use converging lens of short focal length. There are two subcases -

Sub-case I - When final image is formed at near point (D) (least distance of distinct vision = It is the minimum distance at which the object can be seen clearly without straining our eyes).

Then magnification is given by

$$m = 1 + \frac{D}{f}$$

Sub case - II - When final image is formed at infinity

$$m = \frac{D}{f}$$

- Far point is the farthest point from eye at which an object can be seen clearly without straining the eye.

For normal eye, near point is 25 cm and far

point is infinity.

- For relaxed eye viewing, the final image is formed at infinity.
- Maximum magnification is produced at least distance of distinct vision.

2. Compound Microscope

In compound microscope, two converging lenses are placed coaxially at the ends of a tube, one lens is objective lens which has short aperture and short focal length & the other one is eye piece with large aperture and short focal length, but eye piece focal length is greater than objective lens i.e. $f_0 < f_e$

Magnifying power (m)

$m = m_0 \times m_e$

m_e = magnification produced by eye piece

m_0 = magnification produced by objective lens

Based on the final position of image, there are two subcases -

Subcase-I - When final image is formed at near point -

$$m = \frac{v_0}{u_0}\left(1 + \frac{D}{f_e}\right)$$

Subcase-II - When final image is formed at infinity

$$m = \frac{v_0}{u_0} \times \frac{D}{f_e}$$

- If lenses are adjusted in such a way that $v_0 \approx L$ (tube length) then -

→ When final image is formed at near point

$$m = (-)\frac{L}{f_0}\left(1 + \frac{D}{f_e}\right)$$

→ When final image is formed at infinity-

$$m = (-)\frac{L}{f_0} \times \frac{D}{f_e}$$

→ When final image is formed at infinity, then tube length, $L = v_0 + f_e$

3. Refracting Astronomical telescope

→ It is used to see heavenly bodies. There are two lenses. Objective lens with large aperture and large focal length and eye piece with small aperture and short focal length.

→ Based on the position of image there are two subcases.

Subcase - I: When final image is formed at near point (D)

Magnification (m) -

$$m = -\frac{f_0}{f_e}\left(1 + \frac{f_e}{D}\right)$$

and length of telescope

$$L = |f_0| + \frac{f_e D}{f_e + D}$$

Subcase - II: When final image is formed at infinity (∞) -

$$m = (-)\frac{f_0}{f_e}$$

and length of telescope

$$L = |f_0| + |f_e|$$

4. Reflecting astronomical telescope -

→ In this telescope a concave spherical mirror is used as objective which has large light gathering power.

→ The image formed is free from chromatic aberration.

→ It is also free from spherical aberration.

→ $m = \frac{f_0}{f_e}$ (for final image formed at ∞)

Wave Optics

→ **Wave front -** A source of light sends disturbance in all directions. The disturbance reaches to all particles present in the medium and each particle at same distance from source start vibrating in same phase. "The locus of all the particles of the medium which are vibrating in the same phase is called a **wavefront**.

(1) Spherical wavefront -

Produced by point source of light.

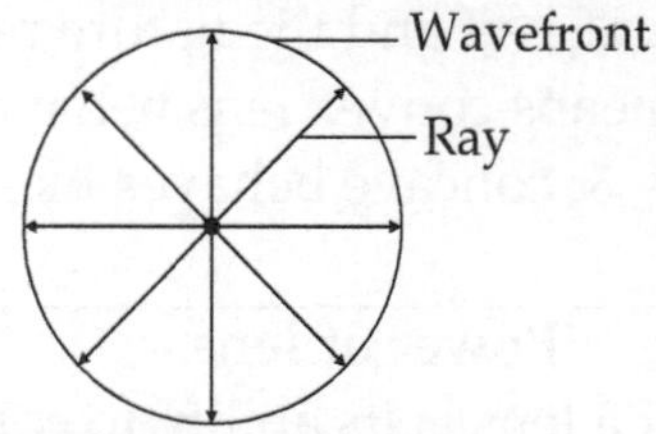

(2) Cylindrical wavefront -

When source of light is linear.

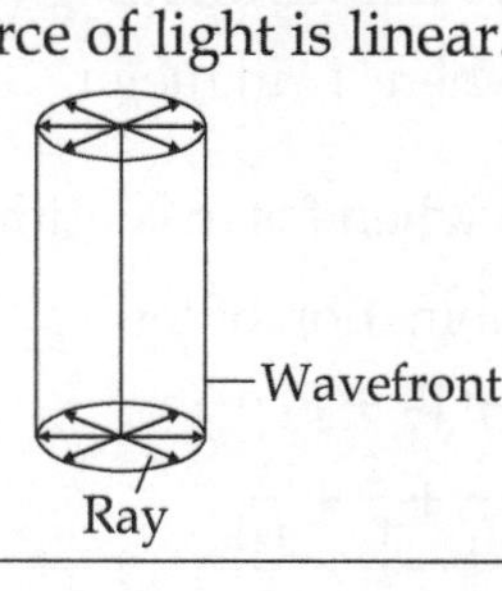

(3) Plane wavefront -
It is a small part of spherical or cylindrical wavefront from a distant light source.

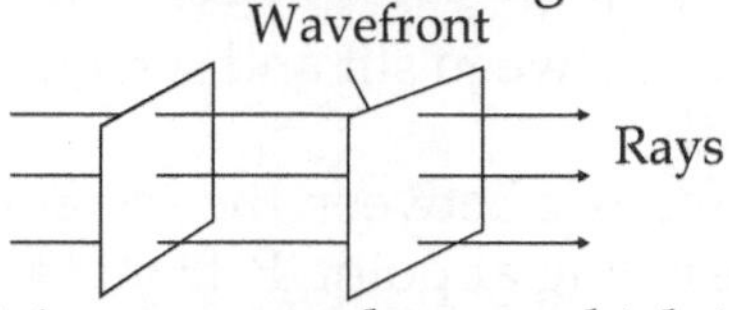

Ray - It is an arrow drawn which is perpendicular to the wavefront and pointing towards the direction of propagation of light.

⇒ Huygens Principles
(1) Each point on the primary wavefront acts like a fresh source of light which sends disturbance in all directions as the primary source.
(2) The new wavefront due to secondary wavelets in the forward direction is called **"Secondary Wavefront".**

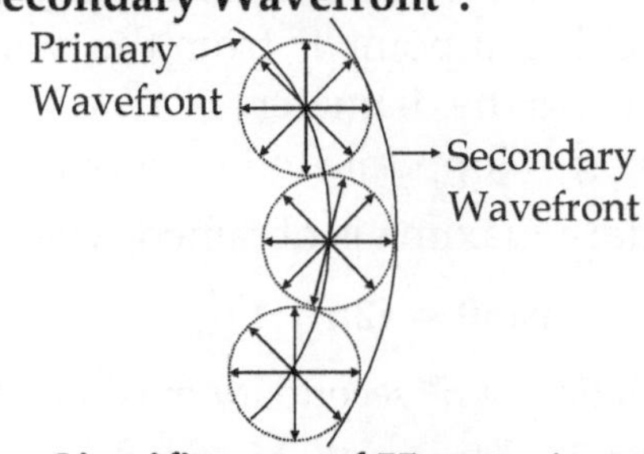

Significance of Huygen's principles
We can explain interference, diffraction and polarisation of light which can not be explained by Ray optics.
These principles verify laws of reflection and refraction, as well.

⇒ Superposition Principle
"When two or more waves of light travel in the same medium and incident at same point, they superimpose on each other and the resultant wave is formed whose displacement is equal to the vector sum of the displacements due to each wave.

$$\vec{y} = \vec{y}_1 + \vec{y}_2 + \vec{y}_3 + \cdots$$

Interference of light
It is the phenomenon of redistribution of energy in a medium due to superposition of waves at a point. At some points intensity of light is maximum which are called constructive interference, while at other points, the intensity of light is minimum which are called "destructive interference".

Conditions for sustained interference -
(1) The two sources of light must emit waves continuously and lie very close to each other.
(2) The light waves must be of same wavelength and frequency.
(3) The waves emitted by sources either have a constant or zero phase difference.
(4) To have good contrast at constructive and destructive interference point, the amplitude of waves must be equal.
(5) The sources of light should be narrow of point sized because broad sources act like a large number of point sources lying close to each other.
(6) Monochromatic light sources are used.

COHERENT SOURCES
These are the sources which emit light of same frequency, same wavelength and having a zero or constant phase difference
Formula of resultant amplitude -

$$R = \sqrt{a^2 + b^2 + 2ab\cos\phi}$$

Here a, and b are amplitudes of two waves respectively. [ϕ = phase difference]
Formula of resultant intensity

$$I_R = I_1 + I_2 + \sqrt{I_1 I_2}\cos\phi$$

For constructive interference -
→ $\cos\phi = +1$, i.e. $\phi = 0, 2\pi, 4\pi \cdots$
→ $\phi = 2n\pi$ (Here $n = 1,2,3 \cdots$)
→ path difference $(x) = n\lambda (n = 1,2,3 \ldots)$
→ Amplitude 'R' $= \sqrt{(a+b)^2} = a + b$
→ Intensity 'I_R' $= \left(\sqrt{I_1} + \sqrt{I_2}\right)^2$

For destructive interference
→ $\cos\phi = -1$
→ $\phi = \pi, 3\pi, 5\pi \ldots$
→ $\phi = (2n-1)\pi$ (Here $n = 1,2,3$)
→ Path difference $(x) = (2n-1)\frac{\lambda}{2}$
(Here $n = 1,2,3$)
→ Amplitude 'R' $= \sqrt{(a-b)^2} = a - b$
→ Intensity 'I_R' $= \left(\sqrt{I_1} - \sqrt{I_2}\right)^2$

Please note that total energy is conserved during Interference

⇒ Young's Double slit Experiment

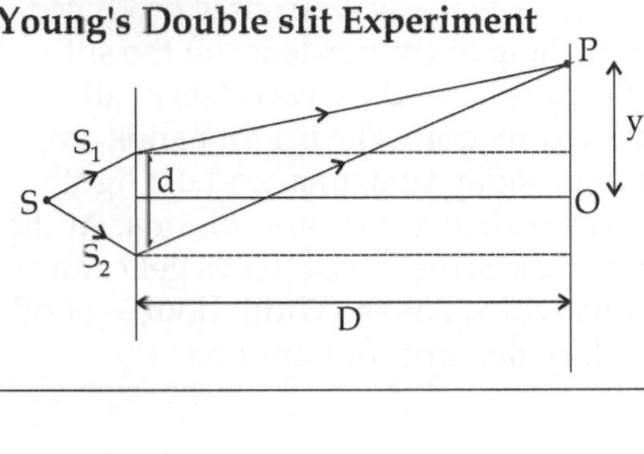

Position of nth bright fringe-

$$y_n = n\frac{\lambda D}{d} \quad n = 1,2,3$$

D = Distance between slits and screen
d = Distance between two slits
λ = wavelength of light used

Position of nth dark fringe

$$y_n = (2n-1)\frac{\lambda D}{2d}$$

- Fringe width for both dark and bright fringes-

$$\beta = \frac{\lambda D}{d}$$

- Angular fringe width,

$$\theta = \frac{\beta}{D} = \frac{\lambda}{d}$$

- Ratio of intensities of light due to two coherent sources

$$\frac{I_1}{I_2} = \frac{a^2}{b^2} = \frac{\omega_1}{\omega_2}$$

Here ω_1 and ω_2 are slit widths.

- Ratio of maximum and minimum intensity:

$$\frac{I_{max}}{I_{min}} = \frac{(a+b)^2}{(a-b)^2} = \frac{(\sqrt{I_1}+\sqrt{I_2})^2}{(\sqrt{I_1}-\sqrt{I_2})^2}$$

- If another medium of refractive index 'μ' and width 't' is placed in front of one of the slit then fringe pattern is shifted by 'n' fringe and shift will be **$(\mu - 1)t = n\lambda$**
- If the whole apparatus is immersed in another medium of refractive index 'μ' then fringe width β' will be-

$$\beta' = \frac{\beta}{\mu}$$

- If both the slits are covered by two refracting slabs of thickness t_1 and t_2 and refractive indices μ_1 and μ_2 then shift is given by - **$(\mu_2 - \mu_1)t = n\lambda$**

Diffraction of light

"The phenomenon of bending of light around the sharp corners of an obstacle or an aperture, which has size of the order of wavelength of light used, due to which light spreads in the region of geometrical shadow is called diffraction of light".It is produced by single slit and based on Huygen's principle. A plane wavefront is produced with the help of convex lens, on the slit which emits secondary wavelets in all possible directions.The bright bands are called secondary maxima, and the light bands are called secondary minima. At the centre of the screen, we always get central maxima which has the width double of other secondary maxima and minima.

Secondary Minima

Distance of nth secondary minima

$$y_n = \frac{n\lambda D}{a} \; [n = 1,2,3 \ldots]$$

D = Distance between slit and screen
a = slit width.

→ Path difference between the secondary wavelets reaching at point 'P' from the two ends of the slit across diameter $x = \pm n\lambda$ (opposite to interference)

→ n^{th} secondary minima is obtained when -

$$a \sin\theta = n\lambda$$

→ Angular position of nth secondary minima -

$$\theta = \frac{y_n}{D} = \frac{n\lambda}{a}$$

Secondary maxima

→ Distance of nth secondary maxima

$$\Rightarrow y_n = (2n+1)\frac{\lambda D}{2a}$$

→ Path difference between the secondary wavelets reaching at point 'P' from the two ends of slit across the diameter

$$x = (2n+1)\frac{\lambda}{2} \; \text{[opposite to interference]}$$

→ n^{th} secondary maxima is obtained when -

$$a\sin\theta = (2n+1)\frac{\lambda}{2}$$

→ Angular position of nth secondary maxima (θ_n)

$$\theta_n = \frac{y_n}{D} = (2n+1)\frac{\lambda}{2a}$$

→ Linear width of secondary maxima or minima

$$\beta = \frac{\lambda D}{a}$$

→ Linear width of central maxima

$$\beta_0 = \frac{2\lambda D}{a}$$

→ In diffraction pattern intensity of secondary maxima goes on decreasing with the increasing order of maxima.

Main differences between Interference and Diffraction

1. Fringe width - In interference pattern are the fringes are of same width but in diffraction the central bright fringe has double width than other secondary maxima or minima.

2. Intensity of fringes - In interference pattern all the bright fringes are of equal intensity but in diffraction, the intensity of secondary maxima (bright fringes) keeps on decreasing with increasing order of maxima.

3. Contrast between bright and dark fringes - There is a good contrast in interference pattern while it is not seen in diffraction.

4. Interference is produced by two different wavefronts coming from two coherent sources while diffraction is produced due to interaction of light coming from different parts of same wavefront.

Unit - VII
Dual Nature of Radiation and Matter

Dual nature of radiation - Interference, diffraction, polarisation can be explained on the basis of wave nature of electromagnetic radiation.
Photo electric effect, Compton effect can be explained on the basis of particle (quantum) nature of radiation.
Rectilinear propagation, reflection, refraction can be explained on the basis of dual nature of radiation.

Work function - The minimum amount of energy required to eject an electron from a metal surface is called work function (w).

Photoelectric effect

The phenomenon of ejection of electrons from a metal surface when light of sufficient frequency falls upon it, is called photoelectric effect.

Properties of Photon -
(a) It's a bundle of energy $E \rightarrow h\upsilon$
(b) It travels with speed of light.
(c) Its rest mass is zero.
(d) momentum $p = \frac{E}{c}$

Alkali metals Li, Na, K show photoelectric effect with visible light whereas metals like Zn, Mg, Ca show photoelectric effect with ultraviolet light.

Electron emission - Electron can come out from the metal surface by the following physical processes:
(i) Thermionic emission: When we heat the metal surface, free electrons attain thermal energy and overcome the attraction force of the nucleus.
(ii) Field emission: When we apply strong electric field of order $10^8 Vm^{-1}$ to a metal, electrons are pulled out by positive terminal.
(iii) Photoelectric emission - When light of a suitable minimum frequency falls on the metal surface, electrons are emitted from it. These electrons are called photo electrons.

Hertz's observations

The phenomenon of photoelectric emission was discovered by Heinrich Hertz in 1887 in his electromagnetic wave experiments. He observed that when the emitter plate was illuminated by U.V. light from an arc lamp there is high voltage spark generated near the detector Loop. This happened because electrons on the metal surface gained sufficient energy from the incident light and escape from the metal surface.

Hallwach's and Lenard's observation

1- Ultraviolet radiation falling on the emitter plate made electrons to come out from the surface of the zinc metal.
2- The emitted electrons are pushed towards the collector plate by the electric field.
3 - They also discovered that when the incident light has frequency lesser than a minimum value, photo electric emission does not take place.
4 - Certain metals like Zn, Cd, Mg etc. responds to only U.V. light which has short wavelength. Some alkali metals such as Li, Na, K, Cs and Rb can show photoelectric effect with visible light.

Einstein's photoelectric equation

During collision between a photon and a particle, the total energy and total momentum are conserved.
When a photon of energy '$h\upsilon$' is absorbed by an electron, some amount of energy (equals to work function 'w') is used up in

liberating the electron free and the remaining (hυ - w) is available to the electron as its Kinetic energy.

$$\therefore \frac{1}{2}mv^2_{max} = h\upsilon - w \qquad [\text{Here } w = h\upsilon_0]$$

$$\text{or } h\upsilon = w + \frac{1}{2}mv^2_{max}$$

$$\text{or } h\upsilon - h\upsilon_0 = \frac{1}{2}mv^2_{max}$$

Experimental study of photoelectric effect - The negative value of collector potential when no electron will reach to collector is called stopping potential.

In this case

$$eV_0 = \frac{1}{2}mv^2_{max}$$

Collector can be maintained at positive or negative potential with respect to emitter.

$$h\upsilon = h\upsilon_0 + \frac{1}{2}mv^2$$

If $h\upsilon = h\upsilon_0$, electrons will be ejected without Kinetic energy.

Conclusions -

(1) The electrons will be ejected only when $h\upsilon = h\upsilon_0$ - (Here υ_0 = threshold frequency)

(2) Emission of electrons starts as soon as light fails on metal surface.

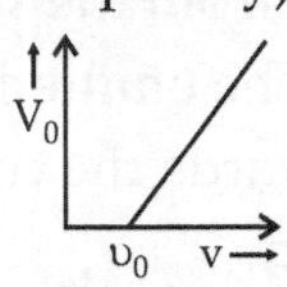

(3) Number of photoelectrons emitted depends upon intensity of incident light.

(4) The maximum kinetic energy with which photoelectrons are emitted from a metal surface depends only upon the frequency of incident light and independent of intensity.

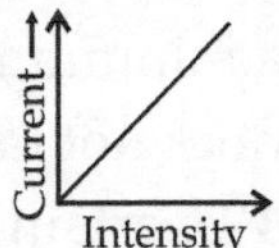

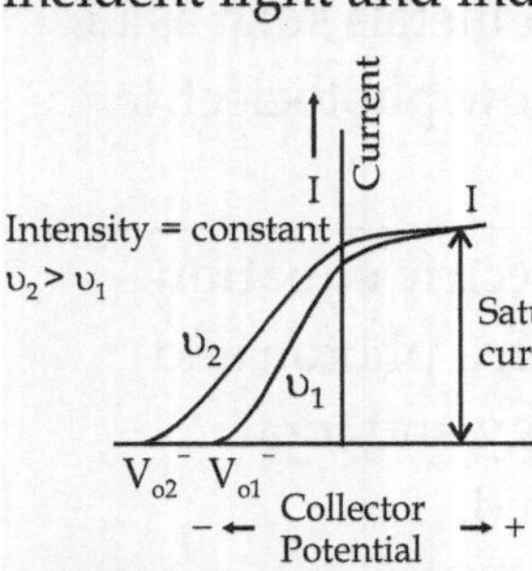

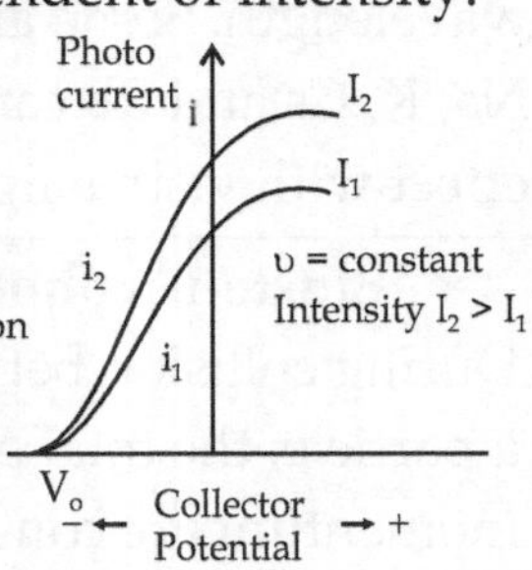

Electron volt - Energy gained by an electron, when it is accelerated through a potential difference of one volt. $1eV = 1.6 \times 10^{-19}$ J

All photons of same energy have same frequency & their energy is independent of intensity of radiation.

Laws of photoelectric emission from Einstein's photoelectric emission -

1. Photoelectrons will possess K.E. when $h\upsilon > h\upsilon_0$.
2. Maximum K.E. increases with the increase in frequency of incident radiations.
3. The number of photo electrons emitted depends upon the intensity of incident radiation.
4. One photon is capable of ejecting only one electron.
5. As the photon is absorbed, the emission of electron takes place instantaneously.

Verification of Einstein's photoelectric equation -

R.A. Millikan performed experiments to prove this equation.

From Einstein's photoelectric equation, we have

$$\frac{1}{2}m\upsilon^2_{max} = h\upsilon - h\upsilon_0$$

$$\text{then } \frac{1}{2}m\upsilon^2_{max} = eV_0$$

If 'e' is the charge on electron and 'V_0' is the stopping potential

$$\text{i.e. } eV_0 = h\upsilon - h\upsilon_0$$

$$\text{Or } V_0 = \frac{h\upsilon}{e} - \frac{h\upsilon_0}{e}$$

Millikan plotted a graph between frequency 'υ' and the stopping potential for a range of frequencies. He found a straight line. By measuring the slope, he also determined the value of 'h' (Planck's cons.). When $\upsilon = \upsilon_0$, $V_0 = 0$, i.e. when light of threshold frequency is incident electrons just come out, no stopping potential is required.

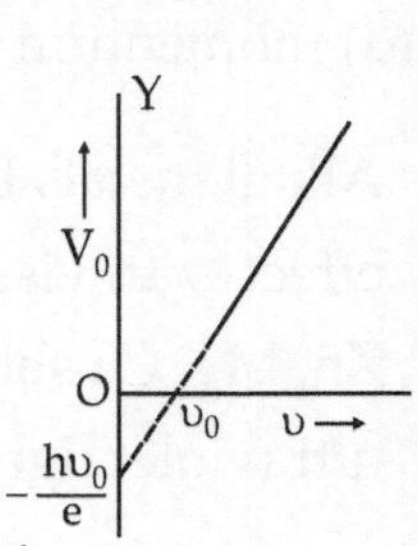

Photoelectric cells - A photoelectric cell is an arrangement which converts light energy into electrical energy. On cathode, a layer of photosensitive material is deposited (Cs on Ag_2O)

There is no time lag between the incident light and emission of electrons and current depends upon intensity of light.

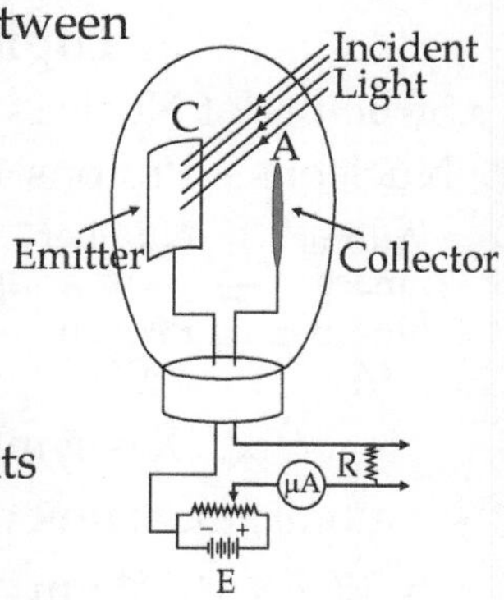

Uses - 1) Automatic switching of street lights and traffic signals.
2) In burglar-alarms.

Unit - VIII
Atoms and Nuclei

- The word '**Atom**' is derived from Greek word "a-tomio" means "uncuttable".
- Static electricity proved that the atom is divisible and made of three subatomic particles - Proton, Neutron and Electron.

Alpha particle scattering experiment - Rutherford's model of an atom:
He performed α-particle scattering experiment using a thin gold foil.

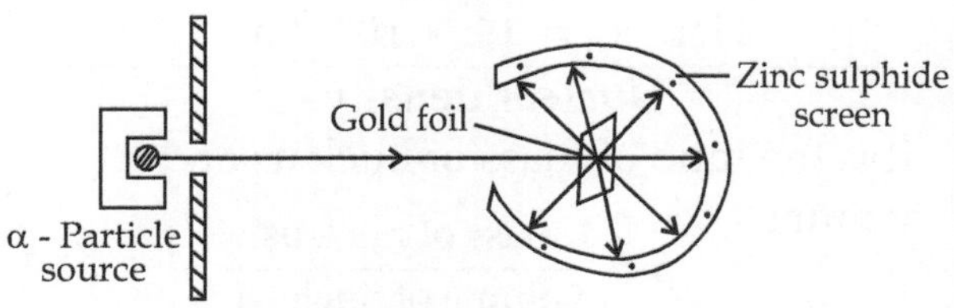

When α-particle approaches gold nucleus its kinetic energy gets converted into electrostatic potential energy. Using this we can calculate distance of closest approach 'd'.

$$K = \frac{1}{2}mv^2 = \frac{k(2e)Ze}{d}$$

$$\therefore d = \frac{2kZe^2}{\frac{1}{2}mv^2}$$

Conclusions -
1) Most of the space inside the atom is empty since many alpha particles passed undeflected.
2) Very few showed large deflections which suggest that there is heavy, dense, positively charged mass present in the atom which he called 'nucleus'.
3) The nucleus is surrounded by negatively charged electrons. This makes an atom electrically neutral.

Bohr's Model for Hydrogen atom
(1) Atom consists of positively charged nucleus in the centre and electrons revolve around it in circular orbits with fixed value of energy called energy levels. The different energy levels are numbered as 1,2,3,4… etc or designated as K, L, M, N-··· etc

$$E_n = -\frac{2\pi^2 me^4 z^2}{n^2h^2}\ \frac{kJ}{mol}$$

$$E_n = -13.6\frac{Z^2}{n^2} = eV/atom$$

For H - atom

$$K = \frac{e^2}{8\pi\varepsilon_o r}$$

$$U = -\frac{e^2}{4\pi\varepsilon_o r}$$

$$E = -\frac{e^2}{8\pi\varepsilon_o r}$$

(2) Energy levels are quantized. The angular momentum in a definite energy level is given by **mvr** $= \frac{nh}{2\pi}$ i.e. only those levels are permissible in which the angular momentum is an integral multiple of $h/2\pi°$
(3) Bohr radius

$$\Rightarrow r_n = 0.529\frac{n^2}{Z}\,\text{Å}$$

(4) Energy is emitted or absorbed when an electron jumps from one level to the other, to the amount of energy released or absorbed is given by
$\rightarrow \Delta E = E_2 - E_1$
(5) The Velocity of electron in different orbits is given by

$$v_n = v_0 \times \frac{Z}{n}$$

Here, $v_0 = 2.188 \times 10^8$ cm/s

Limitations of Bohr's model-
(1) Applicable to only to hydrogen like species. Because Bohr's model does not include forces between electrons.
(2) It cannot explain the relative intensities of radiations in line spectrum of hydrogen like atoms.

Bohr's postulate of angular momentum can be derived from de-Broglie equation -
Electron revolves around the nucleus in circular orbits and behaves like a wave, to complete the wave, the circumference of the orbit must be integral multiple of wavelength associated with that wave.

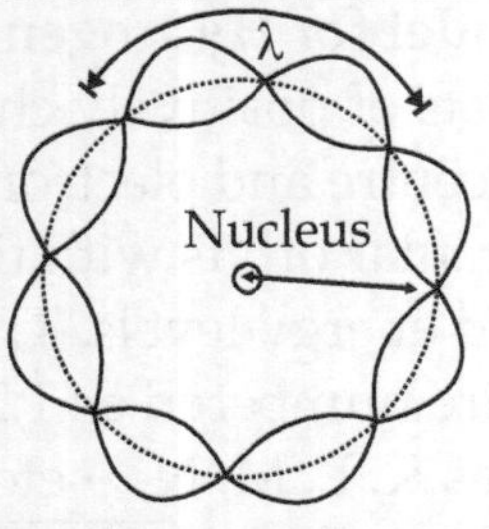

i.e. $2\pi r = n\lambda$ ....(1)

Also $\lambda = \frac{h}{mv}$ ....(2)

substituting 'λ' in eq. (1) we get

$$2\pi r = n\frac{h}{mv} \quad \text{or} \quad mvr = \frac{nh}{2\pi} \quad .... (3)$$

Equation (3) is Bohr's postulate of angular momentum.

Evidence for quantized energy levels in an atom -

(Emission and absorption spectra) When white light from any source is first passed through the solution or vapours of a chemical substance and then analysed by spectroscope we get some dark lines in a continuous spectrum, this is called **"absorption spectrum"**. The missed wavelengths absorbed by electrons and thus electrons get excited to higher energy levels, when they come to their original level, they emit the same wavelengths radiations, when analysed by spectroscope. This is called **"emission spectrum"**. Absorption of only specific wavelengths gives the evidence of quantized energy levels.

Rydberg's formula -

$$\bar{\upsilon} = R\left[\frac{1}{n_1^2} - \frac{1}{n_2^2}\right] \text{cm}^{-1} \qquad R = 109677 \text{ cm}^{-1}$$

$$\bar{\upsilon} = \frac{1}{\lambda} = \text{wave number}$$

when $n_1 = 1$, $n_2 = 2,3,4\cdots$ (Lyman series, U.V. region)

when $n_1 = 2$, $n_2 = 3,4,5\cdots$ (Balmer series, visible region)

when $n_1 = 3$, $n_2 = 4,5,6\cdots$ (Paschen series, I.R. region)

when $n_1 = 4$, $n_2 = 5,6,7\cdots$ (Brackett series, I.R. region)

when $n_1 = 5$, $n_2 = 6,7,8\cdots$ (Pfund dents, I.R. region)

- For H-Like particles

$$\bar{\upsilon} = R\left[\frac{1}{n_1^2} - \frac{1}{n_2^2}\right] Z^2 \text{ cm}^{-1}$$

- Limiting line for any series is when $n_2 = \infty$

$$\bar{\upsilon} = \frac{1}{\lambda} = \frac{\upsilon}{c} \quad \because c = \upsilon\lambda \quad \text{or} \quad \upsilon = c\bar{\upsilon}$$

Topic - Nuclei

- Structure of Nucleus

Nucleons = Protons + electrons

$$\text{Atomic mass Number (A)} = \text{Number of Protons (Z)} + \text{Number of Neutrons (n)}$$

${}^{A}_{Z}X$, X = symbol of element

- 1 atomic mass unit or 1 amu $= 1.67 \times 10^{-24}$ gm

It is the mass of one nucleon.

Also it is $\frac{1}{12}$th the mass of one ${}^{12}_{6}C$ atom.

- It is observed that the volume of nucleus is directly proportional to its mass number.

$$\frac{4}{3}\pi R^3 \propto A$$

R = Radius of Nucleus

A = Atomic mass number

$$\therefore R \propto A^{1/3}$$

$$\text{or } R = R_0 A^{1/3}$$

Here $R_0 = 1.2 \times 10^{-15}$ m

Nuclear density -

It is the ratio of mass of nucleus and its volume.

$$\rho = \frac{\text{mass of nucleus}}{\text{volume of nucleus}}$$

$$\frac{4}{3}\pi R^3 = \frac{4}{3}\pi\left(R_0 A^{1/3}\right)^3$$

$$V = \frac{4}{3}\pi R_0^3 A$$

and mass of nucleus = mA

Here m = mass of one nucleon

$$\therefore \text{Density } (\rho) = \frac{mA}{\frac{4}{3}\pi R_0^3 A}$$

$$\rho = \frac{3m}{4\pi R_0^3} = \text{A constant}$$

Density of every nucleus is a constant.

$$\rho = 2.29 \times 10^{17} \text{ kg/m}^3$$

Mass defect

Difference is sum of masses of nucleons and actual mass of nucleus.

$\Delta m = [Zm_p + (A - Z)m_n)] - [\text{Actual mass of nucleus}]$

Mass energy Relation -

$E = \Delta mc^2$

During nuclear fission and fusion some of the mass is converted into energy.

Nuclear Force

1. Strongest force
2. Short range
3. When distance between the nucleons becomes lesser than 0.8 fermi the nuclear force becomes strongly repulsive.
4. Non central force
5. Spin dependent force, between parallel spin nucleons it's greater.
6. Independent of charge.
7. Doesn't follow inverse square law.
8. Nuclear forces are due to exchange of π-mesons between nucleons, thus they are also called exchange forces.

Binding Energy - (B.E.)

- Binding energy of a nucleus is the energy required to separate the nucleons present inside the nucleus and send them infinite distance apart.
- When nucleons are brought together, some of their mass is converted into energy that is why the actual mass of nucleus is always lesser than the sum of masses of all nucleons present inside the nucleus.

$$\text{B. E.} = \Delta m \text{ (mass of defect) } \times c^2$$

Binding energy per nucleon

$$= \frac{\text{Total B.E.}}{\text{number of nucleons (A)}}$$

Binding energy curve - Its a graph between binding energy per nucleon versus mass number (A).

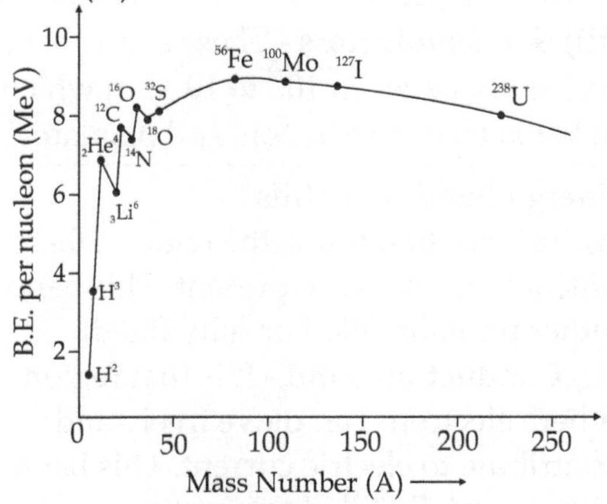

The main features of the curve-

(i) B.E. for middle mass numbers $30 < A < 170$ is almost constant.

(ii) For A = 56 its value is 8.75 MeV and for A = 238 its value is 7.6 MeV.

(iii) B.E. value is lower for both lighter i.e., $A < 30$ and heavier $A > 170$ nuclei.

(iv) more the value of B.E. per nucleon, more stable is the nucleus.

(v) Since binding energy is constant in the range $30 < A < 170$, hence we can say that nuclear force is short-ranged and they exhibit saturation property.

(vi) If a very heavy nucleus A = 240 (which has lower binding energy per nucleon) divides into two atoms with A = 120 nuclei, its nucleons are more tightly bound that means energy is released when a heavy nucleus breaks into two small nuclei. This is the basis of nuclear fusion.

(vii) When two very light nuclei ($A \leqslant 10$) join to form a heavier nucleus, the energy is released & nucleons get more tightly bound. This is the basis of "nuclear fusion". This process is taking place on sun.

(viii) We also see that more energy is released when lighter nuclei join to form heavy nucleus (Nuclear fusion) than a heavy nucleus breaks to form lighter nuclei (Nuclear fission).

Radioactivity

- A.H. Becquerel discovered radioactivity in 1896 by accident.
- Radioactivity is a nuclear phenomenon in which three types of particles released. They are –

(i) α-particles - In which ${}^{4}_{2}He^{2+}$ nucleus is emitted.

(ii) β-particles - They are either electrons or positions (like electrons but with positive charge).

(iii) γ-decay - These are high energy photons.

Nuclear energy

- From binding energy curve we can conclude that both Nuclear fusion and Nuclear fission release energy.
- During nuclear reaction the energy release is the order of Mev.
- Nuclear sources produce million times more energy than a normal exothermic reaction. For example fission of one kilogram of uranium generates 10^{14} J of energy and 1 kg coal gives 10^{7} J.

Nuclear fission

When a heavy nucleus breaks into two lighter nuclei having more binding energy per nucleon than heavy nucleus, energy is released and this process is called nuclear fission.

Example - When a uranium isotope $^{235}_{92}U$ nucleus is bombarded with a high energy neutron, it breaks into two intermediate nuclear fragments.

$$^{1}_{0}n + ^{235}_{92}U \longrightarrow ^{144}_{56}Ba + ^{89}_{36}Kr + 3^{1}_{0}n$$

- The **'Q-value'** is the energy released when nuclear fission takes place.
  For Uranium nucleus, it is 200 MeV per fissioning nucleus.
- **How would you calculate this 'Q value' -** We consider a heavy nucleus with mass number A = 240 having B.E./nucleon about 7.6 MeV, which breaks into two lighter nuclei with A = 120 nuclei of about 8.5 MeV then gain in B.E./nucleon is
  = 8.5 - 7.6 = 0.9 MeV.
  Total number of nucleons before were 240, thus total gain in Binding energy will be
  = 240 × 0.9 = 216 MeV

Nuclear fusion

When two lighter nuclei combine and form one heavy nucleus, it is called nuclear fusion.

Example.

$$^{1}_{1}H + ^{1}_{1}H \longrightarrow {}_{1}H^{2} + e^{+} + \upsilon + 0.42 \text{ mev}$$

It is achieved at a temperature of around 10^{7} K.

- When we increase the temperature, the nucleons get enough kinetic energy to overcome the coulomb repulsive force and this process is called **"thermonuclear fusion"**.
- In stars interior and sun the temperature is of order of 10^{7} K which is suitable for thermonuclear fusion.
- Nuclear fusion reaction on sun is called **"proton-proton cycle"** which takes place in multiple steps.

$^{1}_{1}H + ^{1}_{1}H \rightarrow ^{2}_{1}H + e^{+} + \upsilon + 0.42\text{MeV}$ .... (1)

$e^{+} + e^{-} \rightarrow \gamma + \gamma + 1.02 \text{ MeV}$ .... (2)

$^{2}_{1}H + ^{1}_{1}H \rightarrow ^{3}_{2}He + \gamma + 5.49 \text{ MeV}$ .... (3)

$^{3}_{2}He + ^{3}_{2}He \longrightarrow ^{4}_{2}He + ^{1}_{1}H + ^{1}_{1}H + 12.86 \text{ Mev}$ .... (4)

For 4th step to take place the first three steps occur twice, thus the total energy released when four hydrogen atoms combine to form one $^{4}_{2}He$ atom is 26.7 Mev.

Overall reaction is presented as –

$$4^{1}_{1}H + 2e^{-} \longrightarrow ^{4}_{2}He + 2\upsilon + 6\gamma + 26.7 \text{ MeV}$$

- When Helium nucleus is synthesized, the hydrogen gets depleted and core starts cooling and star begins to collapse due to its own gravity which again increases the temperature of core. Now when temperature increases, helium nucleus starts fusing and form carbon nucleus.

Controlled thermonuclear fusion

- Many countries are trying to replicate thermonuclear fusion process on the earth.
- It requires a temperature of order 10^{8} K and fuel mixture composed of positive ions and electrons which is called plasma.

Unit - IX
Electronic Devices

Semiconductor Electronics: Materials, Devices and Simple Circuits

- Solids are classified into three categories based on their resistivity value -

(i) Conductors - Those solids which have very low resistivity which lies between 10^{-2} - 10^{-8} Ωm.

(ii) Insulators - Those solids which have very high resistivity & it lies between 10^{11} - 10^{19} Ωm.

(iii) Semiconductors - These solids have resistivity between 10^{-5} to 10^{6} Ω m which is in between of conductors and insulators.

Energy bands of solids

(i) Valence band- It is the region where valence electrons are present. This band is either partially filled or fully filled.

(ii) Conduction band - It is that region where electrons can move freely and contribute to electric current. This band is either partially filled or empty.

(iii) Forbidden energy gap or Forbidden band-(E_g)

The energy gap between valence band and conduction band is called forbidden gap.

To shift electrons from valence band to conduction band; energy needs to be supplied. General representation of valence band, conduction band and forbidden energy gap -

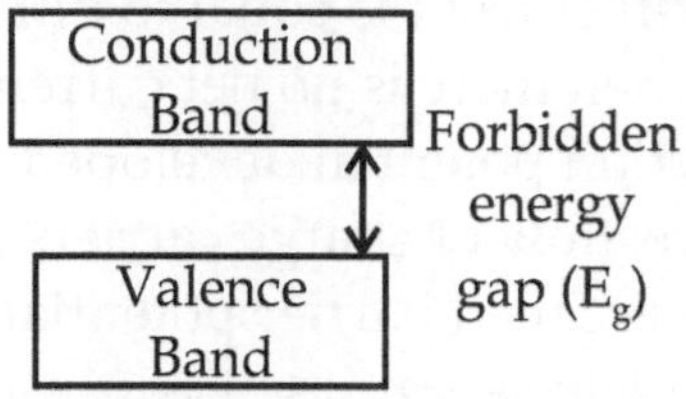

Note - (1) In metals either conduction band and valence band overlap each other, or they have very less energy gap. Here conduction band always have some electrons.

(2) In insulators, valence band is filled but conduction band is empty also they have very high forbidden energy gap.

$E_g > 3eV$

(3) In semiconductors forbidden energy gap is lesser than 3 eV. At absolute zero temperature that means, zero Kelvin, semiconductors act like insulators. When temperature is increased, some electrons in valence band acquire thermal energy and jump to conduction band.

Intrinsic semiconductors

A pure semiconductor which is free from any impurity is called intrinsic semiconductor.

Examples- Group-14 elements such as Germanium (Ge) and silicon (si) are important semiconductors.

We take n_e = number density of electrons and n_h = number density of holes

Note - When heat is absorbed by electrons, the covalent bond is broken and a hole is formed, which is considered equivalent to a positive charge in semiconductors.

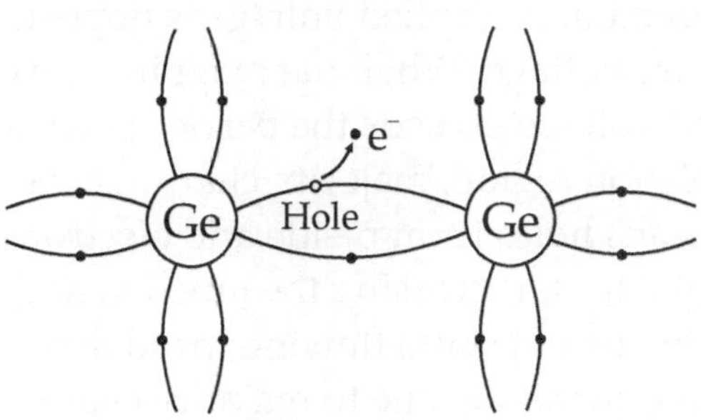

When the covalent bond is broken, the electron is shifted to conduction band and a hole is formed in the valence band.

In intrinsic semiconductors the number density of electron n_e is equal to number density of holes n_h.

$n_e = n_h = n_i$

Here n_i is called intrinsic charge carrier concentration.

At equilibrium - $n_e n_h = n_i^2$

There are two currents flowing in semiconductors when external electric field is applied, electron current I_e and hole current I_h. And total current is given by $I = I_e + I_h$

Extrinsic Semiconductors

- We add some **impurity** of either **group-13 or group-15 elements** to intrinsic semiconductors **to increase their conductivity**, such impure semiconductors are called **extrinsic** or **doped semiconductors**. Based on the nature of dopant (impurity), the extrinsic semiconductors are of the types -

(i) n-type semiconductors –

When a **pentavalent impurity** (group-15 elements) is added to pure semiconductor, then one electron per impurity atom is free to conduct electricity. Since pentavalent atoms are donating electrons, they are called **donor atoms.**

Here electrons are the majority charge carriers and holes are the minority charge carriers. In n-type semiconductors, an **donor energy level (E_D)** is created just below the conduction band

$n_e \gg n_h$

i.e., $I_e \gg I_h$

Please remember you are adding neutral pentavalent atoms to neutral intrinsic semiconductors which means extrinsic semiconductors are also neutral.

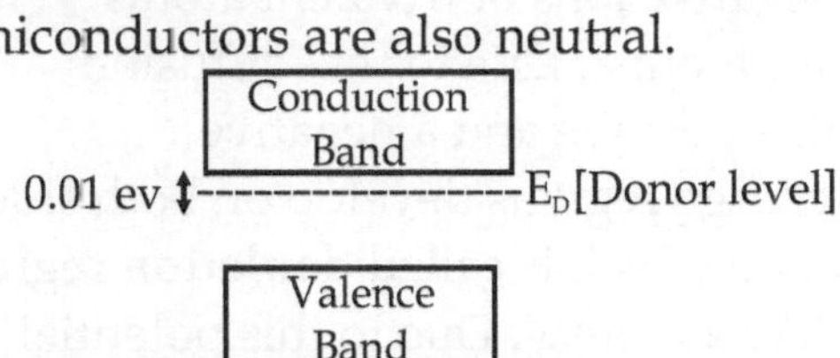

The formula $n_e \times n_h = n_i^2$ is applicable here as well.

(ii) p-type semiconductors - When a **trivalent impurity (group-13 elements)** is added to pure semiconductor (group-14 elements), then one hole per impurity atom is introduced in the lattice of pure semiconductor. Since trivalent atoms are introducing holes which act like electron acceptors, these are called **acceptor atoms**. Here holes are majority charge carriers and $n_h \gg n_e$ that means $I_h \gg I_e$

- p-type semiconductors are also electrically neutral and not positively charged.
- In p-type semiconductors, an **acceptor energy level (E_A)** is created just above the valence band and with very small supply of energy, the electron can jump from valence band to acceptor energy level.

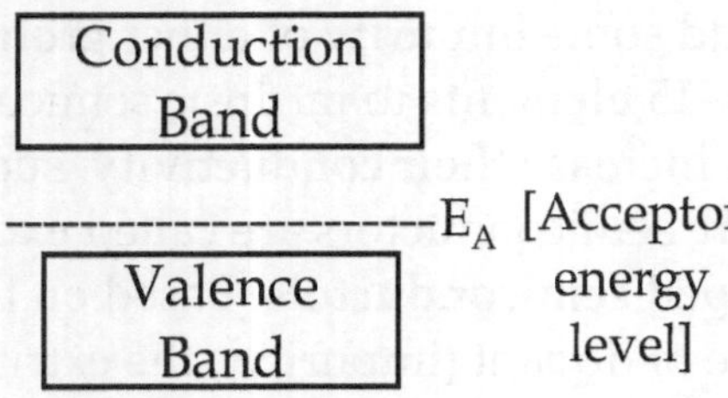

$$n_e \times n_h = n_i^2$$

p-n Junction

- p-n junction or a junction diode is formed when in the single wafer of pure semiconductor, donor impurity is introduced into one side and acceptor impurity is introduced into other side.
- P-side of the wafer is rich in holes and n-side of the wafer is rich in electrons, thus due to concentration gradient they move to opposite sides and a diffusion current starts flowing across the junction.
- When electrons from n-side diffuse to p-side, they leave positive ions of pentavalent atoms (donor atoms) and when holes move from p-side to n-side, they leave negative ions of trivalent atoms (acceptor atoms). Due to this diffusion current, a positive and a negative space-charge regions develop on both sides of junction which is called **depletion region**. It acts like a battery. Due to this potential region electrons from p-side drift towards n-side and holes from n-side drift towards p-side and a drift current gets established.

Initially diffusion current magnitude is more but due to increase in junction potential, with time drift current magnitude increases and after some time an equilibrium is established when drift current is equal to diffusion current then there is no net current.

- The junction potential developed which resists the flow of charge carriers across the junction is called barrier potential.

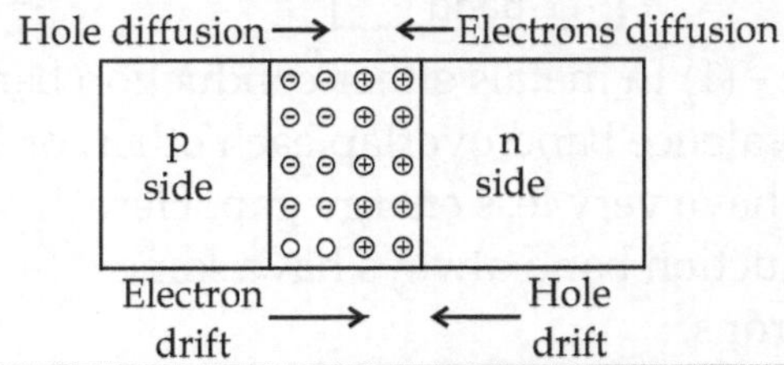

Semiconductor Diode

A semiconductor diode is a p-n junction with metallic contacts at the ends for the application of an external voltage.

It is represented as- 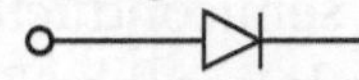

The direction of arrow is in the direction of conventional current when diode is forward biased.

I-V characteristics - These are graphical representations of variation of current flowing through the p-n junction with the applied voltage.

p-n junction diode under forward biasing

- When the **positive terminal** of external battery is connected **to p-side** and **negative terminal** to external battery is connected to **n-side** of the junction diode, the diode is said to be forward biased.

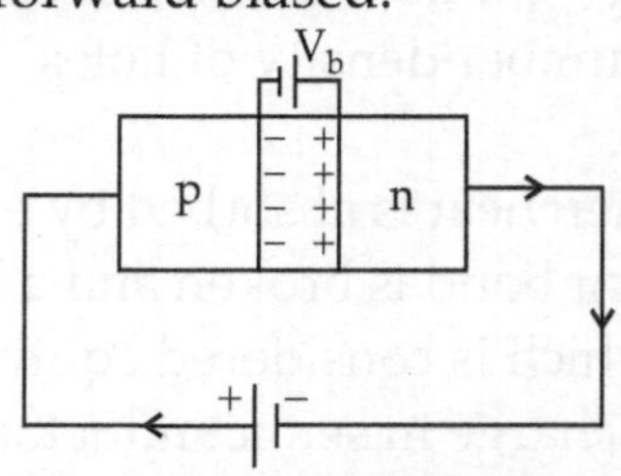

- The direction of applied voltage is opposite to barrier voltage. When the magnitude of applied voltage exceeds the barrier potential of depletion region, majority charge carriers, that means holes from p-side and electrons from n-side start crossing the junction and forward current starts flowing through it. Since the current is due to majority charge carriers, the magnitude of current is usually in milliampere (mA).

- In forward biasing the width of depletion layer decreases.

V_0 = Barrier potential
1 = Barrier potential without battery
2 = Barrier potential with low battery voltage
3 = Barrier potential with high battery voltage

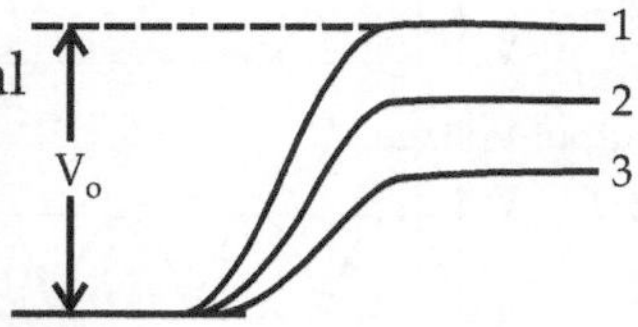

- Barrier potential in Ge is 0.3 V and in Si it is 0.7 V. From the graph we see that small current flows, till the applied forward biased voltage becomes greater than the potential barrier voltage. (Represented as OA).

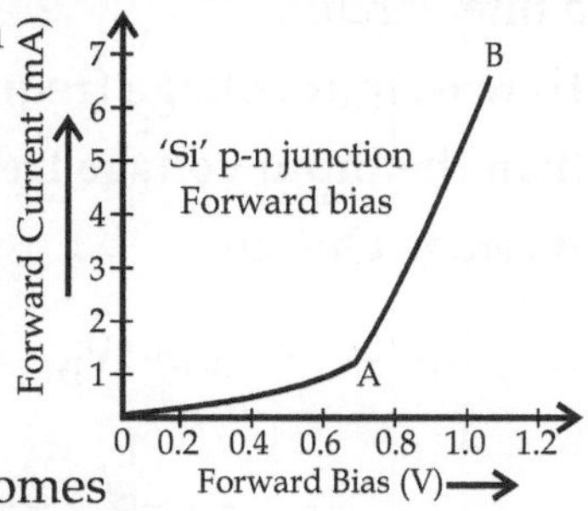

- When external voltage becomes greater than the barrier potential, the forward current increases linearly. At point 'A' external voltage becomes greater than barrier potential and is called **"knee voltage"**.
- In forward biasing the resistance offered by p-n Junction is of low value.

Reverse Biasing

- When the **positive terminal** of external battery is connected to **n-side** of junction diode and **negative terminal** is connected to **p-side** of junction diode, it is called reverse biasing.
- In this case holes from p-section get attracted towards the negative terminal of battery and electrons get attracted towards positive terminal of battery thus in both sides majority charge carriers move away from the junction.
- A very low magnitude current **micro-ampere (μA)** order flows through the junction due to minority charge carriers, it is called **reverse current**.
- If the voltage in reverse biasing is very high, all the covalent bonds near the junction break and a large number of electron hole pairs are liberated and reverse current increases abruptly to high value.
- The maximum reverse biasing voltage, which a diode can tolerate without breakdown is called **"reverse breakdown voltage"** or **"Zener voltage"**.
- In reverse biasing, external battery voltage and barrier potential are connected in same polarity due to which the thickness of depletion layer increases. In this case junction diode offers a high value of resistance.

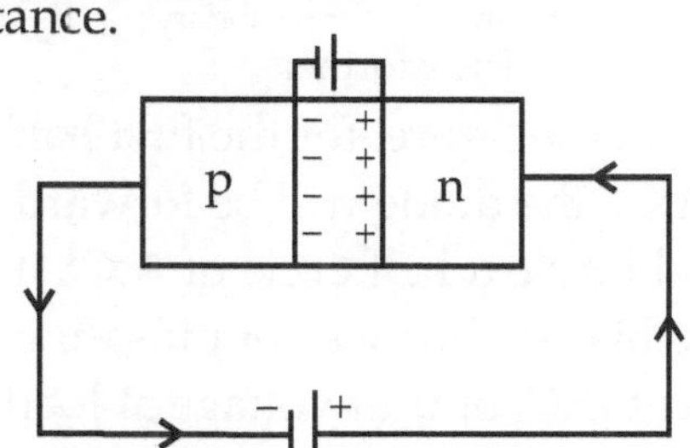

V_0 = Barrier potential under reverse biasing
1 = without external voltage
2 = with external voltage

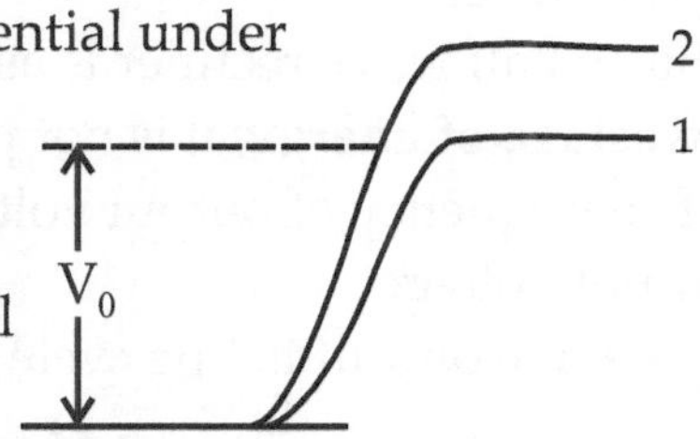

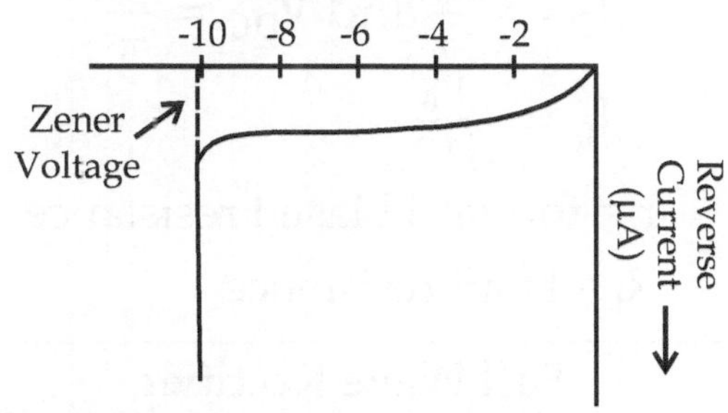

Because of increased thickness of the depletion layer, the p-n junction offers high resistance.

Dynamic resistance (r_d)

It is the ratio of small change in voltage to the small change in current produced in diode.

$$r_d = \frac{\Delta V}{\Delta I}$$

Note - In any circuit, when diode is forward biased, it allows current through it that means circuit is closed. When it is reversed biased, it does not allow the current through it that means it's acting like an open circuit.

Half Wave Rectifier

A Rectifier is a device which converts applied A.C. voltage to D.C. voltage.

Principle - It is based on the principle that when diode is forward biased, it allows current to pass through it and in reverse biased, it does not allow current to pass through it.

Half wave Rectifier

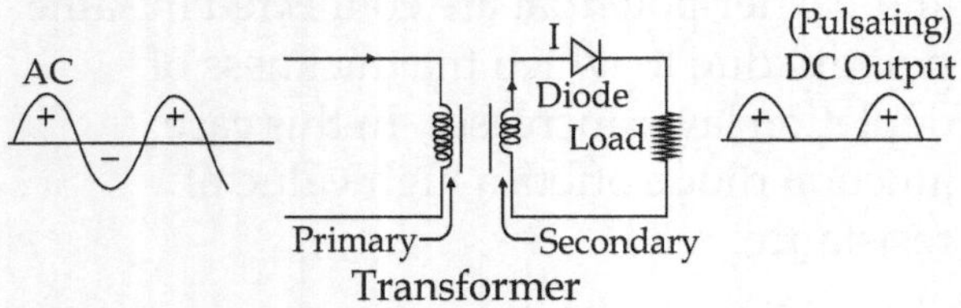

As shown in the figure, for the first half cycle of A.C., the diode will be forward biased and for next half cycle of A.C., it will be reverse biased thus we get pulsating, intermittent D.C. output voltage at load resistance.

Since half wave rectifier is associated with **wastage of energy**, it is not preferred to use.

The frequency of output voltage is equal to input voltage.

Average output in one cycle

$$I_{DC} = \frac{I_0}{\pi} \text{ and } V_{DC} = \frac{V_0}{\pi}$$

$$I_0 = \frac{V_0}{r_f + R_L}$$

r_f = forward biased resistance

R_L = Load resistance

Full Wave Rectifier

- "**A rectifier which rectifies both halves of A.C. input cycle is called a full wave rectifier**". Here we connect two diodes. As shown in the figure.

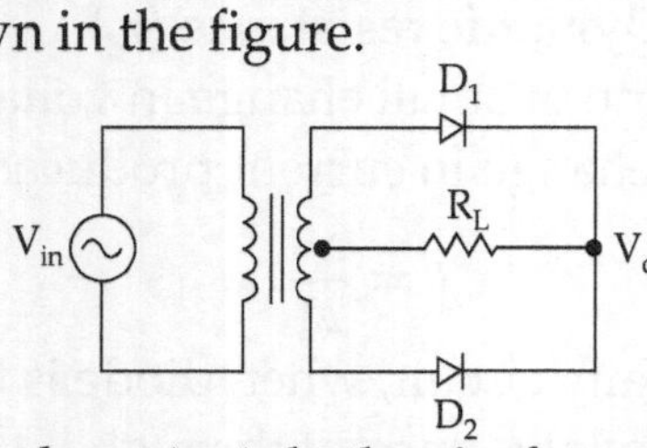

- Based on the principle that **in forward biasing diode offers low resistance and in reverse biasing it offers high resistance.**
- As shown in the figure, for positive half cycle, D_1 is forward biased and D_2 is reverse biased and for negative half cycle, D_2 is forward biased and D_1 is reversed biased. Thus we get the output which is continuous but pulsating.

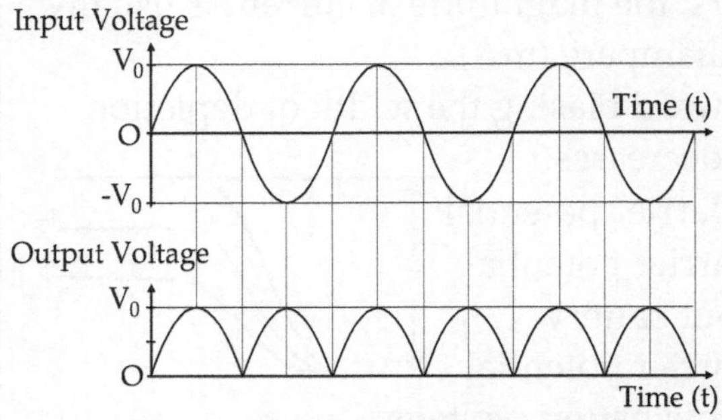

To make the output smooth, we use a filter circuit.

Here output voltage frequency is double than the input voltage frequency.

Average Output -

$$I_{DC} = \frac{2I_0}{\pi} \text{ and } V_{DC} = \frac{2V_0}{\pi}$$

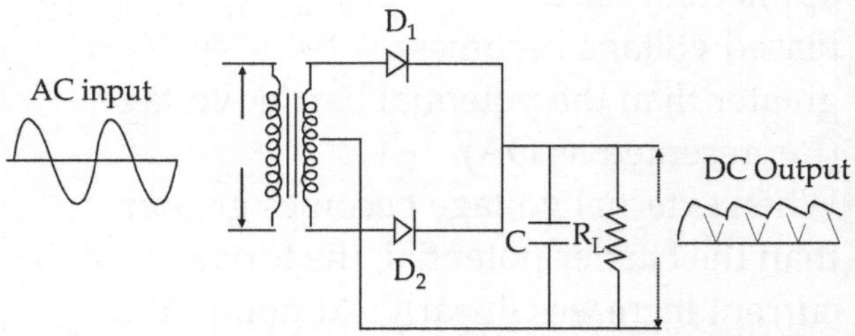

Full wave rectifier with capacitor filter

When the voltage across capacitor for increases, it gets charged to the peak value of voltage.

If there is no external load, it remains charged.

When there is a load, it gets discharged through the load and voltage across it starts falling. In the next half cycle it again gets charged and supply almost constant voltage to the Load.

Note - The rate of fall of voltage across the capacitor is inversely proportional to the product of capacitance (C) and load resistance (R_L).

This product is called time constant (τ).

τ = RC.

Derivations and Proofs

Chapter - 1

Electrostatics

1. Electric Field Intensity due to a point charge

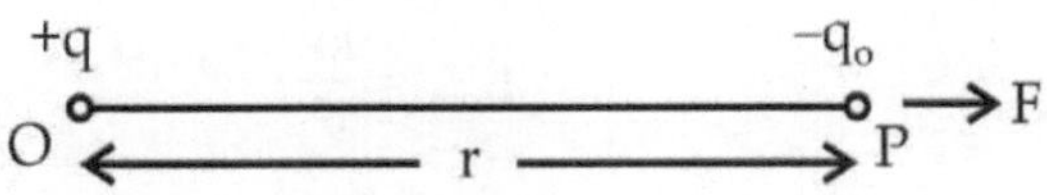

An isolated point-charge is kept at point **O**, in a medium whose dielectric constant (relative permittivity) is **K**. We need to determine the electric field intensity at **P** at distance **r** from charge **+q**. We assume there is a test charge $\mathbf{q_o}$ at **P**. Then according to coulomb's law, the electric force on $\mathbf{q_o}$ will be

$$|F| = \frac{1}{4\pi\varepsilon_0 K}\frac{qq_0}{r^2}\ N$$

And electric field intensity at point **P** will be

$$E = \frac{F}{q_o} = \frac{1}{4\pi\varepsilon_0 K}\frac{q}{r^2}\ \frac{\text{Newton}}{\text{coulomb}}$$

If charge is placed in vacuum (or air) i.e., **K = 1**, then

$$E = \frac{1}{4\pi\varepsilon_0}\frac{q}{r^2} \quad \text{or} \quad E = \frac{kq}{r^2}$$

Here $\mathbf{k} = \frac{1}{4\pi\varepsilon_0}$ Coulomb's force constant

In Vector form, $\vec{E} = \frac{1}{4\pi\varepsilon_0}\frac{q}{r^2}\hat{r}$

Direction of $\vec{E}$ is along $\overrightarrow{OP}$

2. Electric field Intensity at a point on the Axis of a Dipole

An electric dipole **AB** is made of two equal & opposite charges kept **2a** distance apart. From its centre **O** at distance **r** at a point **P**, we want to determine electric field intensity.

Now from **+q** charge, point **P** is at a distance **(r – a)** and from **–q** charge, point **P** is at a distance of **(r + a)**.

$$\therefore E_1 = \frac{kq}{(r+a)^2}\ \text{(towards dipole due to } -q\text{)}$$

$$\text{and } E_2 = \frac{kq}{(r-a)^2}\ \text{(Away from the dipole due to } +q\text{)}$$

$$\because |E_2| > |E_1| \therefore E_{net} = E_2 - E_1$$

$$E_{net} = \frac{kq}{(r-a)^2} - \frac{kq}{(r+a)^2}$$

$$\text{cr } E_{net} = kq\left[\frac{(r+a)^2-(r-a)^2}{(r^2-a^2)^2}\right]$$

Solving this, we get

$$E_{net} = \frac{kq\times 2a\times 2r}{(r^2-a^2)^2}$$

$$\because p = q \times 2a \text{ (Electric dipole moment)}$$

$$\therefore E_{net} = \frac{2kpr}{(r^2-a^2)^2}$$

If $r \gg a$ then $r^2 - a^2 \approx r^2$

$$\therefore E_{net} = \frac{2kpr}{r^4}$$

$$\text{Or } \vec{E}_{net} = \frac{2k\vec{p}}{r^3}$$

Please note both 'E' and 'p' are in the same direction.

3. Electric field Intensity at a point on Equatorial Line of electric Dipole

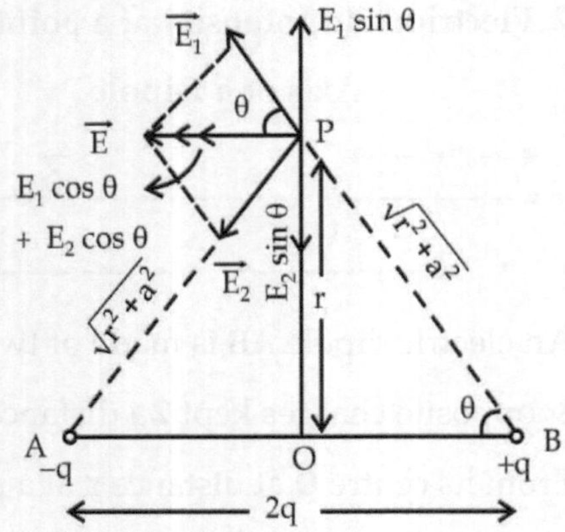

An electric dipole **AB** is made of two equal and opposite charges kept at distance **2a**. We want to determine electric field intensity at point **P** lying on perpendicular bisector of the dipole at a distance **r** from the centre of dipole.

From figure -

$$E_1 = \frac{kq}{(r^2+a^2)} \quad ... (1) \quad \text{(away from +q)}$$

$$\text{And } E_2 = \frac{kq}{(r^2+a^2)} ... (2) \text{ (towards -q)}$$

Now $|\mathbf{E_1}| = |\mathbf{E_2}|$ but directions are different.

We resolve $\mathbf{E_1}$ & $\mathbf{E_2}$ into two perpendicular components.

$\mathbf{E_1 sin\theta}$ and $\mathbf{E_2 sin\theta}$ are equal and opposite hence cancel each other,

but $\mathbf{E_1 cos\theta}$ and $\mathbf{E_2 cos\theta}$ are in same direction, hence get added.

$$\therefore E_{net} = E_1\cos\theta + E_2\cos\theta \quad ...(3)$$

$$E_{net} = \frac{kq}{(r^2+a^2)}\cos\theta + \frac{kq}{(r^2+a^2)}\cos\theta$$

$$\text{Or } E_{net} = \frac{2kq}{(r^2+a^2)}\cos\theta \quad ... (4)$$

$$\text{From figure } \cos\theta = \frac{a}{\sqrt{r^2+a^2}}$$

Then from equation (4)

$$E_{net} = \frac{2kq}{(r^2+a^2)}\frac{a}{\sqrt{r^2+a^2}}$$

$$\because p = q \times 2a$$

$$\therefore E_{net} = \frac{kp}{(r^2+a^2)^{3/2}}$$

If $a \ll r$ then $r^2 + a^2 \approx r^2$

$$\therefore E_{net} = \frac{kP}{r^3}$$

$\vec{P}$ and $\vec{E}_{net}$ are in opposite directions.

Hence

$$\vec{\mathbf{E}} = -\frac{\mathbf{k\vec{P}}}{\mathbf{r^3}}$$

$$\text{Also } \vec{E}_{axial} = -2\vec{E}_{equatorial}$$

4. Torque on a dipole in uniform electric field

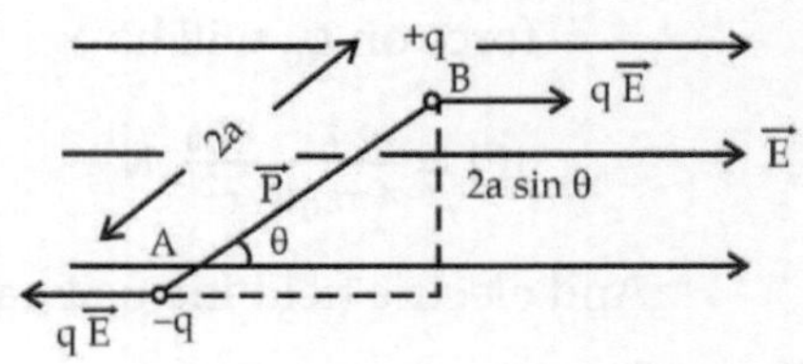

An electric dipole **AB** placed in uniform electric field $\vec{\mathbf{E}}$ and the dipole moment vector is making an angle $\boldsymbol{\theta}$ with the direction of uniform electric field $\vec{\mathbf{E}}$. Due to external electric field $\vec{\mathbf{E}}$, charge $+\mathbf{q}$ experiences force $\mathbf{qE}$ in the direction of electric field & charge $-\mathbf{q}$ experiences force in the opposite direction of electric field.

Two equal and anti-parallel forces acting on a body form a couple and cause dipole moment (Torque), which is given by

$$\tau = \text{ Force } \times \perp \text{ distance b/w two forces}$$

$$\tau = qE \times 2a\sin\theta$$

$$\because p = q \times 2a$$

$$\therefore \tau = pE \sin \theta$$

In Vector form, $\vec{\tau} = \vec{p} \times \vec{E}$

Or $\boldsymbol{\tau} = \mathbf{PE \sin \theta}$

5. Electric field due to an Infinite Line charge

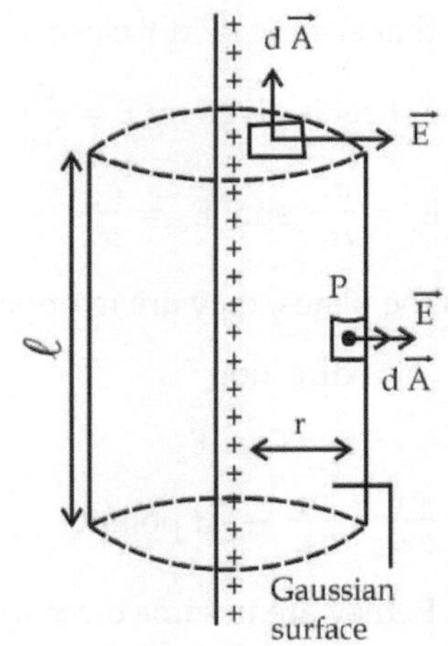

There is a uniformly, positively charged, infinite length wire, having linear charge density **λ** coulomb/metre. We want to find electric field intensity at point **P** at a distance **r** from the wire. We draw a Gaussian surface, cylindrical in shape keeping wire at the axis of the cylinder having radius **r**. Small area element vector $\overrightarrow{\mathbf{dA}}$ makes angle **0°** on the curved surface with electric field but **90°** on the caps of the cylinder.

∴ For curved surface -

Electric flux $d\phi = \vec{E} \cdot d\vec{A} = EdA\cos\ 0° = EdA$ ... (1)

Now flux through entire Gaussian surface

$$\phi = \oint \vec{E} \cdot d\vec{A} = E \times 2\pi r\ell \quad \text{... (2)}$$

Also we know that $\phi = \frac{q}{\varepsilon_0}$

Since linear charge density $\lambda = \frac{q}{\ell}$

$$\therefore q = \lambda\ell \ \&\ \phi = \frac{\lambda\ell}{\varepsilon_0} \quad \text{... (3)}$$

From equations (2) and (3)

$$E \times 2\pi r\ell = \frac{\lambda\ell}{\varepsilon_0}$$

$$\text{or } E = \frac{\lambda}{2\pi\varepsilon_0 r} = \frac{2k\lambda}{r}$$

$$\mathbf{\vec{E} = \frac{\lambda}{2\pi\varepsilon_0 r}\hat{r}}$$

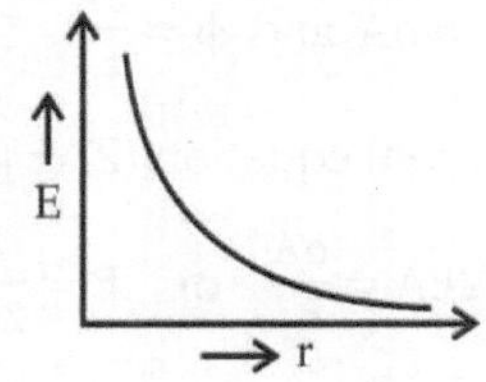

6. Electric field due to a plane Sheet of charge [∞ Area]

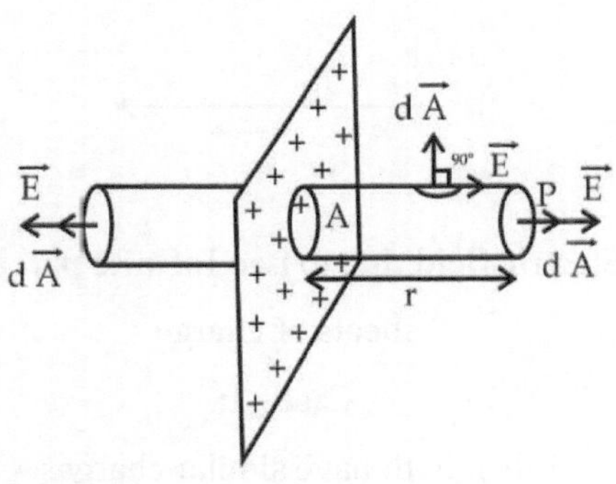

We consider a very thin, infinite in area, a positively charged sheet of surface charge density **σ**. Due to this sheet at point **P,** at a distance **r,** we want to find electric field intensity **E**. We consider a Gaussian surface, cylindrical in shape cutting through the sheet. We consider area **A** for circular caps. As shown in the figure, the area element $\overrightarrow{\mathbf{dA}}$ on the curved surface makes an angle **90°** with the direction of electric field $\vec{\mathbf{E}}$ hence no flux will pass through this.

$\therefore$ Flux through both the caps will be

$$\phi = \oint \vec{E} \cdot d\vec{A} + \oint \vec{E} \cdot d\vec{A} \quad \dots (1)$$

$$\because \phi = \oint E\, dA \cos 0^\circ + \oint E\, dA \cos 0^\circ$$

$$\text{Or } \phi = E \oint dA + E \oint dA$$

$$\phi = EA + EA = 2EA \dots (2)$$

By Gauss theorem $\phi = q/\varepsilon_0$ where q is the charge enclosed by Gaussian cylinder.

$$\because \text{Surface charge density } \sigma = \frac{q}{A}$$

$$\therefore q = \sigma A \text{ and } \phi = \frac{\sigma A}{\varepsilon_0} \quad \dots (3)$$

From equation (2) & (3)

$$2EA = \frac{\sigma A}{\varepsilon_0} \quad \text{or} \quad \mathbf{E} = \frac{\sigma}{2\varepsilon_0}$$

Note: E is independent of **r**.

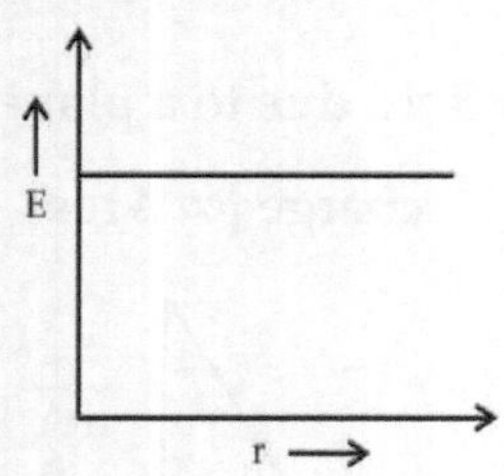

7. Electric field due to two infinite parallel sheets of charge

Case - 1:

When both have similar charges – Two infinite plane sheets, positively charged placed at distance **d.**

Sheet - 1 has surface charge density $\boldsymbol{\sigma_1}$

and

Sheet - 2 has surface charge density $\boldsymbol{\sigma_2}$.

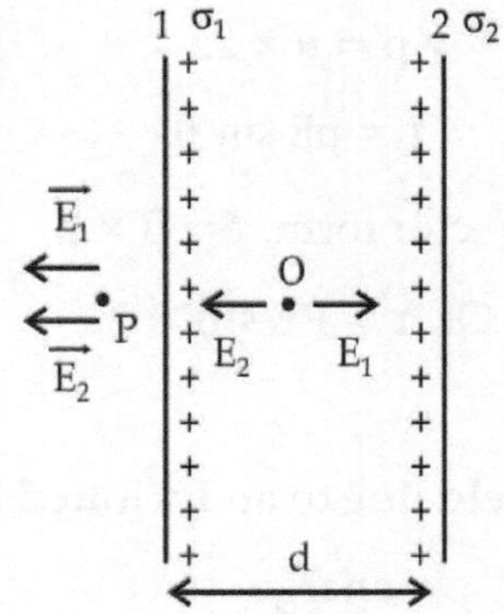

We know that electric field **E** close to an infinite sheet is given by $\mathbf{E} = \frac{\sigma}{2\varepsilon_0}$

$$\therefore E_1 = \frac{\sigma_1}{2\varepsilon_0} \text{ and } E_2 = \frac{\sigma_2}{2\varepsilon_0}$$

In between the plates, they are in opposite direction

$$\therefore E = E_1 - E_2$$

$$E = \frac{\sigma_1}{2\varepsilon_0} - \frac{\sigma_2}{2\varepsilon_0} \quad \text{– At point O}$$

At point **P**, they are in same direction

$$\therefore E = \frac{\sigma_1}{2\varepsilon_0} + \frac{\sigma_2}{2\varepsilon_0}$$

If $\sigma_1 = \sigma_2 = \sigma$ then at Point **P**

$$\mathbf{E} = \frac{\sigma}{\varepsilon_0}$$

Case - 2:

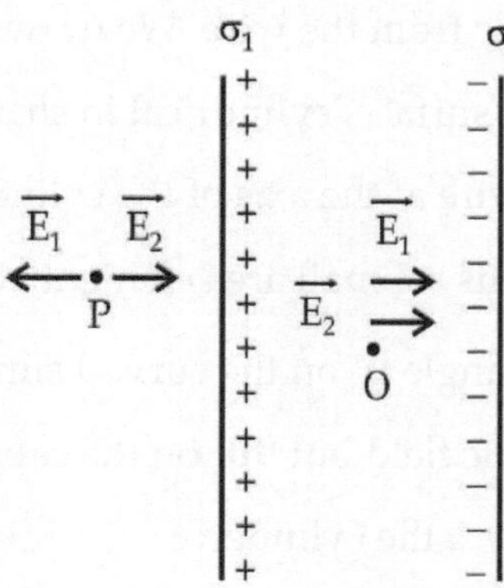

When both the plates are oppositely charged -

At **P**- E_1 & E_2 are in opposite directions.

$$\therefore E = \frac{\sigma_1}{2\varepsilon_0} - \frac{\sigma_2}{2\varepsilon_0}$$

At O – E_1 & E_2 are in same direction.

Hence

$$\mathbf{E = \frac{\sigma_1}{2\varepsilon_0} + \frac{\sigma_2}{2\varepsilon_0}}$$

8. Electric field intensity due to a uniformly charged spherical shell

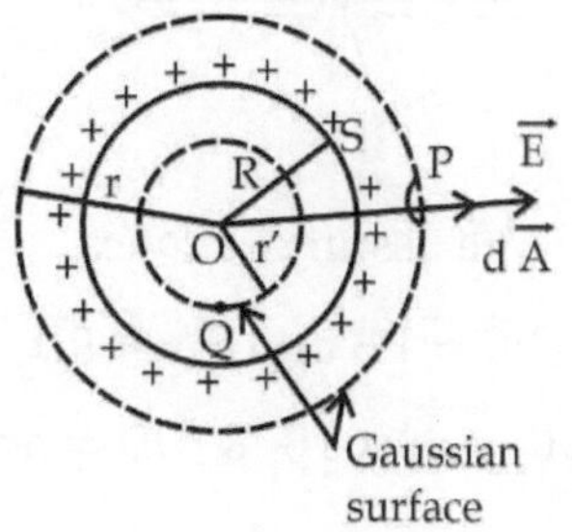

We consider a uniformly charged shell of radius **R** carrying charge **+q** which is uniformly distributed. We need to determine electric field intensity at point **P** (outside), **Q** (inside) & **S** on the surface of shell.

Case – I - Outside the shell

We consider a spherical Gaussian surface enclosing the point **P** (lying on the surface) and makes a sphere of radius **r**. Area element ($\mathbf{d\vec{A}}$) and $\vec{E}$ are in the same direction ($\boldsymbol{\theta = 0°}$).

∴ According to Gauss' theorem

$$\phi = \oint \vec{E} \cdot d\vec{A} = \frac{q}{\varepsilon_0} \quad \dots (1)$$

$$\text{Or } E\oint dA = E \times 4\pi r^2 = \frac{q}{\varepsilon_0}$$

$$\text{or } E = \frac{q}{4\pi\varepsilon_0 r^2} \quad \text{or} \quad \mathbf{E = \frac{kq}{r^2}} \quad \text{Here } r > R$$

Case - II - On the surface (At point **s**)

$$r = R \therefore E = \frac{kq}{R^2} \text{ (max.)}$$

Here Gaussian surface coincides the surface of shell.

$$\text{Surface charge density } (\sigma) = \frac{q}{4\pi R^2}$$

$$\text{then } q = 4\pi R^2 \cdot \sigma$$

$$\therefore E = \frac{\mathbf{4\pi R^2 \sigma}}{\mathbf{4\pi\varepsilon_0 R^2}} = \frac{\boldsymbol{\sigma}}{\boldsymbol{\varepsilon_0}}$$

Case - III- Inside the shell (At point Q) -

The Gaussian surface does not enclose any charge for **r < R**.

$$\therefore \phi = \oint \vec{E} \cdot d\vec{A} = \frac{q}{\varepsilon_0} = 0$$

Or **E = 0**

Variation of E with distance from the centre of a charged spherical shell.

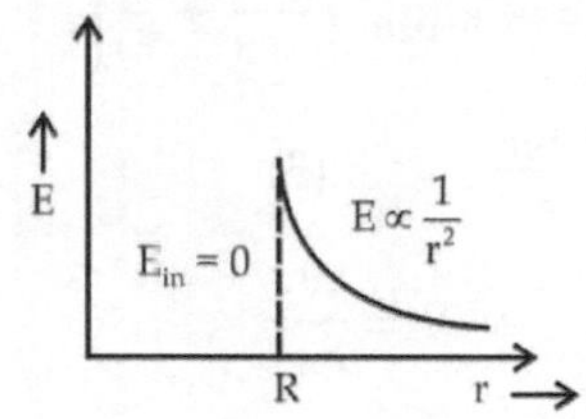

Chapter - 2

Electric Potential and Capacitance

9. Electrostatic potential due to a point charge

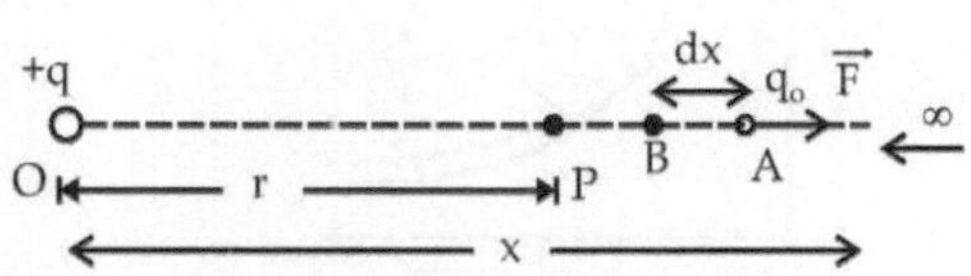

We consider a point charge of **+q** coulomb is placed at point **O**.

We want to determine electric potential due to this charge at point **P**.

There is a test charge $\mathbf{q_o}$ brought from ∞ to point **A** which is at distance **x** from **+q**. The force experienced by $\mathbf{q_o}$ due to charge

+q is $F = \frac{kqq_0}{x^2}$ .....(1) (in air or vacuum)

Please note $\vec{F}$ is away from **O**. Now if q_o is moved by small distance **dx** towards **O** then the work done by electric field can be written as $dW = \vec{F} \cdot d\vec{x} = Fdx \cos 180°$

$$\therefore dW = -Fdx \quad .... (2)$$

Total work done in moving q_o from ∞ to point **P** will be.

$$W = \int_{\infty}^{r} -F\,dx = \int_{\infty}^{r} -\frac{kqq_0}{x^2}dx$$

$$\text{or } W = -kqq_0 \int_{\infty}^{r} \frac{dx}{x^2} = -kqq_0 \left[-\frac{1}{x}\right]_{\infty}^{r}$$

$$\text{or } W = -kqq_0 \left[-\frac{1}{r} + \frac{1}{\infty}\right]$$

$$\text{Or } W = \frac{kqq_0}{r} \quad .....(3)$$

$$\because V = \frac{W}{q_0}$$

$$\therefore V = \frac{kqq_0}{r} \times \frac{1}{q_0}$$

$$\text{Or } \mathbf{V = \frac{kq}{r}}$$

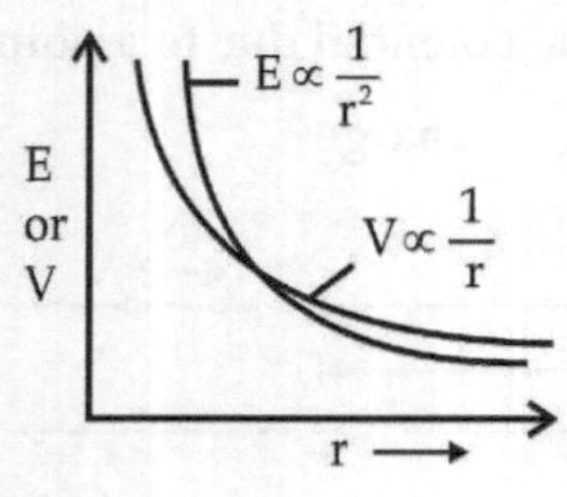

10. Electric potential due to a dipole at any point

An electric dipole **AB** is shown in the figure. Dipole length is **2a**. We want to calculate electric potential at **P**. Which is at a distance **r** from the centre of dipole and the position vector is making angle **θ** with the direction of dipole moment ($\vec{p}$).

Now we drop perpendiculars **BC** & **AD** on the line joining **O** and **P**.

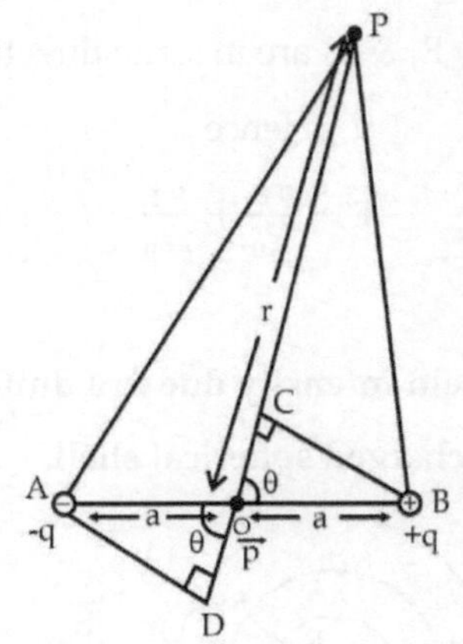

We assume, $\mathbf{r \gg 2a}$

$$\therefore PC \approx PB \text{ and } PA \approx PD$$

$$\text{Now } OC = a\cos\theta \;\&\; OD = a\cos\theta$$

$$\therefore PC = r - a\cos\theta \quad \text{and} \quad PD = r + a\cos\theta$$

$$\text{or } PB = r - a\cos\theta \quad ... (1)$$

$$PA = r + a\cos\theta \quad ... (2)$$

Electric potential due to **+q** charge at **P**

$$V_1 = \frac{kq}{PB} = \frac{kq}{(r - a\cos\theta)} \quad ... (3)$$

Similarly electric potential due to **−q** charge at **P**

$$V_2 = \frac{-kq}{PA} = \frac{-kq}{(r + a\cos\theta)} \quad ... (4)$$

Since electric potential is a scalar quantity, we can calculate the total potential at P as

$$V = V_1 + V_2 = kq\left[\frac{1}{(r - a\cos\theta)} + \frac{1}{r + a\cos\theta}\right]$$

$$\text{or } V = kq\left[\frac{(r + a\cos\theta) - (r - a\cos\theta)}{r^2 - a^2\cos^2\theta}\right]$$

$$\text{or } V = \frac{kq \times 2a\cos\theta}{(r^2 - a^2\cos^2\theta)}$$

$$\therefore \mathbf{V = \frac{kP\cos\theta}{r^2 - a^2\cos^2\theta}} \quad ... (5)$$

$$[\because p = q \times 2a \text{ Dipole moment}]$$

$$\because r \gg a \therefore r^2 - a^2\cos^2\theta \approx r^2$$

Then from equation (5) -

$$\mathbf{V = \frac{kP\cos\theta}{r^2}} \quad ... (6)$$

Case - I - When point **P** lies on the axis i.e.

$$\theta = 0°$$

then from equation -(6)

$$V = \frac{kP}{r^2}$$

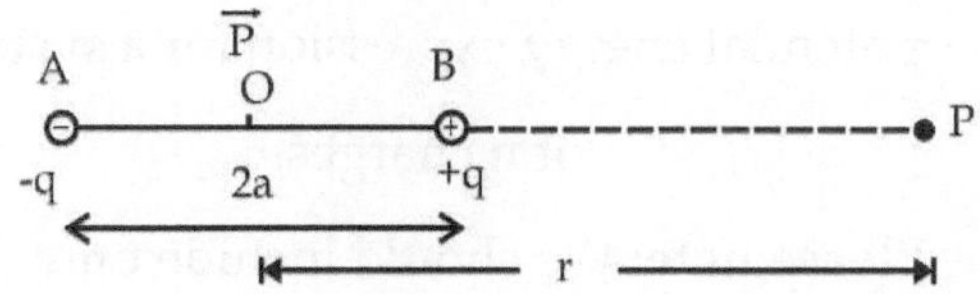

Case - II - When point **P** lies on equatorial line of the dipole, i.e., $\theta = 90°$

then from equation **(6)**

$$V = \frac{kP\cos 90°}{r^2} = 0$$

Note: No work is done in moving the charge along equatorial line of the dipole, because V = 0 along the equatorial line

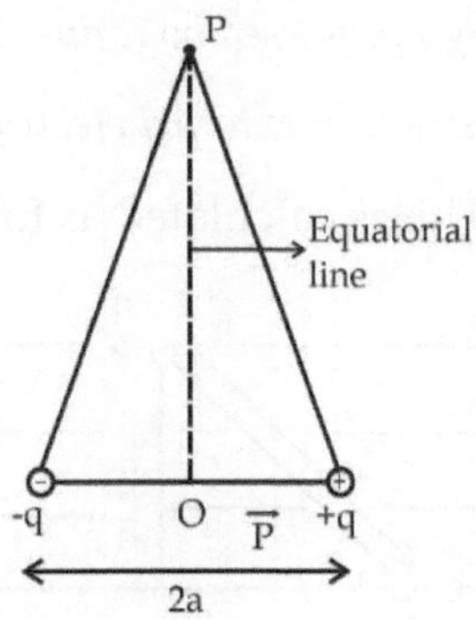

11. Electric potential due to a group of point - charges

Since electric potential is a scalar quantity hence the net potential at any point due to a group of charges is algebraic sum of the potential of each charge.

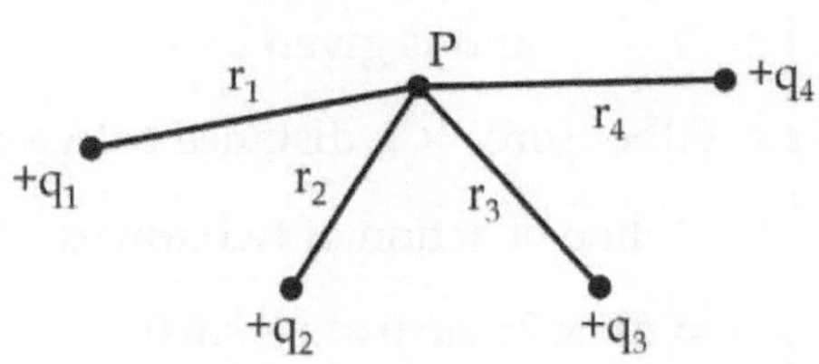

$$V_p = \frac{kq_1}{r_1} + \frac{kq_2}{r_2} + \frac{kq_3}{r_3} + \frac{kq_4}{r_4}$$

If there are **n** point charges, then due to them the total potential at **P** will be -

$$V = k \sum_{i=1}^{i=n} \frac{q_i}{r_i}$$

12. Electrostatic potential energy due to a system of two point charges.

Definition - "*Electrostatic potential energy of a system of point charges is the amount of work done in bringing those charges from ∞ to their respective positions*".

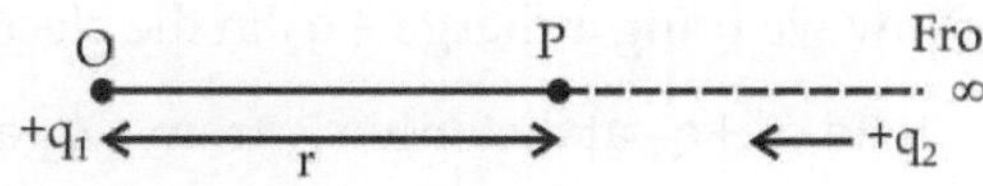

As shown in the figure, a charge $+\mathbf{q_1}$ is placed at **O** and another charge $+\mathbf{q_2}$ is brought from ∞ to a point **P** in the electric field of charge $+\mathbf{q_1}$ and as per the definition, the work done in bringing the charge $+\mathbf{q_2}$ at point **P** is stored as potential energy of the system.

Electric potential due to $+\mathbf{q_1}$ is given by

$$V = \frac{kq_1}{r} \quad \text{... (1)}$$

And work done in bringing the charge $+\mathbf{q_2}$ from ∞ to **P** will be

W = Electric potential at P × Charge q_2

$$\therefore W = V \times q_2 = \frac{kq_1q_2}{r} \quad \text{... (2)}$$

This is stored as the potential energy of the system of two charges

$$\therefore U = W = \frac{kq_1q_2}{r} \quad \text{... (3)}$$

Note- When two charges of same sign are

brought together, they repel each other hence work is done against electrostatic force of repulsion due to which potential energy of the system increases [i.e., positive].

13. Electrostatic potential energy due to a system of n point charges.

When we bring the first charge $+\mathbf{q_1}$ from ∞ to a point in space where there is no electric field, no work is done. Suppose the point where charge is placed is **O**.

$$W_1 = 0 \quad \dots (1)$$

Now we bring a charge $+\mathbf{q_2}$ in the electric field of $+\mathbf{q_1}$ at distance $\mathbf{r_{12}}$ from $+\mathbf{q_1}$ at point **P** hence, the work done will be

$$W_2 = \text{Potential due to } q_1 \text{ at P} \times q_2$$

$$\therefore W_2 = \frac{kq_1}{r_{12}} \times q_2 = \frac{kq_1q_2}{r_1} \quad \dots (2)$$

Now, we bring a charge $+\mathbf{q_3}$ at point **Q** which is at distance $\mathbf{r_{13}}$ from $+\mathbf{q_1}$ and $\mathbf{r_{23}}$ from $+\mathbf{q_2}$.

At **Q** work has to be done against electrostatic forces of both $\mathbf{q_1}$ and $\mathbf{q_2}$.

$$W_3 = \text{Potential at Q due to } q_1 \,\&\, q_2 \times q_3$$

$$\therefore W_3 = \left[\frac{kq_1}{r_{13}} + \frac{kq_2}{r_{23}}\right] \times q_3 \quad \dots (3)$$

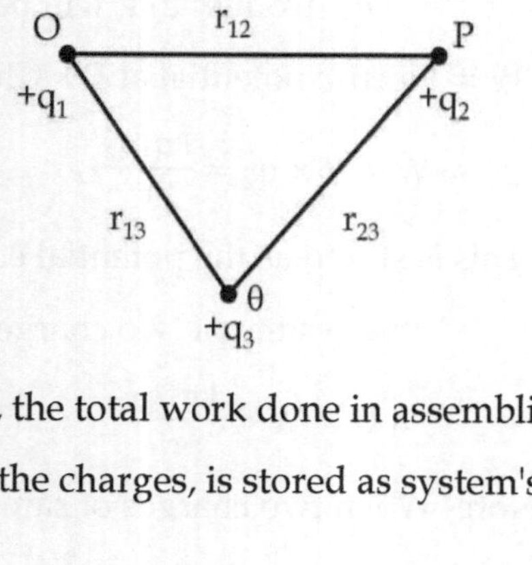

Now, the total work done in assembling all the charges, is stored as system's potential energy.

$$U = W_1 + W_2 + W_3$$

$$U = 0 + \frac{kq_1q_2}{r_{12}} + \frac{kq_1q_3}{r_{13}} + \frac{kq_2q_3}{r_{23}}$$

We can proceed similarly and write potential energy expression for a system of **n** charges,

Please note: We should include only one term for each pair of charges -

$$\therefore \mathbf{U = k \sum_{\substack{\text{all}\\ \text{possible}\\ \text{pairs}}} \frac{q_a q_b}{r_{ab}}}$$

14. Electrostatic potential energy of a dipole in external uniform electric field.

"*Electrostatic potential energy of a dipole is the energy possessed by it due to its peculiar position in external electric field*".

This is calculated as follows.

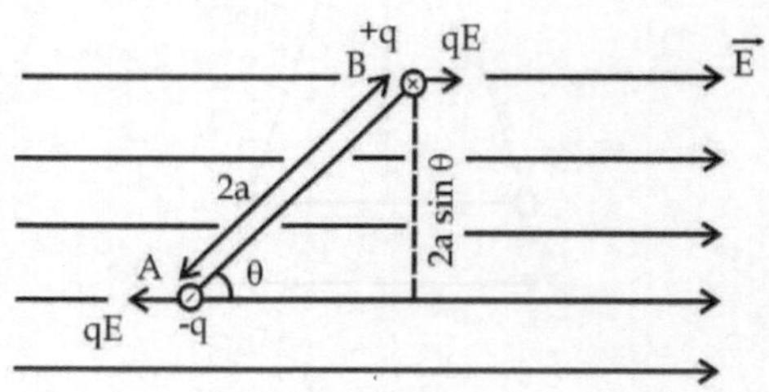

As shown in the figure, a dipole **AB** is placed in external electric field $\vec{\mathbf{E}}$, [uniform], dipole is making an angle $\boldsymbol{\theta}$ with the direction of external electric field. Two equal, and antiparallel forces acting on the dipole cause rotational effect ($\boldsymbol{\tau}$) and is given by -

τ = Either force $\times \perp$ distance between the line of action of two forces

$$\therefore \tau = qE \times 2a \sin\theta = pE \sin\theta \quad \dots (1)$$

This torque tries to align the dipole along

the direction of **E**.

Now we work against this **τ** by rotating the dipole through small angle **dθ**

$$\therefore dW = \tau d\theta$$

$$\text{or } dW = PE \sin\theta\, d\theta \quad \dots (2)$$

Total work done in rotating the dipole through angle θ_1 to θ_2 will be

$$W = \int_{\theta_1}^{\theta_2} PE \sin\theta\, d\theta = PE\,[-\cos\theta]_{\theta_1}^{\theta_2}$$

$$\text{or } \mathbf{W = -PE\,[\cos\theta_2 - \cos\theta_1]} \quad \dots (3)$$

This work done is stored in the form of electrostatic potential energy of the dipole.

Case – I:

When $\theta_1 = 0°$ and $\theta_2 = \theta$

$$\text{then } W = -PE\,[\cos\theta - \cos 0°]$$

$$\mathbf{W = -PE\,[\cos\theta - 1]} \quad \dots (4)$$

Case – II:

When $\theta_1 = 90°$ and $\theta_2 = \theta$, then from equation (3)

$$W = -PE\,[\cos\theta - \cos 90°]$$

$$\mathbf{U = W = -PE\cos\theta = -\vec{P}\cdot\vec{E}} \quad \dots (5)$$

Note - $\boldsymbol{\theta_1 = 0°}$ means P.E. of the system is negative and minimum and $\boldsymbol{\theta = 90°}$ means P.E. of the dipole is zero, and $\boldsymbol{\theta = 180°}$ means P.E. of dipole is positive and maximum.

15. Relation between $\vec{E}$ and V. (Electric Potential Gradient)

Consider a point charge **+q** placed at **O**. A test charge $+\mathbf{q_0}$ is placed at point **P** as shown in the figure.

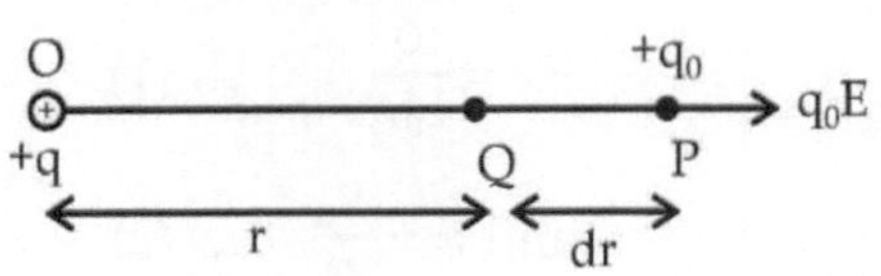

$\mathbf{q_0}$ experiences force due to **+q** charge which is $\mathbf{q_0E}$ away from **+q**. Now if we want to shift $\mathbf{q_0}$ from **P** to **Q** we need to work against $\mathbf{q_0E}$ which is given by

$$dW = \vec{F}.d\vec{r}$$

$$\text{or } dW = F\,dr \cos 180°$$

$$dW = -q_0 E dr$$

$$\text{or } \frac{dW}{q_0} = -Edr \quad \dots (1)$$

We assume potential at **Q** due to **+q** is **V** and at **P** its **V – dV**

$$\therefore \frac{dW}{q_0} = V - (V - dV) = dV \quad \dots (2)$$

from equation (1) and (2)

$$-Edr = dV \text{ or } \mathbf{E = -\frac{dV}{dr}} \quad \dots (3)$$

Here $\frac{dV}{dr}$ is called potential gradient.

Negative sign shows that potential decreases in the direction of $\vec{E}$.

16. Capacitance of an isolated spherical conductor.

Definition – *Capacitance of a conductor is its ability to store charge.*

We consider an isolated spherical conductor having radius **r**. When we give it any arbitrary charge **+q** is will uniformly spread on the outer surface of the conductor whether the sphere is solid or hollow. Now the potential at any point on the surface will be

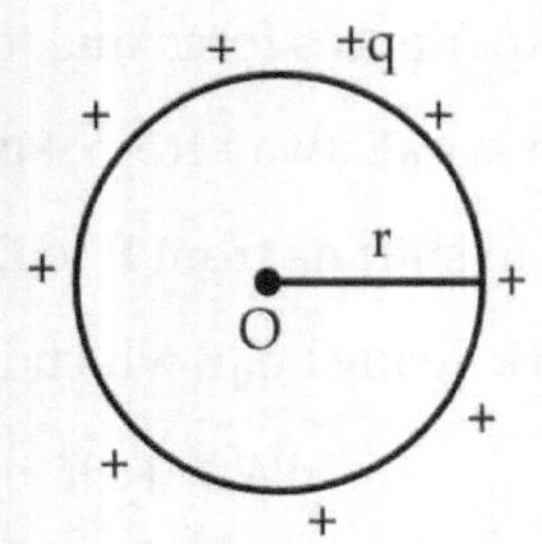

$$V = \frac{kq}{r} \quad \dots (1)$$

Since capacitance (C) of the capacitor is given by -

$$\because C = \frac{q}{V}$$

$\therefore$ From equation (1) $C = \frac{q}{kq} \times r$

Or $C = \frac{r}{k}$ or $4\pi\varepsilon_0 r$.

$$\therefore \mathbf{C = 4\pi\varepsilon_0 r}$$

17. Capacitance of parallel plate capacitor.

Definition – *A capacitor is an arrangement for storing large amount of charge and thus electrical energy is a very small space.*

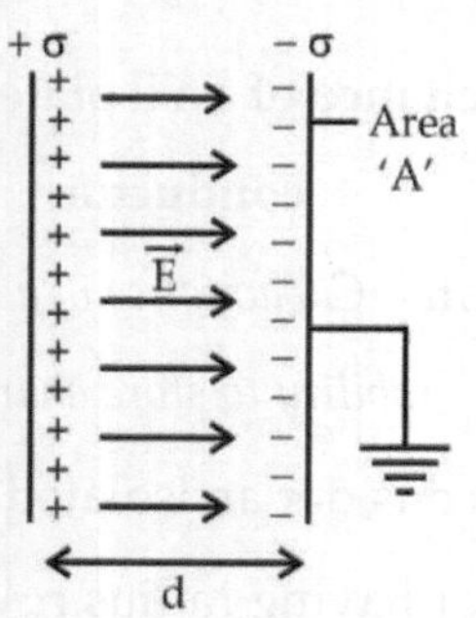

A parallel plate capacitor is composed of two plates of same area **A**, placed at distance **d** apart have surface charge densities **+σ** and **−σ**. Plate with **−σ** is earthed as shown in the figure

$$\because +\sigma = \frac{+Q}{A} \quad \text{and} \quad -\sigma = \frac{-Q}{A}$$

also **E** between the two plates $= \frac{\sigma}{\varepsilon_0}$

$$\text{or } E = \frac{Q}{A\varepsilon_0} \quad \dots (1)$$

Electric field is uniform between the two plates and if the potential difference between the two plates is **V** then

$$V = E \times d = \frac{Q}{A\varepsilon_0} \times d \dots (2)$$

$\because C = \frac{Q}{V}$ substituting **V** value from equation (2) we get

$$C = \frac{Q}{Qd} \times \varepsilon_0 A = \frac{\varepsilon_0 A}{d}$$

$$\therefore \mathbf{C = \frac{\varepsilon_0 A}{d}}$$

Here the medium between the plates is air or there is vacuum.

18. Capacitance of parallel plate capacitor with dielectric slab.

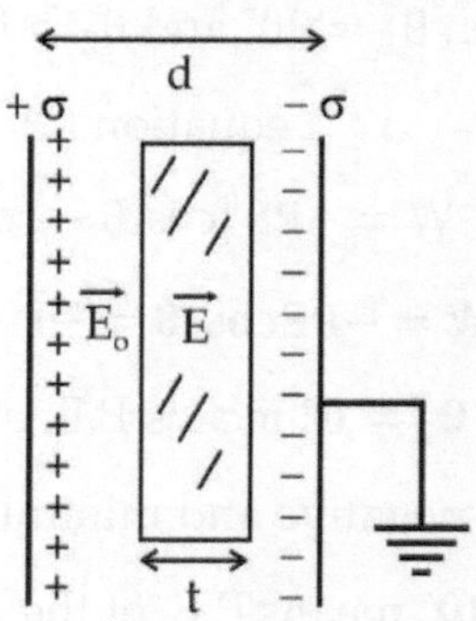

We consider a parallel plate capacitor having surface charge densities **+σ** and **−σ** placed at **d** distance apart.

A dielectric slab is inserted between the plates of thickness **t**. Now **(d − t)** -space between the plates will have vacuum/air. Thus **(d − t)** space has electric field $\mathbf{E_0}$ and **t** space has electric field **E**, then-

$$E_0 = \frac{Q}{\varepsilon_0 A} \quad \dots (1)$$

$$\text{and } E = \frac{E_0}{K} \quad \dots (2)$$

[K = dielectric constant of slab material]

And total potential between the plates will be

$$V = E_0(d-t) + Et$$

$$\text{or } V = \frac{Q}{\varepsilon_0 A}(d-t) + \frac{Q}{\varepsilon_0 AK}t$$

$$\text{or } V = \frac{Q}{\varepsilon_0 A}\left[(d-t) + \frac{t}{K}\right] \quad \ldots (3)$$

And capacitance of the capacitor will be

$$C = \frac{Q}{V} = \frac{Q}{\frac{Q}{\varepsilon_0 A}\left[d-t+\frac{t}{K}\right]}$$

$$\text{or } C = \frac{\varepsilon_0 A}{d-t+\frac{t}{K}} \quad \ldots (4)$$

If t = d then

$$C = \frac{\varepsilon_0 A}{d/K}$$

$$\text{Or } \mathbf{C} = \frac{\mathbf{K\varepsilon_0 A}}{\mathbf{d}} \quad \ldots.(5)$$

$\because$ K > 1, inserting dielectric slab increases the capacitance.

19. Capacitance of parallel plate capacitor using conducting slab

A parallel plate capacitor is shown here having same area **A**, charge densities **+σ** and **−σ**, kept at distance **d**.

A conducting slab of thickness **t** is inserted between the plates.

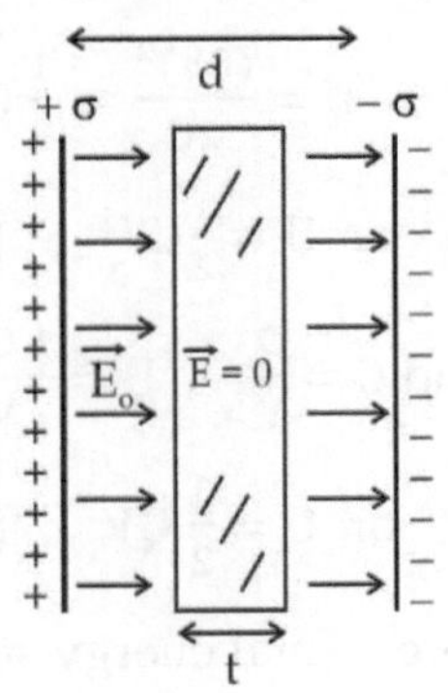

As we know, electric field inside the conductor is zero thus the electric field $\mathbf{E_0}$ is present in (**d − t**) space between the plates. Thus electric potential between the plates will be

$$V = E_0(d-t) \text{ or } V = \frac{\sigma}{\varepsilon_0}(d-t) \left[\because E_0 = \frac{\sigma}{\varepsilon_0}\right]$$

$$\text{or } V = \frac{Q}{\varepsilon_0 A}(d-t) \quad \ldots.(1) \quad \left[\because \sigma = \frac{Q}{A}\right]$$

$$\text{As } C = \frac{Q}{V}$$

From equation (1) we can write -

$$\therefore C = \frac{Q}{\frac{Q}{\varepsilon_0 A}(d-t)}$$

$$\text{or } C = \frac{\varepsilon_0 A}{(d-t)} = \frac{\varepsilon_0 A}{d\left(1-\frac{t}{d}\right)}$$

$$\text{or } \mathbf{C} = \frac{\mathbf{C_0}}{\left(\mathbf{1}-\frac{\mathbf{t}}{\mathbf{d}}\right)} \quad \text{Here } C_0 = \frac{\varepsilon_0 A}{d}$$

Please note: $\therefore C > C_0$ i.e., capacitance will increase, if we put a conducting slab between the plates.

20. Capacitors in series.

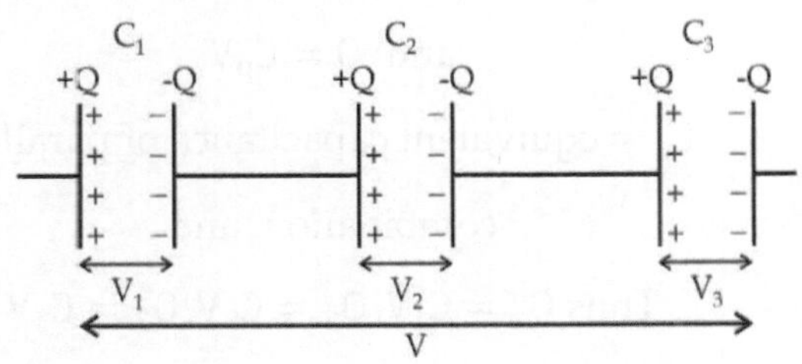

In series combination, the charge on each capacitor is same and electric potential is divided.

$$\text{From figure } V_1 = \frac{Q}{C_1},\ V_2 = \frac{Q}{C_2},\ V_3 = \frac{Q}{C_3}$$

$$\text{Total potential } V = V_1 + V_2 + V_3$$

$$\therefore V = \frac{Q}{C_1} + \frac{Q}{C_2} + \frac{Q}{C_3}$$

$$\text{Here } V = \frac{Q}{C_s}$$

C_s = equivalent capacitance of series combination

$$\therefore \frac{Q}{C_s} = \frac{Q}{C_1} + \frac{Q}{C_2} + \frac{Q}{C_3}$$

$$\text{or } \frac{1}{C_s} = \frac{1}{C_1} + \frac{1}{C_2} + \frac{1}{C_3}$$

Please Note- Total capacitance in series is less than the least one.

21. Capacitors in Parallel.

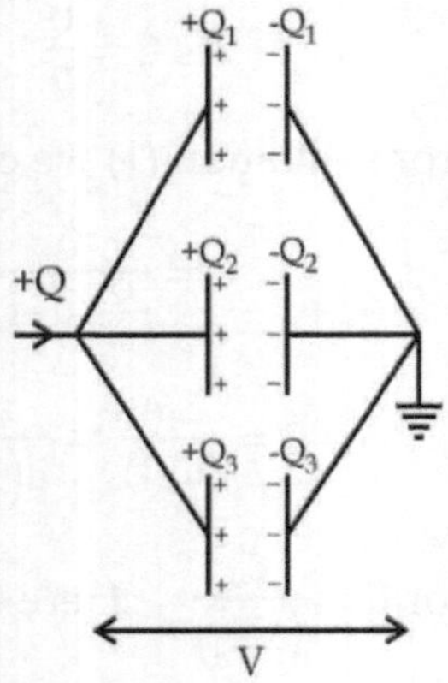

In parallel combination electric potential across each capacitor is same, but charges will be different.

But Total charge

$$Q = Q_1 + Q_2 + Q_3$$

$$\text{and } Q = C_pV$$

C_p = equivalent capacitance of parallel combination, and

$$\text{Thus } Q_1 = C_1V, Q_2 = C_2V, Q_3 = C_3V$$

$$\therefore C_pV = C_1V + C_2V + C_3V$$

$$\text{Or } \mathbf{C_p = C_1 + C_2 + C_3}$$

22. Energy stored in charged capacitor

When we charge a capacitor by using a battery, work is done by charging battery at the cost of chemical energy. Due to charging, the potential of the plates increases.

"The total work done in charging the capacitor is stored in the form of electrical potential energy."

Consider a parallel plate capacitor connected to a battery and getting charged. We assume that when small amount of charge **dq** is given to the capacitor the p.d. between the plates remains **V** only.

∴ Small amount of work done by the battery in sending charge **dq** will be

$$dW = V\,dq \quad \text{or} \quad dW = \frac{q}{C}dq \qquad \ldots (1)$$

And total work done in giving charge Q will be

$$W = \int_{q=0}^{q=Q} \frac{q}{C}dq = \frac{1}{C}\left[\frac{q^2}{2}\right]_0^Q$$

$$\text{or } W = \frac{Q^2}{2C}$$

"This work done is stored in the form of electrostatic P.E."

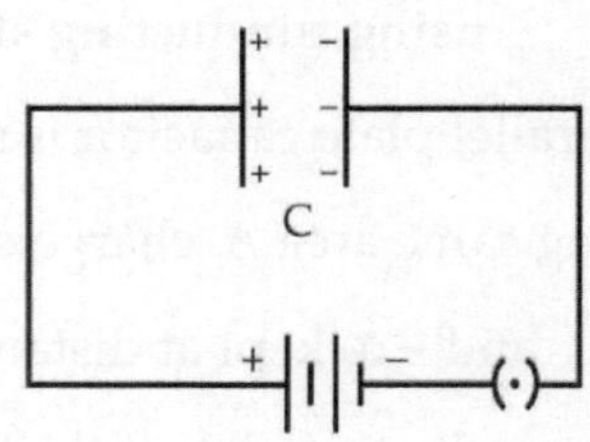

$$\therefore U = W = \frac{Q^2}{2C} \ldots (2) \qquad \because Q = CV$$

$$\therefore U = \frac{(CV)^2}{2C} = \frac{1}{2}CV^2$$

$$\text{or } U = \frac{1}{2}CV^2 \ldots (3)$$

$$\text{and } C = \frac{Q}{V} \quad \therefore U = \frac{1}{2}\frac{Q}{V} \times V^2$$

$$\text{or } \mathbf{U = \frac{1}{2}QV} \ldots (4)$$

Note - Total energy stored in series/parallel combination

$= U_1 + U_2 + U_3 \cdots$

23. Common potential of two capacitors when connected in parallel.

When two capacitors are charged to a different potentials are connected such that their positive plates are joined together and their negative plates are joined together, then their is flow of charge from the capacitor at higher potential to the capacitor at lower potential until both acquire a common potential.

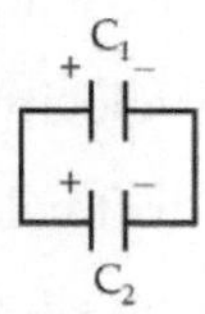

During this process, charge is conserved.

That means

Total charge before sharing = Total charge after sharing

If $Q_1 = C_1V_1$ & $Q_2 = C_2V_2$

& total charge before sharing

$$Q = Q_1 + Q_2 = C_1V_1 + C_2V_2 \qquad \ldots (1)$$

If **V** is the common potential then total charge after sharing will be

$$Q = C_1V + C_2V = (C_1 + C_2)V \qquad \ldots (2)$$

from eq. (1) & (2)

$$(C_1 + C_2)V = C_1V_1 + C_2V_2$$

$$\text{or } \mathbf{V} = \frac{C_1V_1+C_2V_2}{(C_1+C_2)} \qquad \ldots (3)$$

$$\therefore V = \frac{\text{Total charge}}{\text{Total capacitance}}$$

24. Loss of energy in sharing charges by capacitors connected in parallel

When two capacitors are connected in parallel, they share charges until they acquire common potential. Charge is not lost during this process, but energy is lost which can be proven as follows -

Since we know that, after sharing charges the common potential is acquired which is given by

$$\mathbf{V} = \frac{C_1V_1+C_2V_2}{(C_1+C_2)}$$

and total electrostatic potential energy before sharing charges

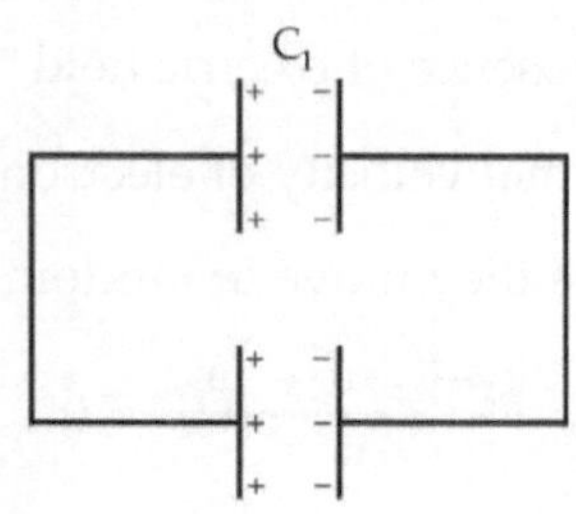

$$U_1 = \frac{1}{2}C_1V_1^2 + \frac{1}{2}C_2V_2^2 \qquad \ldots (1)$$

and total energy after sharing charges

$$U_2 = \frac{1}{2}C_pV^2 \rightarrow C_p = \text{Total capacitance}$$

V = common potential

$$\therefore U_2 = \frac{1}{2}(C_1 + C_2)\left[\frac{C_1V_1+C_2V_2}{(C_1+C_2)}\right]^2$$

$$\text{or } U_2 = \frac{(C_1V_1+C_2V_2)^2}{2(C_1+C_2)} \qquad \ldots (2)$$

subtracting U_1 from U_2 we get-

$$U_1 - U_2 = \frac{1}{2}C_1V_1^2 + \frac{1}{2}C_2V_2^2 - \frac{(C_1V_1+C_2V_2)^2}{2(C_1+C_2)}$$

Solving this we get

$$\Delta\mathbf{U} = \frac{C_1C_2(V_1-V_2)^2}{2(C_1+C_2)} \qquad \ldots (3)$$

Expression-(3) is a positive quantity hence

we can say that during sharing charges, energy is lost.

$\because U_1 > U_2$.

Chapter - 3

Current Electricity

25. Drift velocity of electrons when potential difference is applied across the conductor

Definition – "*It is the average velocity with which the free electrons in a conductor get drifted towards the positive end of the conductor when external potential difference is applied.*"

In the absence of electric field, the average thermal velocity of electrons is zero because they move in random directions.

i.e. $\frac{\vec{u}_1+\vec{u}_2+\vec{u}_3+\cdots u_n}{n} = \vec{0}$ ... (1)

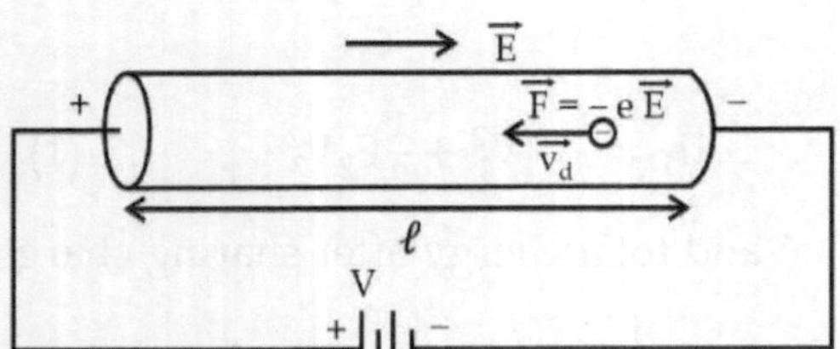

When potential difference is applied across the two ends of the conductor each electron will experience a force which is given by

$$\vec{F} = -e\vec{E} \quad \ldots (2)$$

This force is in the direction opposite to electric field because electrons are negatively charged.

If **m** is the mass of electron, then acceleration will be -

$$\vec{a} = -\frac{e\vec{E}}{m} \quad \ldots (3)$$

Due to this they start accelerating but for a short time because they experience collisions. "The time for which they get accelerated before colliding with positive ions of metal is called **relaxation time**".

We assume an electron having initial thermal velocity $\vec{\mathbf{u}}_1$ accelerates for a time $\mathbf{t_1}$ then it attains a velocity $\mathbf{v_1}$, which is given by -

$$\vec{v}_1 = \vec{u}_1 + \vec{a}t_1 \quad \ldots (4)$$

Similarly $\vec{v}_2 = \vec{u}_2 + \vec{a}t_2$ ... (5)

& $\vec{v}_n = \vec{u}_n + \vec{a}t_n$ ... (6)

and average velocity of free electrons (drift velocity)

$$\vec{v}_d = \frac{\vec{v}_1+\vec{v}_2+\vec{v}_3+\cdots\vec{v}_n}{n}$$

$$\vec{v}_d = \frac{(\vec{u_1}+\vec{a}t_1)+(\vec{u_2}+\vec{a}t_2)+\cdots(\vec{u_n}+\vec{a}t_n)}{n}$$

$$\text{or } \vec{v}_d = \frac{\vec{u}_1+\vec{u}_2+\cdots\vec{u}_n}{n} + \vec{a}\frac{(t_1+t_2+\cdots+t_n)}{n}$$

Now,

$$\frac{t_1+t_2+\cdots t_n}{n} = \text{average relaxation time } (\tau)$$

$$\because \frac{\vec{u}_1+\vec{u}_2+\cdots\vec{u}_n}{n} = 0$$

(Average thermal velocity)

$$\therefore \vec{v}_d = 0 + \vec{a}\tau = \vec{a}\tau \text{ or } \vec{\mathbf{v}}_\mathbf{d} = \frac{-\mathbf{e}\vec{\mathbf{E}}\tau}{\mathbf{m}}$$

(Negative sign means that the electrons are drifted in the opposite direction o electric field)

26. Relation between drift velocity (v_d) and electric current (I)

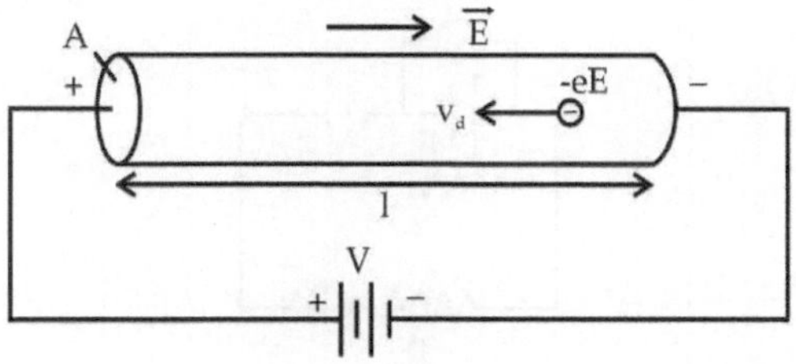

As shown in the figure, a segment of conductor is considered having length ℓ and cross sectional area **A**. When external electric field is applied they are drifted towards positive terminal due to electric force $[\vec{F} = -e\vec{E}]$.

Now volume of conductor = $A \times \ell$ ... (1)

If n represents number density then

$$n = \frac{N}{V}$$

N ⟶ Number of electrons

V ⟶ Volume of conductor

∴ Total number of free electrons in volume $\boldsymbol{A\ell}$ will be = $nA\ell$ ... (2)

If charge on electron is taken **e** then total charge on all the electrons present in volume $A\ell$ will be

$$q = nA\ell \times e \quad \dots (3)$$

Now the time taken by the free electrons to cross the length ℓ of the conductor will be -

$$t = \frac{\ell}{v_d} \dots (4) \quad [v_d = \text{drift velocity}]$$

$$\text{And } I = \frac{q}{t}$$

∴ from eq. (3) & (4)

$$I = \frac{nA\ell e}{\ell/v_d} = neAv_d$$

$$\therefore I = neAv_d \quad \dots (5)$$

$$\because v_d = \frac{eE\tau}{m}$$

$$\therefore \mathbf{I} = \frac{\mathbf{nAe^2\tau E}}{\mathbf{m}} \quad \dots (6)$$

27. Electron Mobility (μ)

*It is the drift velocity ($\boldsymbol{v_d}$) acquired by the electrons per unit strength of electric field (**E**) applied.*

$$\therefore \mu = \frac{v_d}{E} \quad \dots (1)$$

$$\because v_d = \frac{eE\tau}{m} \quad \dots (2)$$

$$\therefore \mu = \frac{\mathbf{eE\tau}}{\mathbf{mE}} \text{ or } \mu = \frac{\mathbf{e\tau}}{\mathbf{m}} \quad \dots (3)$$

28. Deduction of Ohm's Law.

We know that drift velocity

$$v_d = \frac{eE}{m}\tau \quad \dots (1)$$

If ℓ is the length of the conductor and external field **E** is applied then -

$$E = \frac{V}{\ell} \quad \therefore v_d = \frac{e\tau}{m} \times \frac{V}{\ell}$$

Also $I = neAv_d$

Substituting $\mathbf{v_d}$ from eq (2)

$$I = Ane\left[\frac{eV}{m\ell}\tau\right] = \left[\frac{Ane^2\tau}{m\ell}\right]V \quad \dots (2)$$

or $\frac{V}{I} = \frac{m\ell}{Ane^2\tau} = R$ → A constant for a given conductor

∴ V = IR   Ohm's law

$$\text{Here } R = \frac{m}{ne^2\tau} \times \frac{\ell}{A} \quad \dots (3)$$

In this equation term $\frac{m}{ne^2\tau}$ is completely depending upon the nature of conductor and it is called resistivity **ρ**.

$$\therefore \boldsymbol{\rho} = \frac{\mathbf{m}}{\mathbf{ne^2\tau}} \quad \dots (4)$$

29. Relation between Current density and Drift velocity

*"Current density **J** is the amount of current flowing per unit area having area vector in the direction of current".*

$$\vec{J} = \frac{I}{\vec{A}} \quad \ldots (1)$$

$$\because I = neAv_d$$

$$\therefore J = \frac{neAv_d}{A} = nev_d$$

$$\mathbf{J = nev_d} \quad \ldots\ldots (2)$$

It is a vector quantity.

30. Microscopic form of Ohm's Law or vector form of ohm's Law (Relation between J, σ and E)

$$I = neAv_d = neA\left[\frac{eE\tau}{m}\right]$$

$$\text{Or } I = \frac{ne^2\tau E}{m}A \quad \ldots (1)$$

$$\text{Or } \frac{I}{A} = \frac{ne^2\tau E}{m} = \frac{E}{m/ne^2\tau} \quad \ldots (2)$$

$$\because J = \frac{I}{A}$$

$$\therefore J = \frac{E}{m/ne^2\tau} \ldots (3)$$

$$\text{Now } \rho = \frac{m}{ne^2\tau}$$

$$\therefore J = \frac{E}{\rho} \quad \ldots (4)$$

$$\because \frac{1}{\rho} = \sigma \text{ conductivity}$$

$$\therefore \mathbf{J = \sigma E}$$

Or $\vec{J} = \sigma\vec{E}$ → vector form or microscopic form of Ohm's Law

Note: Conductance $G = \frac{1}{R}$ and

Conductivity $(\sigma) = \frac{1}{\rho}$

31. Relation between emf, internal resistance and terminal voltage

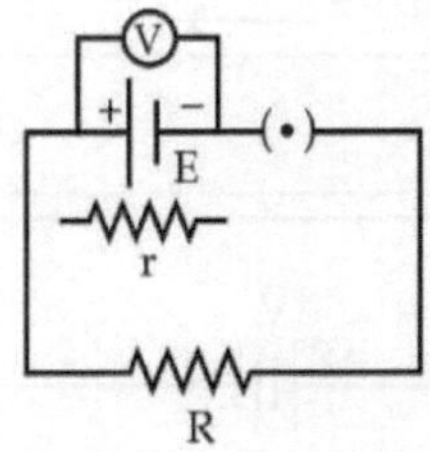

*When the cell is not in use then the potential difference across the terminals of the cell is called emf and represented by **E**.*

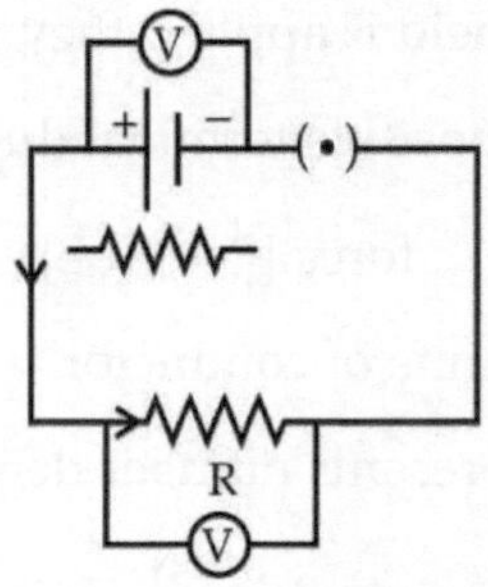

Terminal voltage – *When the cell is in use then some potential drop takes place inside the cell itself due to its internal resistance **r** and available voltage to the circuit decreases which is called terminal voltage.*

It is denoted by -

$$\mathbf{V = E - Ir} \quad \ldots.(1)$$

$$\because V = IR$$

$$\therefore IR = E - Ir$$

$$\text{or } Ir = E - V$$

$$\text{or } r = \frac{E-V}{I}$$

Please note, While charging the cell equation (1) becomes

$$\mathbf{V = E + Ir} \quad \ldots.(2)$$

32. Cells connected in series

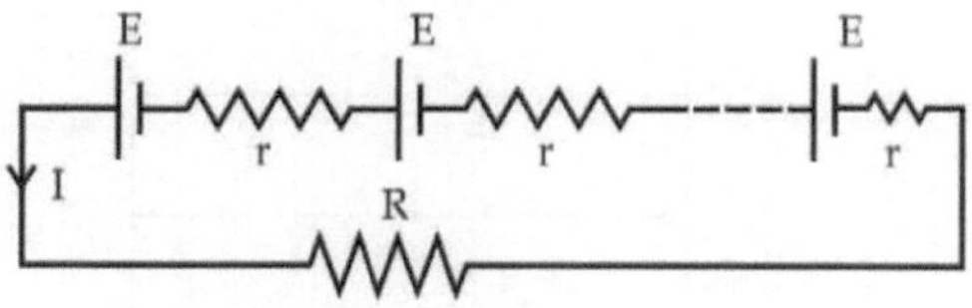

Case – I: n cells connected in series

Here n number of cells are connected in series each having emf E_ and internal resistance **r** also connected to an external resistor **R**,

then $E_{total} = nE$ ... (1)

and $r_{eq} = nr$, $R_{eff} = nr + R$ ... (2)

and $\mathbf{I} = \frac{\mathbf{nE}}{\mathbf{nr+R}}$... (3)

<u>**Subcase-1**</u>

When R ≫ nr

then $nr + R \approx R$ and

$$\mathbf{I} = \frac{\mathbf{nE}}{\mathbf{R}}$$

Here current is **n** times than the current due to single cell.

<u>**Subcase-2**</u>

When R ≪ nr

then $nr + R \approx nr$ and

$$I = \frac{E}{r}$$

Current is same as obtained by a single cell i.e., here no advantage of connecting **n** cells in series.

Note: When external resistance is much greater than equivalent internal resistance, we must connect cells in series.

Case – II: Two cells in series with different emf values

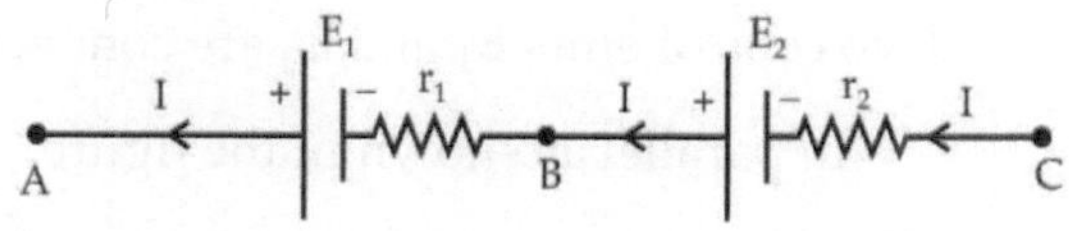

Two cells having emfs $\mathbf{E_1}$ & $\mathbf{E_2}$ & internal resistances $\mathbf{r_1}$ & $\mathbf{r_2}$ respectively.

As shown in the figure, they are connected in series. Now we can say that

$V_A - V_B = E_1 - Ir_1$ ... (1)

& $V_B - V_C = E_2 - Ir_2$ ... (2)

Now adding equations (1) and (2)

We get $V_A - V_c = (E_1 - Ir_1) + (E_2 - Ir_2)$

or $V_A - V_C = (E_1 + E_2) - I(r_1 + r_2)$ ... (3)

Potential difference between **A** and **C** can be represented as $V_{AC} = V_A - V_C$

$\therefore V_{AC} = E_{eq} - Ir_{eq}$ ... (4)

Comparing eq. (3) and (4) we get

$E_{eq} = E_1 + E_2$ & $r_{eq} = r_1 + r_2$

If the polarity of one of the cell is reversed then

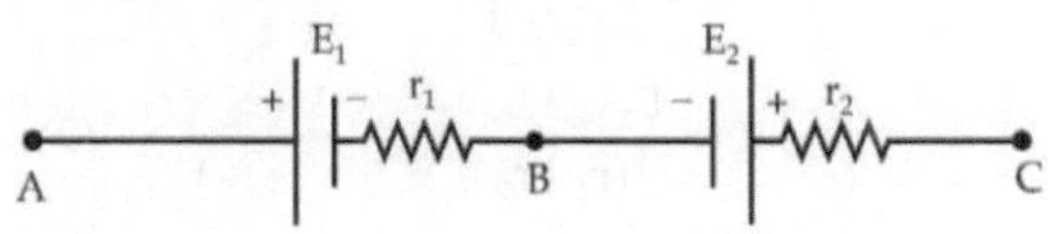

Then $V_A - V_C = (E_1 - E_2) - I(r_1 + r_2)$

$\mathbf{E_{eq} = E_1 - E_2}$ **and** $\mathbf{r_{eq} = r_1 + r_2}$

33. Cells connected in parallel

<u>**Case – I:**</u>

When two cells with different emfs are connected in parallel

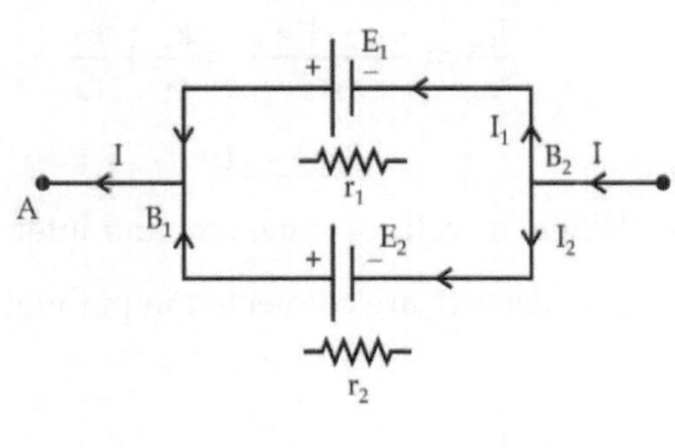

Two cells of emfs $\mathbf{E_1}$ and $\mathbf{E_2}$ are connected in parallel as shown in the figure.

They have internal resistances $\mathbf{r_1}$ and $\mathbf{r_2}$ respectively.

From figure $I = I_1 + I_2$ ... (1)

Now $V_{B_1} - V_{B_2}$ can be written as

$$V = V_{B_1} - V_{B_2} = E_1 - I_1r_1 \quad ... (2)$$

$$\& \; V = V_{B_1} - V_{B_2} = E_2 - I_2r_2 \quad ... (3)$$

From equation (2) $I_1 = \frac{E_1 - V}{r_1}$ ... (4)

$$\& \; I_2 = \frac{E_2 - V}{r_2} \quad ... (5)$$

substituting eq. (4) & (5) in equation -(1) we get

$$I = \left[\frac{E_1 - V}{r_1}\right] + \left[\frac{E_2 - V}{r_2}\right] = \left[\frac{E_1}{r_1} + \frac{E_2}{r_2}\right] - V\left[\frac{1}{r_1} + \frac{1}{r_2}\right]$$

$$\text{or } I = \frac{E_1r_2 + E_2r_1}{r_1r_2} - V\left[\frac{r_1 + r_2}{r_1r_2}\right] \quad ... (6)$$

$$\text{or } V\left[\frac{r_1 + r_2}{r_1r_2}\right] = \frac{E_1r_2 + E_2r_1}{r_1r_2} - I$$

$$\text{or } V = \frac{E_1r_2 + E_2r_1}{(r_1 + r_2)} - \frac{Ir_1r_2}{(r_1 + r_2)} \quad ... (7)$$

Since $V = E_{eq} - Ir_{eq}$ ...... (8)

Hence comparing equations (7) and (8) we get

$$E_{eq} = \frac{E_1r_2 + E_2r_1}{r_1 + r_2} = \frac{(E_1/r_1) + (E_2/r_2)}{(1/r_1) + (1/r_2)} \quad ... (9)$$

$$\& \; r_{eq} = \frac{r_1r_2}{r_1 + r_2} \text{ or } \frac{1}{r_{eq}} = \frac{1}{r_1} + \frac{1}{r_2} \quad ... (10)$$

Dividing eq. (9) by (10) we get

$$\frac{\mathbf{E_{eq}}}{\mathbf{r_{eq}}} = \frac{\mathbf{E_1r_2 + E_2r_1}}{\mathbf{r_1r_2}} = \frac{\mathbf{E_1}}{\mathbf{r_1}} + \frac{\mathbf{E_2}}{\mathbf{r_2}}$$

Case – II:

When 'n' cells of same emf and internal resistance are connected in parallel.

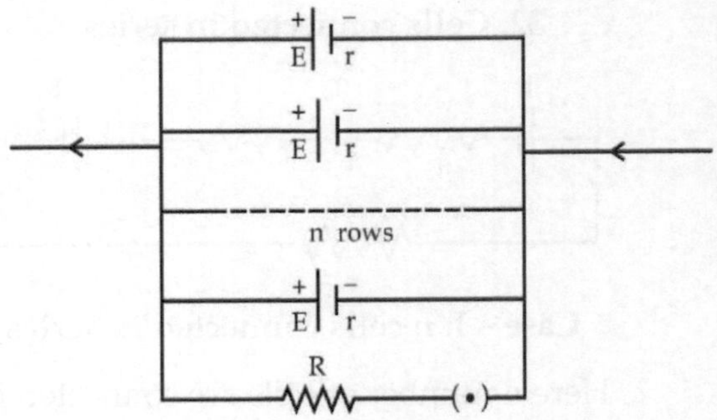

When **n** cells are connected in parallel and they all are connected to external resistance **R** as shown in the figure, then total internal resistance can be calculated as

$$\frac{1}{r_{eq}} = \frac{1}{r} + \frac{1}{r} \cdots \text{n times}$$

$$\therefore \; \frac{1}{r_{eq}} = \frac{n}{r} \text{ or } r_{eq} = \frac{r}{n} \quad ... (1)$$

And total resistance of the circuit

$$R_{eq} = \frac{r}{n} + R \quad ... (2)$$

Since all cells are identical, hence effective emf will be **E** only and total current will be

$$I = \frac{E}{R_{eq}} = \frac{E}{\left(\frac{r}{n} + R\right)}$$

$$\text{or } I = \frac{nE}{r + nR} \quad ... (3)$$

Subcase – I: When $R \ll \frac{r}{n}$, then $I = \frac{nE}{r}$

current becomes n times with respect to single cell.

Subcase – II: When $r \ll R$, then $I = \frac{E}{R}$

same as current due to single cell.

Note: When external resistance is smaller than equivalent internal resistance, we must connect cells in parallel.

34. Wheatstone Bridge

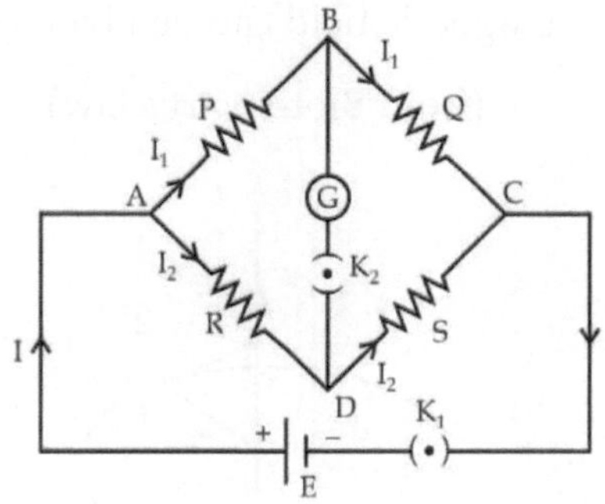

Principle: When four resistances **P, Q, R** and **S** are connected to form a bridge as shown in figure and after pressing keys $\mathbf{K_1}$ and $\mathbf{K_2}$ if no current flows through galvanometer arm then bridge is in balanced state and in this condition -

$$\frac{\mathbf{P}}{\mathbf{Q}} = \frac{\mathbf{R}}{\mathbf{S}}$$

Proof – When no current flows through **BD** arm, then for closed loop **ABDA** using Kirchhoff's second Law -

$$-I_1P + I_2R = 0$$

$$\text{or } I_1P = I_2R \quad \dots (1)$$

Similarly for closed loop **BCDB**

$$-I_1Q + I_2S = 0$$

$$\text{or } I_1Q = I_2S \quad \dots (2)$$

Dividing eq. (1) & (2), we get

$$\frac{I_1P}{I_1Q} = \frac{I_2R}{I_2S}$$

$$\frac{P}{Q} = \frac{R}{S} \text{ Hence proved}$$

Chapter - 4

Magnetic Effects of Electric current

Biot - Savart's Law-

According to this law, the magnetic field (**dB**) at a point **P** due to a small current element **I**$\boldsymbol{d\ell}$ is given by

$$dB = \frac{\mu_0}{4\pi}\frac{Id\ell \sin\theta}{r^2}$$

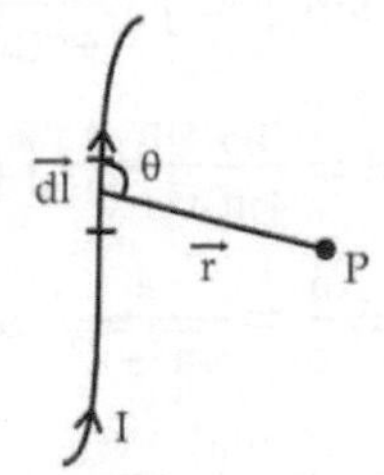

35. Magnetic field at a point on the axis of a circular coil carrying current

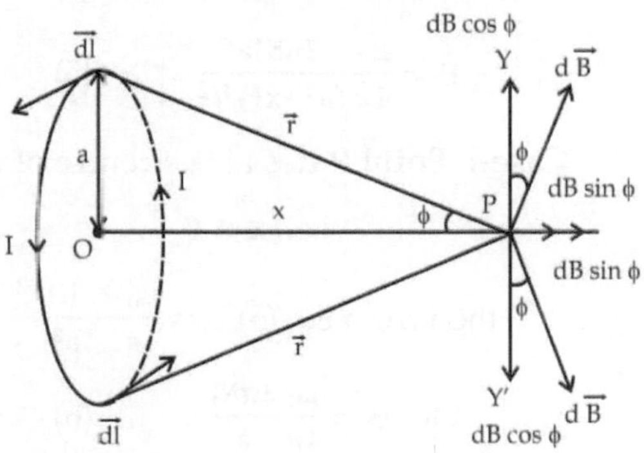

A circular current carrying loop is shown in the figure having current **I** and radius **a**. We want to calculate magnetic field at point **P** lying on its axis at distance **x** from the centre **O**. Distance between current element $\boldsymbol{d\ell}$ and point **P** is denoted by **r**.

$$\therefore\ r = \sqrt{a^2 + x^2} \qquad \dots (1)$$

$d\vec{B}, d\vec{\ell}$ and $\vec{r}$ are mutually perpendicular to each other.

$$\therefore\ dB = \frac{\mu_0 Id\ell \sin 90^\circ}{4\pi r^2}$$

$$\text{or } dB = \frac{\mu_0}{4\pi}\frac{Id\ell}{r^2} \qquad \dots (1)$$

When we resolve $\mathbf{d\vec{B}}$ into two rectangular components as shown in the figure, we get dB cos ϕ components get cancelled & dB sin ϕ are in same direction hence get added. Thus total magnetic field B at point P is given by

$$B = \int dB\sin\phi = \int \frac{\mu_0 I d\ell \sin\phi}{4\pi(a^2+x^2)} \quad \ldots (2)$$

$$\text{or } B = \frac{\mu_0 I \sin\phi}{4\pi(a^2+x^2)}\int d\ell \quad \ldots (3)$$

$$\because \sin\phi = \frac{a}{r} = \frac{a}{\sqrt{a^2+x^2}} \text{ and } \int d\ell = 2\pi a$$

$$\therefore B = \frac{\mu_0 I}{4\pi(a^2+x^2)} \times \frac{a}{\sqrt{a^2+x^2}} \times 2\pi a$$

$$\text{or } B = \frac{\mu_0}{4\pi}\frac{2\pi I a^2}{(a^2+x^2)^{3/2}} \quad \ldots (4)$$

If there are N turns in the loop then -

$$\mathbf{B} = \frac{\mu_0}{4\pi}\frac{2\pi N I a^2}{(a^2+x^2)^{3/2}} \quad \ldots (5)$$

Case-I: Point P lies at the centre of coil i.e., $x = 0$

then from eq. (5) $B = \frac{\mu_0}{4\pi}\frac{2\pi N I a^2}{a^3}$

$$\text{Or } \mathbf{B} = \frac{\mu_0}{4\pi}\frac{2\pi N I}{a} \quad \ldots (6)$$

Case-II: If $x \gg a$,

then $a^2 + x^2 \approx x^2$, then from eq-(5)

$$B = \frac{\mu_0}{4\pi}\frac{2\pi N I a^2}{x^3} = \frac{\mu_0}{4\pi}\frac{2NIA}{x^3} \quad [\text{Here } A = \pi a^2]$$

Now, NIA = Magnetic dipole moment of the loop (M)

$$\therefore B = \frac{\mu_0}{4\pi}\frac{2M}{x^3} \quad \ldots (7)$$

36. Ampere's circuital law and its proof

According to this law,

The line integral of $\vec{B}$ around any closed path is equal to μ_0 times the current enclosed by the closed path.

$$\oint \vec{\mathbf{B}} \cdot d\vec{\boldsymbol{\ell}} = \mu_0 \mathbf{I}$$

Proof:- We consider a long straight conductor **X − Y** carrying current **I**.

Due to this conductor, at a point **P,** which is at a distance **r** from the conductor, magnetic field can be given by (from Biot-swart's law)

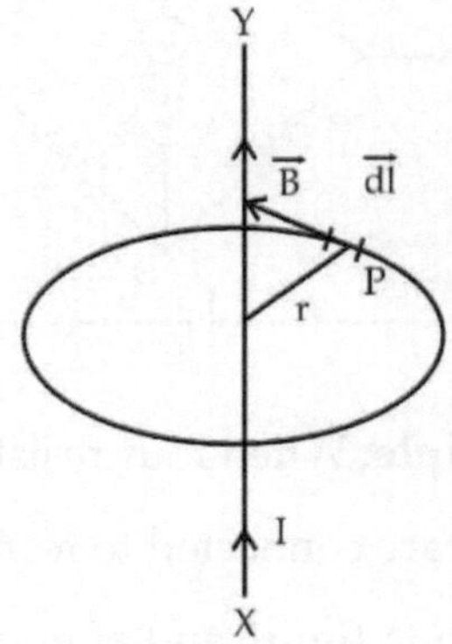

$$B = \frac{\mu_0 I}{2\pi r} \quad \ldots (1)$$

We draw a circle keeping wire at the centre of circle, of radius **r**. We can see that on this circle $\mathbf{d\vec{\ell}}$ and $\vec{\mathbf{B}}$ both are in same direction.

$\therefore$ The line integral of $\vec{\mathbf{B}}$ will be represented as

$$\oint \vec{B}.\vec{d\ell} = \oint B \cdot d\ell \cos 0^\circ$$

$$\text{or } \oint B d\ell = B\oint d\ell = B \times 2\pi r \quad \ldots (2)$$

from eq. (1) we can say that

$$\oint \vec{B} \cdot d\vec{\ell} = \frac{\mu_0 I}{2\pi r} \times 2\pi r = \mu_0 I$$

$$\therefore \oint \vec{\mathbf{B}} \cdot d\vec{\boldsymbol{\ell}} = \mu_0 \mathbf{I}$$

This is Ampere's circuital Law.

37. Magnetic field of a long straight thin wire

[Application of Ampere's circuital Law]

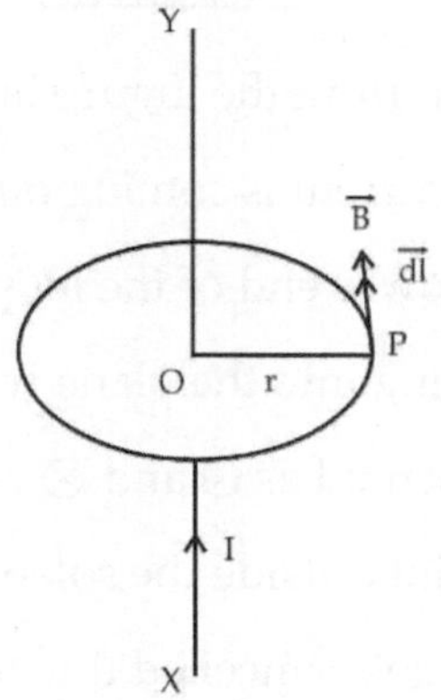

A current carrying conductor **XY** shown here, from this here at point **P**, at a distance **r** we want to find magnetic field. We draw a circle enclosing point **P** and direction of magnetic field is tangent to this circle at every point then from ampere's circuital law -

$$\oint \vec{B} \cdot d\vec{\ell} = \mu_0 I \quad \dots (1)$$

$\because$ $\vec{B}$ and $\vec{d\ell}$ are along same direction $(\theta = 0°)$

$$\therefore \oint B.d\ell = \mu_0 I \quad \dots (2)$$

$$\therefore B \times 2\pi r = \mu_0 I$$

$$\text{or } \mathbf{B} = \frac{\mu_0 I}{2\pi r} \quad \dots (3)$$

38. Magnetic field due to a current carrying thick circular cylinder

[Application of Ampere's circuital Law]

Here a segment of long, thick circular wire of radius **R** is shown.

Current **I** is flowing through entire cross section of wire.

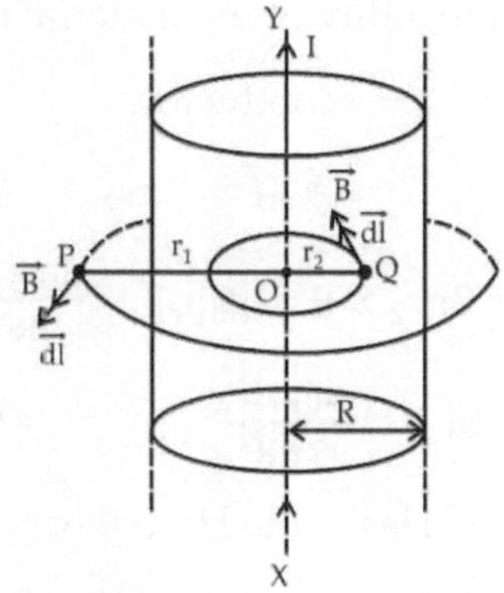

We want to calculate magnetic field at a point lying outside the wire **P**, on the surface and at a point **Q** inside the wire.

Case - I: Point P is lying outside the wire $(r_1 > R)$

We draw Amperian loop of radius $\mathbf{r_1}$ enclosing point **P**. Amperian loop centre and axis of the cylindrical wire coinciding.

$\vec{B}$ and $\vec{d\ell}$ are in same direction, hence applying Ampere's circuital law

$$\oint \vec{B} \cdot d\vec{\ell} = \mu_0 I \text{ or } \oint B d\ell \cos 0° = \mu_0 I$$

$$\text{or } \oint B\, d\ell = \mu_0 I \text{ or } B \times 2\pi r_1 = \mu_0 I$$

$$\therefore \mathbf{B} = \frac{\mu_0 I}{2\pi r_1} \quad \text{Here } B \propto \frac{1}{r_1}$$

Case - II: At a point Q lying inside the wire $(r_2 < R_1)$

Here we assume that the current is uniformly distributed throughout the cross section of the conductor then assume **I′** current is flowing through area πr_2^2 then

$$I' = \frac{I}{\pi R^2} \times \pi r_2^2$$

$$\therefore I' = \frac{I r_2^2}{R^2} \quad \dots (1)$$

Now applying Ampere's circuital law, we get

$$\oint \vec{B} \cdot d\vec{\ell} = \mu_0 \mu_r I'$$

Here μ_r is the relative magnetic

permeability of the material of the conductor.

$$\because \oint d\vec{\ell} = 2\pi r_2$$

$$\therefore 2\pi r_2 \times B = \mu_0\mu_r I' = \frac{\mu_0\mu_r I r_2^2}{R^2}$$

$$\text{or } \mathbf{B} = \frac{\mu_0\mu_r I r_2}{2\pi R^2} \qquad \ldots (2)$$

Please note Here $B \propto r_2$

Case – III: At a point on the surface of the wire i.e., r = R

From equation (2)

$$B = \frac{\mu_0\mu_r IR}{2\pi R^2} = \frac{\mu_0\mu_r I}{2\pi R}$$

$$\therefore \mathbf{B} = \frac{\mu_0\mu_r I}{2\pi R}$$

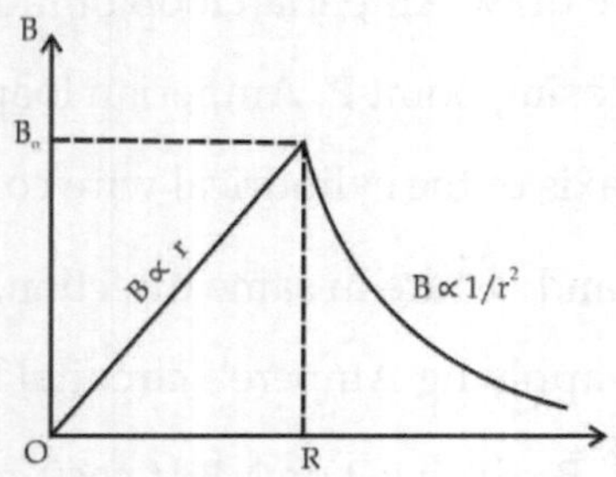

Graphical representation

39. Magnetic field due to a long solenoid (Application of Ampere's circuital Law)

A solenoid is a coil in which wire is closely wound in the form of a helix and its length is very large as compared to its diameter.

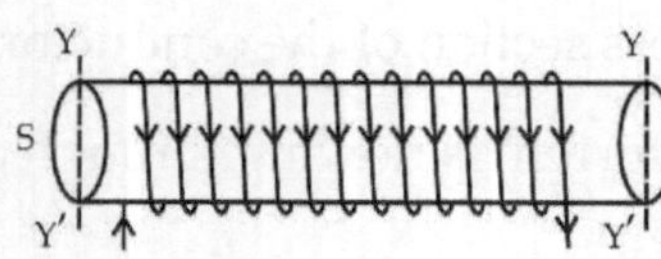

Fig - 1

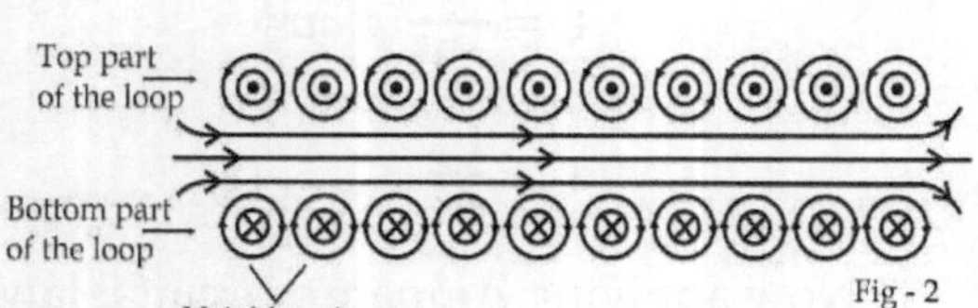

Fig - 2

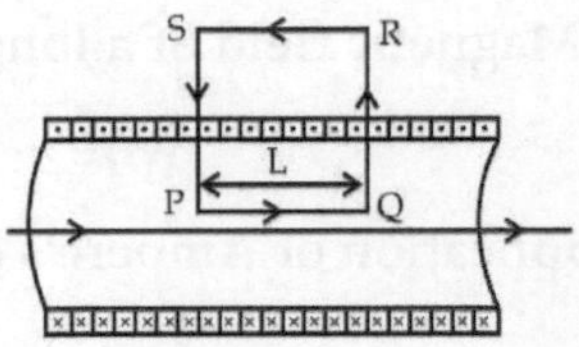

Fig - 3

As clear from the figure in each loop at the top current is coming out of the plane & in lower end of the loop current in going into the plane which are represented as ⊙ and ⊗ respectively.

At a point outside the solenoid magnetic field gets cancelled due to opposite direction of magnetic field in the neighbouring loop. (Figure-2)

At a point inside the solenoid magnetic field is in the same direction due to upper and lower end of each loop hence added.

We consider a rectangular loop (amperian loop) **PQRS** as shown in the figure.

Here we take number density

$$n = \frac{\text{Number of turns total (N)}}{\text{Total length of solenoid (L)}}$$

$\therefore$ Total no. of turns will be = nL

Now line integral of $\vec{\mathbf{B}}$ over closed path **PQRS** is given by

$$\oint_{PQRS} \vec{B} \cdot \vec{d\ell} = \int_P^Q \vec{B} \cdot d\vec{\ell} + \int_Q^R \vec{B} \cdot \vec{d\ell} + \int_R^S \vec{B} \cdot d\vec{\ell} + \int_S^P \vec{B} \cdot d\vec{\ell} \qquad \ldots (1)$$

For PQ segment $\vec{d\ell}$ and $\vec{B}$ are in same direction

$$\therefore \int_P^Q \vec{B} \cdot \vec{d\ell} = \int_P^Q B d\ell \cos 0^\circ$$

$$= \int_P^Q B d\ell \qquad \ldots (2)$$

Also $\int_Q^R \vec{B} \cdot \vec{d\ell} = \int_Q^R B d\ell \cos 90^\circ = 0 \ldots (3)$

$\int_R^S \vec{B} \cdot \vec{d\ell}$ = zero because outside the solenoid magnetic field is zero. ... (4)

$$\& \int_S^P \vec{B} \cdot \vec{d\ell} = \int_S^P B d\ell \cos 90^\circ = 0 \ \ldots (5)$$

$$\oint_{PQRS} \vec{B} \cdot \vec{d\ell} = \int_p^Q B d\ell = BL \ \ldots (6)$$

From Ampere's circuital law

$$\oint \vec{B} \cdot \vec{d\ell} = \mu_0 \times \text{total current through PQRS}$$

$$= \mu_0 \times \text{no. of turns} \times \text{current}$$

$$\oint \vec{B} \cdot \vec{d\ell} = \mu_0 nL \times I$$

$$\text{or } BL = \mu_0 nI \times L$$

$$\text{or } \mathbf{B = \mu_0 nI} \ \ldots (7)$$

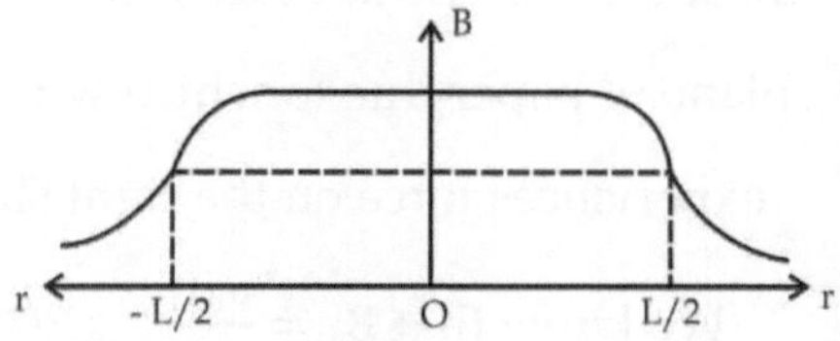

Note - Magnetic field inside the solenoid is uniform and outside the solenoid it is almost zero.

40. Motion of a charged particle in uniform magnetic field.

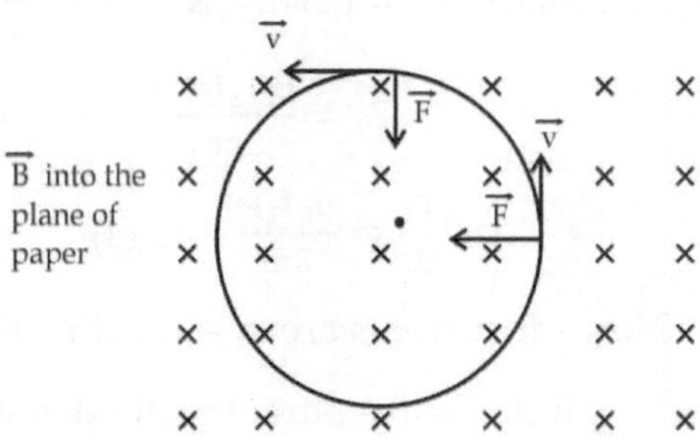

Assume a charged particle entering an external magnetic field (into the plane of paper) having velocity $\vec{v}$ then it gets deflected due to magnetic force which is given by

$$\vec{F} = q(\vec{v} \times \vec{B}) \text{ or } F = qvB\sin\theta$$

If $\theta = 90^\circ$ then, this force provides centripetal force to the charged particle hence angular velocity & time period can be calculated as follows -

$$qvB = \frac{mv^2}{r} \text{ or } v = \frac{qBr}{m}$$

$$\because v = r\omega \text{ or } \omega = \frac{v}{r} = \frac{qBr}{m \times r}$$

$$\therefore \omega = \frac{qB}{m}$$

and time period

$$T = \frac{2\pi}{\omega}$$

$$\text{Or } \mathbf{T = \frac{2\pi m}{qB}}$$

Another case: If it enters at an angle other than **90°** then-

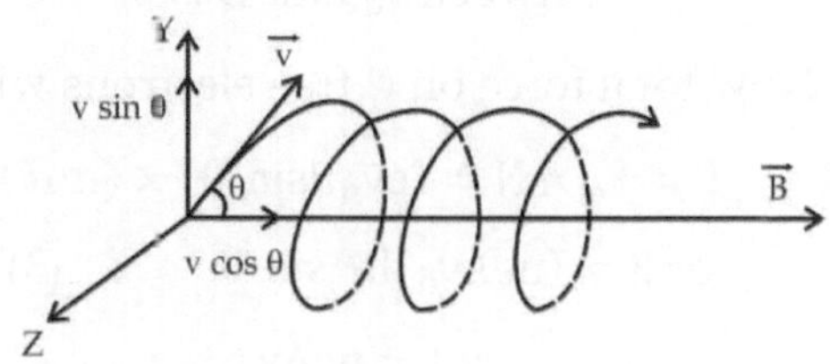

Due to **v sin θ** component it acquires circular motion & due to **vcos θ** it moves linearly, thus it acquires helical path

Now the linear distance covered by the charged particle in time equal to one rotation (T) is called pitch of helix (d).

$$d = v\cos\theta \times T = v\cos\theta \, \frac{2\pi m}{qB}$$

$$\text{and } \boldsymbol{\omega} = \frac{\mathbf{v\sin\theta}}{\mathbf{r}}$$

41. Force on a current carrying conductor placed in external magnetic field

A conductor of length ℓ, carrying current **I** is placed in external magnetic field **B** into the plane of paper as shown in figure. We assume number of electrons per unit volume = **n**. (number density) then total no. of electrons in the length ℓ of the conductor is -

$$N = nA\ell \quad \dots (1)$$

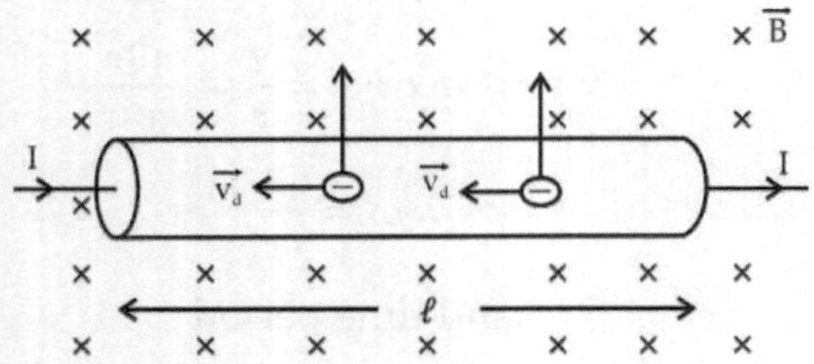

Now force experienced by each free electron will be

$$F_e = ev_d B \sin\theta \quad \dots (2)$$

We assume the angle between $\vec{v}_d$ and $\vec{B}$ is θ.

Now total force on **N** free electrons will be

$$F = F_e \times N = (ev_d B \sin\theta) \times (nA\ell)$$

$$\text{or } F = (neAv_d)B\ell \sin\theta \quad \dots (3)$$

$$\because I = neAv_d$$

$$\therefore F = IB\ell \sin\theta \quad \dots (4)$$

In vector form

$$\vec{F} = I\vec{\ell} \times \vec{B} \quad \dots (5)$$

42. Force between two current carrying parallel conductors [Definition of one Ampere]

Two parallel current carrying conductors shown in the figures. Distance between them is **r**. The magnetic field due to wire **AB** is B_1 which is into the plane (on the right side) hence wire **CD** experiences force towards left side [Flemming's left hand rule].

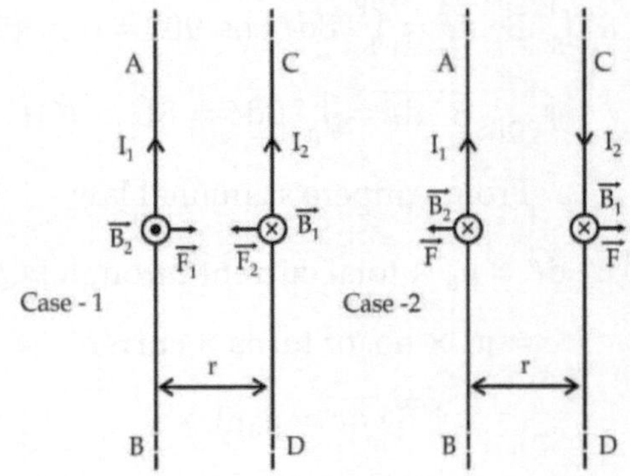

Similarly magnetic field due to wire **CD** at distance **r** on the left side is out of the plane of paper due to which wire **AB** experiences force on the right side.

$$\text{We know that } B_1 = \frac{\mu_0 I_1}{2\pi r} \quad .....(1)$$

$$\text{Thus } F_2 = I_2 B_1 L \sin 90° \quad .....(2)$$

$$\text{or } F_2 = I_2 B_1 L = I_2 \frac{\mu_0 I_1}{2\pi r} \times L$$

or force per unit length will be -

$$\frac{F_2}{L} = \frac{\mu_0}{2\pi}\frac{I_1 I_2}{r} \quad .....(3)$$

$$\text{Similarly } F_1 = I_1 B_2 L \quad \& \quad B_2 = \frac{\mu_0 I_2}{2\pi r}$$

$$\therefore F_1 = \frac{\mu_0 I_1 I_2 L}{2\pi r}$$

$$\text{or } \frac{F_1}{L} = \frac{\mu_0 I_1 I_2}{2\pi r} \quad ....(4)$$

Note - If currents are in same directions, parallel wires attract each other & if currents are in opposite directions they repel each other.

Definition of ampere – Relation (3) or (4) is used to define one ampere.

If $I_1 = I_2 = 1\text{ A}$ & $r = 1\text{ m}$ then

$$F = \frac{\mu_0}{4\pi}\frac{2I_1 I_2}{r} = 10^{-7} \times \frac{2\times1\times1}{1}$$

$$\text{Or } F = 2 \times 10^{-7}\ \text{Nm}^{-1}$$

$$[\text{Here } \frac{\mu_0}{4\pi} = 10^{-7}]$$

Thus *we can say that one ampere is that amount*

of current which is flowing through two parallel linear conductors placed in air/vacuum at a distance of one meter from each other due to which the conductors attract/repel each other with a force of $2 \times 10^{-7} N$ per meter of their length.

43. Torque acting on a current carrying loop placed in an external magnetic field

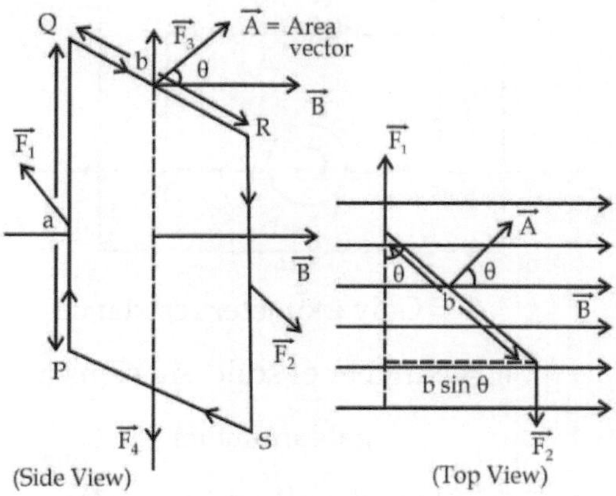

A closed rectangular loop **PQRS** is shown in figure. Sides **PQ** & **RS** are having lengths **a** and sides **QR** and **SP** having lengths **b**. Sides **PQ** and **RS** are perpendicular to the direction of magnetic field, hence experience force which is given by

$$\vec{F_1} = \vec{F_2} = I(\vec{a} \times \vec{B}) \quad \dots (1)$$

$\mathbf{F_1}$ & $\mathbf{F_2}$ are equal and opposite also parallel to each other, hence causing rotational effect, but forces $\mathbf{F_3}$ & $\mathbf{F_4}$ are having same line of action, equal and opposite thus cancel each other.

Now τ = Either force × ⊥ distance

$$\therefore \tau = IaB \times b\sin\theta \quad \dots (2)$$

Now $a \times b = A$ (area of the loop)

$$\therefore \tau = IAB \sin\theta \quad \dots (3)$$

If there are N turns in the loop then

$$\boldsymbol{\tau = NIAB \sin\theta} \quad \dots (4)$$

Term **NIA** is called magnetic dipole moment denoted by symbol **M**.

$$\therefore \vec{M} = NI\vec{A}$$

$$\therefore |\tau| = MB\sin\theta$$

$$\text{or } \vec{\tau} = \vec{M} \times \vec{B}$$

44. Moving coil Galvanometer

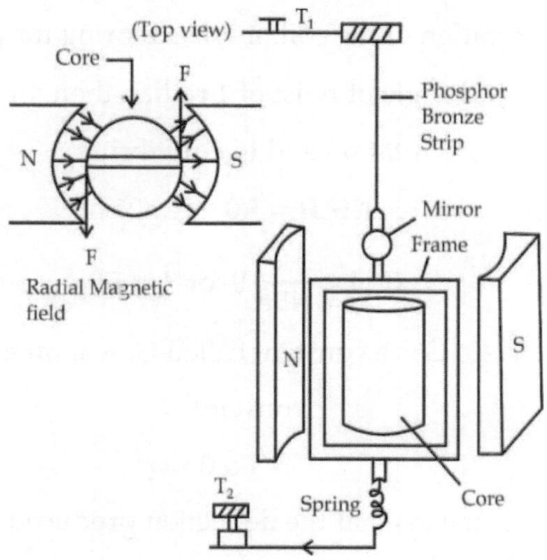

Moving coil galvanometer is an instrument which is used to detect and measure small amount of electric current.

Principle - *When a current carrying coil is placed in a uniform magnetic field, it experiences a torque.*

Radial field: We make a special arrangement in construction of galvanometer by using cylindrical core and curved magnetic poles. Due to this arrangement, the magnetic field and area vector of coil always perpendicular to each other.

Thus due to radial magnetic field

$$\tau = NIBA\sin 90°$$

$$\tau = \mathbf{NIAB} \quad \dots (1)$$

Working: When current is passed, coil rotates due to the formation of couple (pair of forces equal, opposite and parallel to each other). As coil rotates, the suspension strip and spring are twisted & an elastic restoring torque is developed in them which opposes the rotation of coil. This torque is proportional to the angle of rotation of the coil. If **k** is restoring torque per angle of twist of 1 radian then for θ rotation and in equilibrium.

$$\mathbf{NIAB = k\theta} \quad \dots (2)$$

$$\text{Or } I = \frac{k}{NBA}\theta \text{ or } I = G\theta$$

Here G is a constant called Galvanometer constant.

$$\therefore \mathbf{I \propto \theta}$$

It means that the deflection produced is directly proportional to the current flowing through the galvanometer i.e., galvanometer can have a linear scale. (This is another advantage of radial magnetic field)

Current sensitivity: It is the deflection produced in galvanometer when a unit current flows through it.

$$\mathbf{I_s = \frac{\theta}{I} = \frac{NBA}{k}}$$

$$\text{units} = \frac{\text{rad}}{\text{A}}$$

Voltage sensitivity: It is deflection produced in galvanometer when a unit voltage is applied across the terminals of galvanometer.

$$\mathbf{V_S = \frac{\theta}{V} = \frac{\theta}{IR} = \frac{nBA}{kR} = \frac{I_S}{R}}$$

$$\text{Units} = \text{rad V}^{-1}$$

45. Conversion of Galvanometer into Ammeter.

*We can convert a galvanometer into ammeter by using a low resistance wire in parallel called **SHUNT**.*

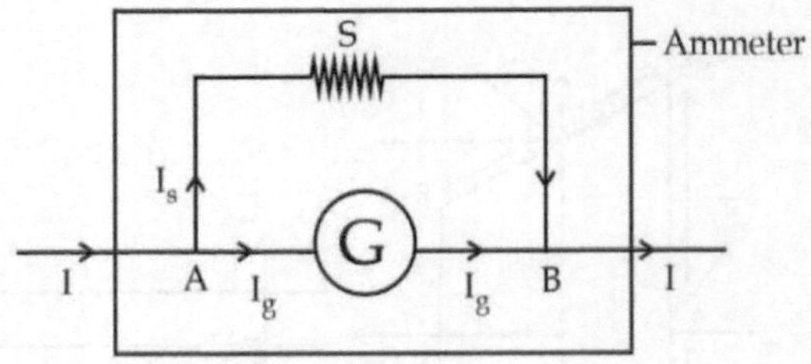

G = Galvanometer resistance

n = Number of scale divisions in galvanometer

$$\text{Now } V_A - V_B = I_g G \dots (1)$$

$$\text{Also } V_A - V_B = (I - I_g)S \quad \dots (2)$$

From equations (1) and (2)

$$\mathbf{S = \frac{I_g}{(I - I_g)} G} \quad \dots (3)$$

Effective resistance of ammeter will be

$$\mathbf{R_p = \frac{G \times S}{G + s}} \quad \dots (4)$$

Note: (1) If the ammeter range has to be increased **n** times its original range, the value of shunt resistance should be

$$\mathbf{S = \frac{G}{n-1}}$$

(2) Ammeter is connected in series.

46. Conversion of Galvanometer into Voltmeter

We can convert a Galvanometer into Voltmeter by connecting high resistance in series.

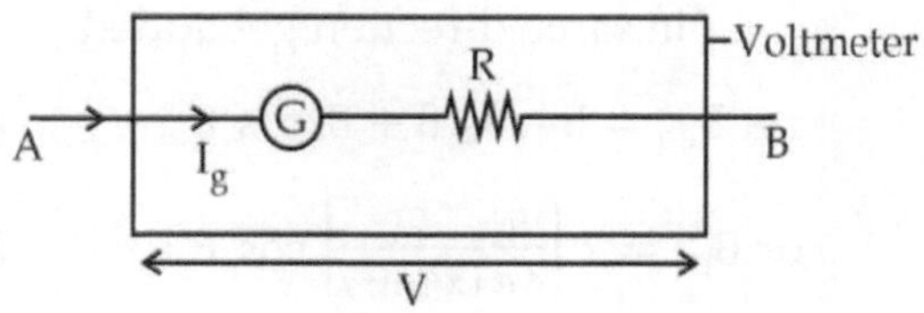

G = Galvanometer resistance

n = number of scale divisions

$$I_g = \frac{V}{G+R}$$

$$\text{Or } G + R = \frac{V}{I_g}$$

$$\text{or } R = \frac{V}{I_g} - G$$

Effective resistance $R_s = R + G$

Note: (1) If the voltmeter range is to be increased to **n** times its original value, the series resistance need to be connected will be

$$\mathbf{R = (n - 1)G}$$

(2) Resistance of Ideal voltmeter is infinity.

(3) Voltmeter is always connected in parallel.

Chapter - 5
Magnetism and Matter

47. Bar magnet and magnetic field intensity due to bar magnet along its axial line

- A bar magnet is magnetic dipole having two poles called as North pole and South pole.
- We use **m** to represent pole strength. It is analogous to a charge in electric dipole.
- **+m** corresponds to North pole and **−m** corresponds to South pole.

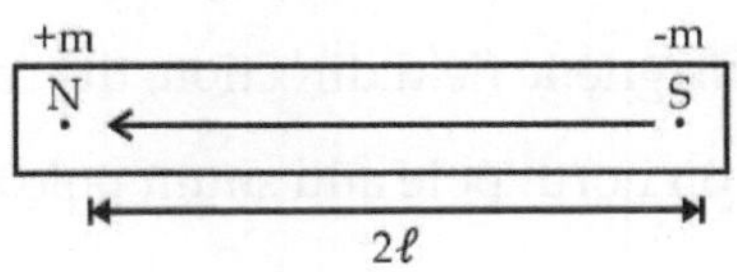

- Magnetic dipole moment of a bar magnet is the magnetic pole strength multiplied with the magnetic length of bar magnet.

$$\vec{M} = m \times 2\vec{\ell} \quad \dots (1)$$

$\mathbf{2\vec{\ell}}$ is directed from South pole to North pole.

- Magnetic field due to a bar magnet at a point on its axis is derived in a similar way as we derived for electric field due to electric dipole on axial line.

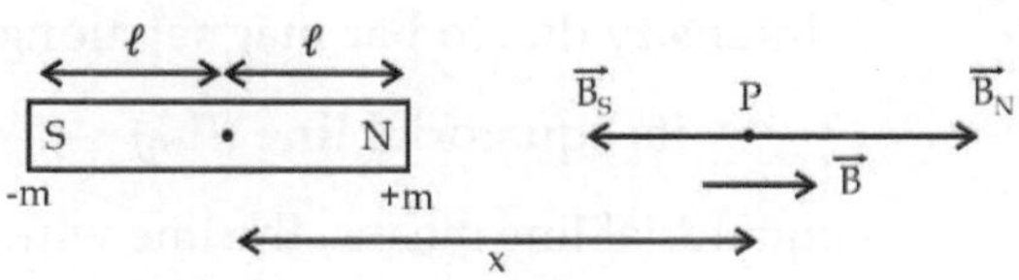

Magnetic field due to north pole at point P (way from North)

$$B_N = \frac{\mu_0}{4\pi}\frac{m}{(x-\ell)^2} \text{ (away from north)} \dots (2)$$

Magnetic field due to south pole at point P-

$$B_S = \frac{\mu_0}{4\pi}\frac{m}{(x+\ell)^2} \dots (3) \text{ (towards south)}$$

Net magnetic field strength can be written as

$$B_{axis} = B_N - B_s \quad (\because B_N > B_s)$$

$$B_a = \frac{\mu_0}{4\pi}\frac{m}{(x-\ell)^2} - \frac{\mu_0}{4\pi}\frac{m}{(x+\ell)^2}$$

$$\text{or } B_a = \frac{\mu_0 m}{4\pi}\left[\frac{1}{(x-\ell)^2} - \frac{1}{(x+\ell)^2}\right]$$

$$B_a = \frac{\mu_0 m}{4\pi}\left[\frac{(x+\ell)^2-(x-\ell)^2}{(x^2-\ell^2)^2}\right]$$

$$B_a = \frac{\mu_0 m}{4\pi}\left[\frac{4x\ell}{(x^2-\ell^2)^2}\right]$$

$$B_a = \frac{\mu_0}{4\pi}\frac{m\times 2\ell\times 2x}{(x^2-\ell^2)^2}$$

$$B_a = \frac{\mu_0}{4\pi}\frac{2Mx}{(x^2-l^2)^2}$$

$$\because M = m \times 2\ell$$

(magnetic moment of the magnet)

If $x \gg \ell$ then $x^2 - \ell^2 \approx x^2$

$$\therefore B_a = \frac{\mu_0}{4\pi}\frac{2Mx}{x^4} = \frac{\mu_0}{4\pi}\frac{2M}{x^3}$$

$$\mathbf{B_a = \frac{\mu_0}{4\pi}\frac{2M}{x^3}}$$

Note - It is similar to electric field due to electric dipole on axial line

$$E_a = \frac{2kp}{x^3}$$

48. Bar magnet and magnetic field intensity due to bar magnet along its equatorial line (B_{eq})

Equatorial line means, the line which divides the bar magnet into two halves perpendicular to its axis.

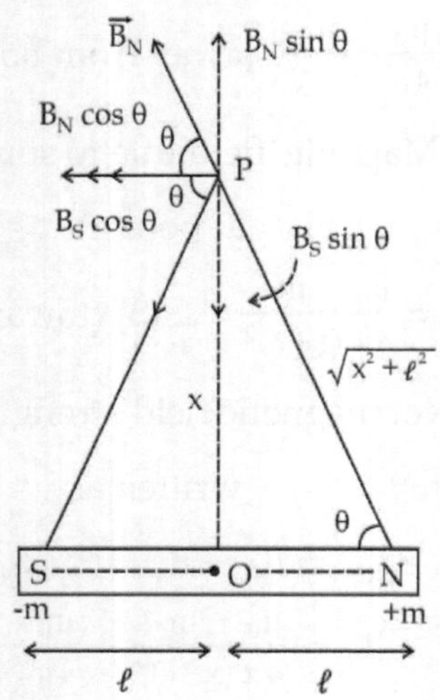

In the figure, the magnetic field direction due to both poles, is shown.

The magnetic field strength due to both poles can be written as

$$B_N = B_S = \frac{\mu_0}{4\pi}\frac{m}{\left(\sqrt{x^2+\ell^2}\right)^2}$$

$$\text{or } B_N = B_S = \frac{\mu_0}{4\pi}\frac{m}{(x^2+\ell^2)} \quad \text{... (1)}$$

The vertical components of $\mathbf{B_s}$ and $\mathbf{B_N}$ cancel each other and horizontal components are being in same direction get added.

$$\text{i.e. } B_{eq} = B_N\cos\theta + B_s\cos\theta \quad \text{... (2)}$$

$$\text{or } B_{eq} = 2\left[\frac{\mu_0}{4\pi}\frac{m}{(x^2+l^2)}\right]\cos\theta \quad \text{... (3)}$$

From the figure $\cos\theta = \frac{\ell}{\sqrt{x^2+\ell^2}}$

substituting this in eq. (3), we get

$$B_{eq} = 2\left[\frac{\mu_0}{4\pi}\frac{m}{(x^2+\ell^2)}\right]\frac{\ell}{(x^2+\ell^2)^{1/2}}$$

$$\text{or } B_{eq} = \frac{\mu_0}{4\pi}\frac{2m\ell}{(x^2+\ell^2)^{3/2}} \quad (\because 2m\ell = M)$$

$$\therefore B_{eq} = \frac{\mu_0}{4\pi}\frac{M}{(x^2+\ell^2)^{3/2}} \quad \text{... (4)}$$

If $x >> \ell$, then $x^2 + \ell^2 \approx x^2$

$$\therefore B_{eq} = \frac{\mu_0}{4\pi}\frac{M}{x^3} \quad \text{... (5)}$$

In vector form

$$\vec{\mathbf{B}}_{\mathbf{eq}} = (-)\,\frac{\mu_0}{4\pi}\frac{\vec{\mathbf{M}}}{\mathbf{x^3}}$$

Net magnetic field intensity on perpendicular bisector is opposite to the magnetic dipole moment vector. ($\vec{\mathbf{M}}$)

49. Torque on a bar magnet placed in a uniform magnetic field

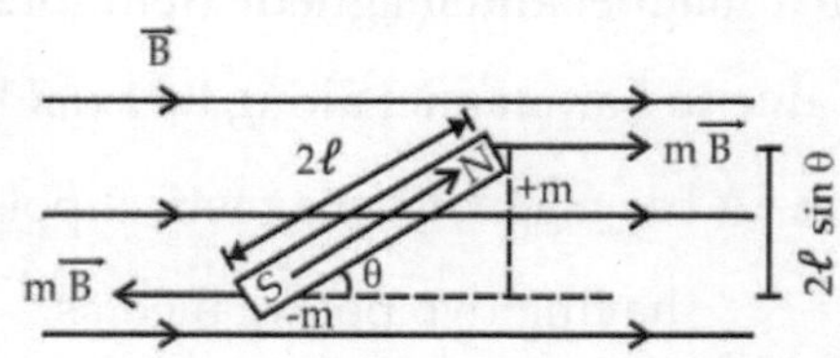

A bar magnet is placed in uniform magnetic field. Force on North and South poles due to external magnetic field is $\mathbf{m\vec{B}}$, as shown in the figure. Since the magnet axis is making an angle $\boldsymbol{\theta}$ with external magnetic field direction, the force acting on north pole and south pole are equal, and antiparallel, thus net force acting on the magnetic dipole is zero but two equal antiparallel forces generate a torque.

τ = Either force × ⊥ distance between line of action of two forces

$$\therefore \tau = mB \times 2\ell \sin \theta$$

$$\text{Or } \tau = (2m\ell)B\sin \theta$$

$$\therefore \tau = MB\sin \theta$$

$$\text{Or } \vec{\boldsymbol{\tau}} = \vec{\mathbf{M}} \times \vec{\mathbf{B}}$$

50. Potential energy of a magnetic dipole in external magnetic field

Potential energy of a magnetic dipole in external magnetic field is due to its particular position in the field.

When a magnet of magnetic dipole moment **M** is held in external magnetic field $\vec{\mathbf{B}}$ making an angle $\boldsymbol{\theta}$, then the torque acting on it is given by

$$\tau = MB\sin \theta \quad \ldots (1)$$

This torque tends to align the dipole in the direction of the field. When external work is done in rotating the dipole in external magnetic field against this torque, then this work is stored in the form of potential energy. Now small amount of work done in rotating the dipole through a small angle $\mathbf{d\theta}$ against the restoring torque is

$$dW = \tau d\theta \quad \ldots (2)$$

$$\text{or } dW = MB \sin \theta \, d\theta$$

& total work done in moving the dipole from angle θ_1 to an angle θ_2 will be -

$$W = \int_{\theta_1}^{\theta_2} MB \sin \theta d\theta$$

$$\text{or } W = MB \, [-\cos \theta]_{\theta_1}^{\theta_2}$$

$$W = -MB \, [\cos \theta_2 - \cos \theta_1]$$

This work done is stored as potential energy of the dipole

$$U = -MB \, [\cos \theta_2 - \cos \theta_1] \ldots (3)$$

When $\theta_1 = 90°$ and $\theta_2 = \theta$, then

$$U = -MB \, (\cos \theta - \cos 90°)$$

$$\text{or } U = -MB\cos \theta \quad \ldots (4)$$

In vector form -

$$\mathbf{U} = -\vec{\mathbf{M}} \cdot \vec{\mathbf{B}} \quad \ldots (5)$$

Case - I: When $\theta = 90°$

$$\because U = -MB\cos \theta = -MB\cos 90°$$

$$\therefore U = 0.$$

This position is considered as reference point.

Case - II: When $\theta = 0°$

the $U = -MB \cos 0° = -MB$

$U = -MB$ minimum value of potential energy.

This position is stable equilibrium.

Case - III: When $\theta = 180°$

$$U = -MB \cos 180° = MB$$

maximum value of potential energy.

This position is of unstable equilibrium.

51. Magnetic dipole moment of a revolving electron

In an atom electron revolves around the nucleus. It behaves like a current carrying loop. Since we know that a current carrying loop acts like a magnet i.e. it will have magnetic dipole moment. Suppose it completes one revolution in time period **T** and charge on electron is **e** then current

$$I = \frac{e}{T} \ \& \ \omega = \frac{2\pi}{T}$$

$\therefore T = \frac{2\pi}{\omega}$ thus current formula can be written as -

$$I = \frac{e\omega}{2\pi} \quad \dots (1)$$

Area of orbit of electron is

$$A = \pi r^2$$

& Magnetic moment of a current carrying loop is given by -

$$M = I \times A$$

$$\text{or } M = \frac{e\omega}{2\pi} \times \pi r^2$$

$$\text{or } M = \frac{e\omega r^2}{2} \quad \dots (2)$$

According to Bohr's theory, only those orbits are possible in which angular momentum of electron (mvr) is integral multiple of $\frac{h}{2\pi}$.

$$\text{i.e. } mvr = \frac{nh}{2\pi}$$

$$\because v = r\omega \ \therefore m \times r\omega \times r = \frac{nh}{2\pi}$$

$$\text{or } \omega r^2 = \frac{nh}{2\pi m} \quad \dots (3)$$

substituting above value in equation (2), we get

$$M = \frac{e}{2} \times \frac{nh}{2\pi m} = n\frac{eh}{4\pi m}$$

$$\text{i.e. } M = n(\mu_B) \quad \dots (4)$$

$$\text{Here } \boldsymbol{\mu_B} = \frac{\mathbf{eh}}{\mathbf{4\pi m}} \quad \dots (5)$$

$\boldsymbol{\mu_B}$ is the basic value of magnetic dipole moment of a revolving electron.

The higher values are integral multiple of this base value.

The value of $\boldsymbol{\mu_B} = \frac{\mathbf{eh}}{\mathbf{4\pi m}}$ is called **"Bohr Magneton".**

Its value is

$$\mu_B = \frac{1.6\times10^{-19}\times6.6\times10^{-34}}{4\times\frac{22}{7}\times9\times10^{-31}}$$

$$\mu_B = 9.27 \times 10^{-24} \text{ ampere metre}$$

52. Bar magnet as an equivalent solenoid

When we keep compass needle in the vicinity of a solenoid or bar magnet, we observe that the deflection pattern of the needle is similar in both. Using compass needle when we draw magnetic field lines we get same pattern. When a solenoid is cut into smaller solenoid, each segment behaves like a magnetic dipole (a smaller bar magnet) with weaker magnetic field. This is similar to the behaviour observed when bar magnet is cut into pieces.

Magnetic field due to a current carrying solenoid/coil having **N** turns is given by-

$$B_a = \frac{\mu_0}{4\pi}\frac{2\pi NIa^2}{(a^2+x^2)^{3/2}} \text{ (on its axis)}$$

a = Radius of solenoid

x = Distance of point of observation on axial line from the centre.

When $x \gg a$ then $a^2 + x^2 \approx x^2$

then magnetic field on axial line will be

$$B_a = \frac{\mu_0}{4\pi}\frac{2\pi NIa^2}{x^3}$$

$\because$ Magnetic dipole moment

$$M = NI\pi a^2 \text{ (i.e. } M = NIA)$$

$$\therefore B_a = \frac{\mu_0}{4\pi}\frac{2M}{x^3}$$

This is similar to magnetic field due to a bar magnet on the axial line of it.

i.e. $B_a = \frac{\mu_0}{4\pi}\frac{2M}{x^3}$

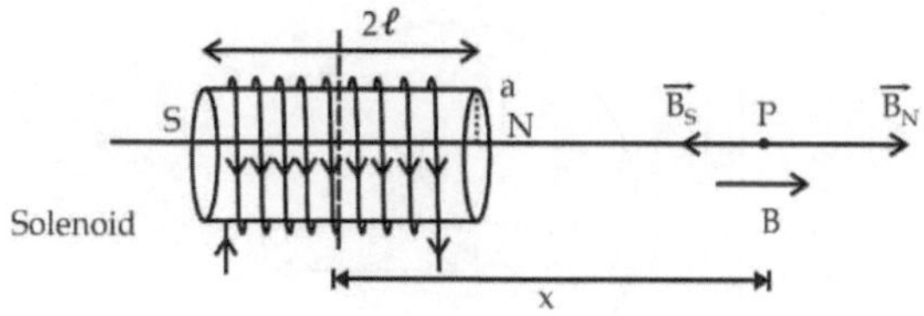

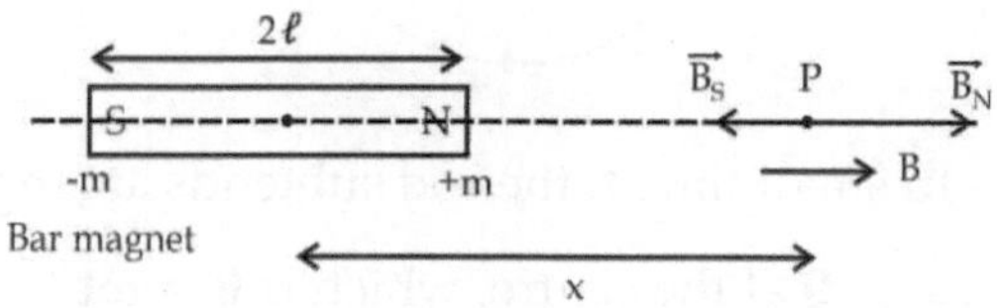

Note- The dissimilarity between bar magnet and solenoid is, outside the solenoid, the magnetic field is negligible i.e. on equatorial line we will not get magnetic field due to current carrying solenoid but we will get due to a bar magnet.

53. Relation between magnetic permeability (μ_r) and susceptibility (χ)

Magnetic susceptibility (χ) represents the response of a magnetic material to external magnetic field or its ease to get magnetised.

$$\because \mu_r = \frac{\mu_m}{\mu_0}$$

then $\mu_r = 1 + \chi$ [Here χ = (Chi)]

Also χ is inversely proportional to temperature.

$$\chi \propto \frac{1}{T}$$

- For paramagnetic materials

  $\mu_r > 1$ (But small)

  And χ is small and positive.

- For Diamagnetic materials-

  $\mu_r < 1$

  And χ is always negative.

- For ferromagnetic materials-

  $\mu_r \gg 1$ & $\chi \gg 1$

Curie's Law

$$\chi = \frac{C}{T}$$

Here C = Curie's constant

Curie Temperature - Temperature at which ferromagnetic material loses its domain structure and start acting like paramagnetic material is called curie temperature.

54. Magnetic flux and Gauss theorem

Gauss' theorem for electrostatics states that the net flux linked with the Gaussian surface is net charge enclosed by it times $\frac{1}{\varepsilon_0}$.

$$\text{i.e., } \phi_E = \oint \vec{E} \cdot \vec{ds} = \frac{q}{\varepsilon_0}$$

But in a magnet, monopole does not exist. There will always be a magnetic dipole however small a magnet be. Also magnetic field lines form a closed curve due to the dipole. If we draw a Gaussian surface enclosing a bar magnet then the number of field lines entering the surface will be equal to the number of field lines leaving the surface thus net magnetic flux through closed surface will be zero.

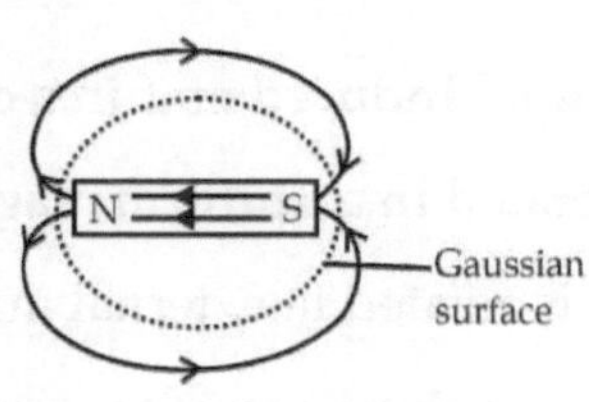

$$\therefore \phi_B = \oint \vec{B} \cdot \vec{ds} = 0$$

Chapter - 6

Electromagnetic Induction

55. Motional Electromotive force

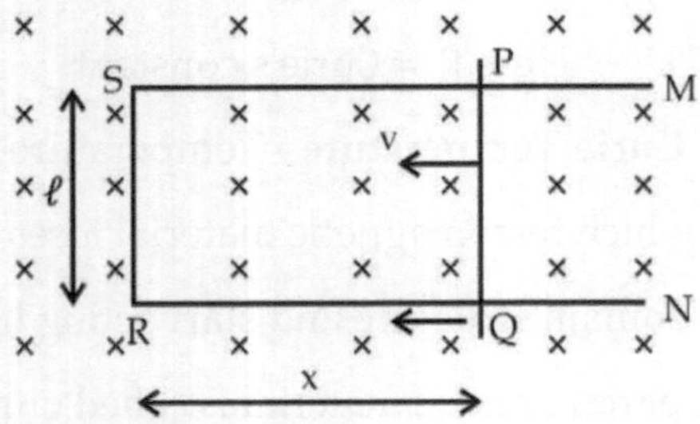

Consider conductor **PQ** moving in magnetic field with velocity **v** over a conducting rectangular frame **MSRN** towards left side. Here **RS** = ℓ and **RQ = x.**

The magnetic flux ϕ_B enclosed by the loop **PQRS** can be written as

$$\phi_B = B \cdot A = B \times x \times \ell \qquad \ldots (1)$$

Since conductor is moving towards left hand side **x** value is changing with time

And $v = (-)\frac{dx}{dt}$ (Because x value is decreasing with time)

The flux magnitude is decreasing with time which induces emf in **PQ** conductor.

$$\therefore \varepsilon = (-)\frac{d\phi_B}{dt} = (-)\frac{d}{dt}(Bx\ell) \qquad \ldots (2)$$

$$\text{or } \varepsilon = B\ell\left((-)\frac{dx}{dt}\right) = Bv\ell$$

$$\therefore \boldsymbol{\varepsilon = Bv\ell} \qquad \ldots (3)$$

56. Motional Induced emf in a conducting rod rotated in a uniform magnetic field

A rod **PQ** is rotated in external magnetic field **(B)** into the plane of the paper. The angular velocity is **ω** and time taken to complete one revolution is **T**.

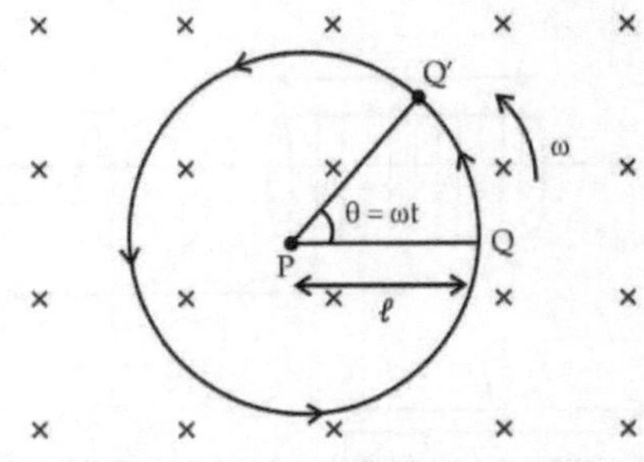

In small time **t**, the rod subtends an angle **θ** at the centre, which is **θ = ωt.**

Now area swept by the metal rod is

$$A = \frac{1}{2} \times \ell \times QQ' \qquad \ldots (1)$$

$$\because \theta = \frac{QQ'}{\ell} \quad \therefore QQ' = \ell\theta$$

$$\text{Thus } A = \frac{1}{2} \times \ell \times \ell\theta = \frac{1}{2}\ell^2\theta \qquad \ldots (2)$$

& Magnetic flux linked with it -

$$\phi = BA\cos\theta \qquad \text{Here } \theta = 0^\circ$$

(Because the area vector and magnetic field are in the same direction)

$$\phi = B \times \frac{1}{2}\ell^2\theta \qquad \ldots (3)$$

And Induced emf will be

$$|\varepsilon| = \frac{d\phi}{dt} = \frac{d}{dt}\left(B \times \frac{1}{2}\ell^2\theta\right)$$

$$\text{or } |\varepsilon| = \frac{1}{2}B\ell^2\frac{d\theta}{dt}$$

$$\therefore \boldsymbol{\varepsilon = \frac{1}{2}B\omega\ell^2} \qquad \ldots\ldots(4)$$

57. Power dissipation in a conductor moved in external magnetic field

As shown in the figure when we move an arm of rectangular conductor of length $\boldsymbol{\ell}$ in a uniform external magnetic field **(B)** which is perpendicular to the plane & directed into the plane, motional emf is induced in the conductor **PQ** which is given by -

$$\varepsilon = Bv\ell \quad \ldots (1)$$

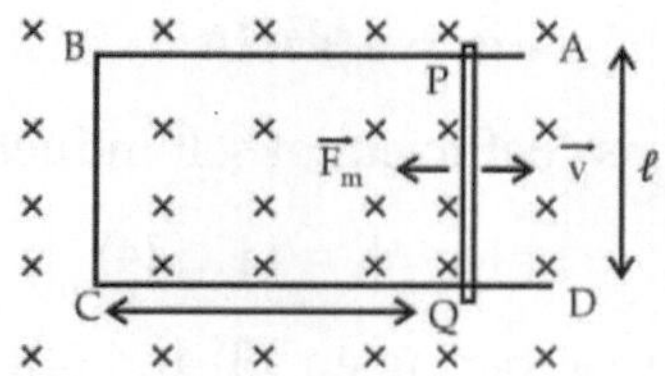

Lets take resistance of arm **PQ** is **R,** then current induced in the conductor will be-

$$I = \frac{\varepsilon}{R} = \frac{Bv\ell}{R} \quad \ldots (2)$$

Now, the force acting on the conductor **PQ** which is moving in magnetic field will be

$$F = BI\ell$$

$$\text{or } \mathbf{F = B \times \ell \times \frac{Bv\ell}{R} = \frac{B^2\ell^2v}{R}} \quad \ldots (3)$$

Direction of force is always opposite to the direction of motion of the conductor.

Hence power required to push the conductor with velocity (**v**) is given by

$$P = F \times v = \frac{B^2l^2v}{R} \times v$$

$$\mathbf{P = \frac{B^2\ell^2v^2}{R}} \quad \ldots (4)$$

Also this mechanical work is dissipated in the form of heat energy.

$$H = P \times t = \frac{B^2\ell^2v^2}{R} \times t$$

Note: We can find the charge induced in the conductor due to change in magnetic flux.

$$\because \varepsilon = IR = \frac{d\phi}{dt}$$

$$\text{Or } \frac{dQ}{dt} \times R = \frac{d\phi}{dt}$$

$$\text{Or } dQ = \frac{d\phi}{R} \quad \ldots (5)$$

58. Energy stored in an inductor

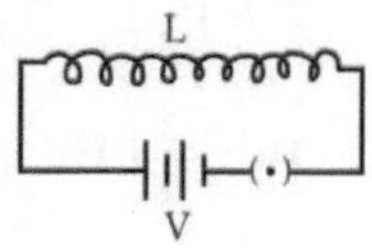

As shown in the figure, a source of emf **E** is connected to an inductor of self inductance **L.** When key is pressed the current starts increasing in the circuit upto a steady value. When current is increasing in the circuit, an opposing emf is induced which opposes the growth of current in the circuit. To maintain this current through the circuit, source emf has to spend energy (in sending the current), and this energy is stored in the inductor in the form of magnetic field.

At **t = 0**, the current in the inductor is zero and as we press the key let's assume rate of growth of current is $= \frac{dI}{dt}$

then emf induced will be

$$\varepsilon = L\frac{dI}{dt} \text{ (in magnitude)} \quad \ldots (1)$$

If the source sends a constant current **I** through the inductor for a small time **dt**, then small amount of work done by the source will be -

$$dW = \varepsilon Idt = \left(L\frac{dI}{dt}\right) Idt$$

$$\text{or } dW = LIdI \quad \ldots (2)$$

The total amount of work done by the source emf till the current reaches to its maximum value i.e. from I = 0 to I = I will be -

$$w = \int_0^I LIdI = L\int_0^I IdI$$

$$w = L\left[\frac{I^2}{2}\right]_0^I = \frac{1}{2}LI^2$$

$$\therefore \mathbf{W = \frac{1}{2}LI^2} \quad \dots (3)$$

Remember: The energy stored in the inductor at the expense of energy of source emf is in the form of magnetic field. If an AC source is connected instead of DC then during one half of the cycle, energy is stored in the inductor and during the next half cycle, it is returned to the source. Because of this reason, the energy and average electric power spent in inductor (in AC circuit) is zero.

59. Self Inductance of a long solenoid

Definition - *Self induction is that property of the coil due to which, it opposes change in the magnitude of current flowing through it by inducing an opposing emf in itself.*

We know that magnetic field inside the solenoid is given by

$$B = \mu_0 nI \quad \dots(1) \quad \left[\text{Here } n = \frac{N}{\ell}\right]$$

N = Total number of turns

ℓ = length of solenoid

I = current flowing through solenoid

Now flux linked with each turn of solenoid

$$\phi = B \times A = \mu_0 nI \times A \quad \dots (2)$$

A = Area of each turn

Total flux through entire solenoid

ϕ = Flux from each turn × Area of Total number of turns

$$\therefore \phi = \mu_0 nI \times A \times n \times \ell$$

Here total number of turns $N = n \times \ell$

$$\therefore \phi = \mu_0 n^2 \ell IA \quad \dots (3)$$

If L is coefficient of self inductance, then

$$\phi = LI \dots (4)$$

Comparing eq. (3) and (4), we get

$$LI = \mu_0 n^2 \ell IA$$

$$\text{or } L = \mu_0 n^2 \ell A \quad \dots (5)$$

$$\because n = \frac{N}{\ell}$$

$$\therefore L = \frac{\mu_0 N^2 \ell A}{\ell^2}$$

$$\text{or } \mathbf{L = \frac{\mu_0 N^2 A}{\ell}} \quad \dots (6)$$

If the core is filled with any other magnetic material, then $\mu = \mu_0\mu_r$ & thus -

$$\mathbf{L = \frac{\mu_0 \mu_r N^2 A}{\ell}} \quad \dots\dots (7)$$

60. Mutual inductance of two long coaxial solenoids

"Mutual induction is that property of two coils due to which each of them opposes the change in current in other coil by inducing an opposing emf in itself".

Lets assume there are two coils in close association with each other as shown in the figure.

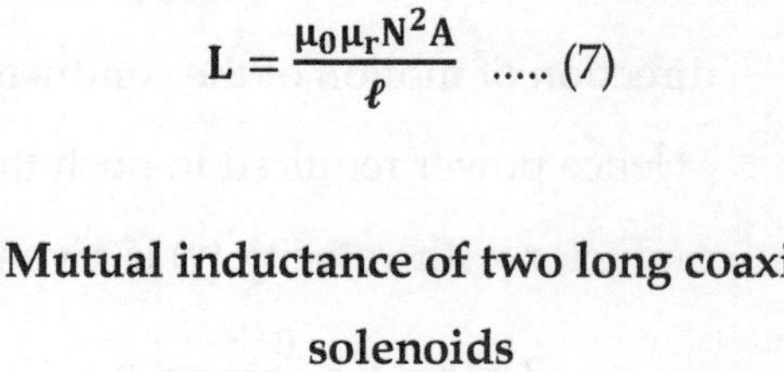

S_1 = solenoid - 1 (inner coil)

S_2 = Solenoid - 2 (Outer coil)

I_1 = current in S_1

I_2 = current in S_2

n_1 = number density in S_1.

n_2 = number density in S_2

ℓ = length of both solenoids

A_1 = Area of cross section of S_1

A_2 = Area of cross section of S_2

When current $\mathbf{I_2}$ starts flowing through $\mathbf{S_2}$, there will be a magnetic flux linked with $\mathbf{S_1}$ which is given by

$$\phi_1 = M_{12}I_2 \quad \dots (1)$$

M_{12} = coefficient of mutual inductance of $\mathbf{S_1}$ with respect to $\mathbf{S_2}$

Now magnetic field of $\mathbf{S_2}$ due to current flowing through it is

$$B_2 = \mu_0 n_2 I_2 \quad \dots (2)$$

$\therefore$ flux associated with $\mathbf{S_1}$ will be

$$\phi_1 = B_2 A_1 N_1$$

Here $N_1 = n_1 \times \ell$ (Total turns in $\mathbf{S_1}$)

$$\text{or } \phi_1 = \mu_0 n_2 I_2 \times A_1 \times n_1 \times \ell \quad \dots (3)$$

From equation (1) & (3)

$$M_{12}I_2 = \mu_0 n_1 n_2 A_1 \ell \times I_2$$

$$\text{or } \mathbf{M_{12} = \mu_0 n_1 n_2 A_1 \ell} \quad \dots (4)$$

Now, the flux associated with $\mathbf{S_2}$ due to magnetic field of $\mathbf{S_1}$ will be

$$\phi_2 = B_1 A_1 N_2 \quad \dots (5)$$

Please note that here we have taken $\mathbf{A_1}$ and not $\mathbf{A_2}$ because outside the solenoid magnetic field is zero thus in $\mathbf{S_2}$ magnetic flux will be present in $\mathbf{A_1}$ area only.

$$\& \ \phi_2 = M_{21}I_1 \quad \dots (6)$$

$$\& \ B_1 = \mu_0 n_1 I_1 \quad \dots (7)$$

Equating equations (5) & (6)

$$M_{21}I_1 = B_1 A_1 N_2$$

Substituting B_1 value

$$M_{21}I_1 = \mu_0 n_1 I_1 \times A_1 \times n_2 \times \ell$$

$$\text{or } \mathbf{M_{21} = \mu_0 n_1 n_2 A_1 \ell} \quad \dots (8)$$

This equation is same as equation (4)

$$\text{Thus } M_{12} = M_{21} \quad \dots (9)$$

Chapter - 7

Alternating Current

61. Mean value or average value of alternating current

"Mean value of AC is that value of steady current, which sends same amount of charge through a given circuit in half cycle of AC as sent by AC through the same circuit in same time".

We know that -

$$I = I_0 \sin \omega t \quad \dots (1)$$

Assuming for a small interval **dt** the current remains constant,

thus small amount of charge sent by source in this interval

$$dq = Idt \quad \dots (2)$$

If **q** is the total charge sent in first half cycle i.e. from $\mathbf{0 \rightarrow T/2}$

$$\text{Then } q = \int_0^{T/2} Idt$$

And from equation - (1)

$$q = \int_0^{T/2} I_0 \sin \omega t \, dt$$

$$\text{or } q = I_0 \left[-\frac{\cos \omega t}{\omega}\right]_0^{T/2}$$

$$\text{or } q = -\frac{I_0}{\omega}\left[\cos \omega \frac{T}{2} - \cos 0°\right]$$

$$\left[\because \omega \frac{T}{2} = \frac{2\pi}{T} \times \frac{T}{2} = \pi\right]$$

$$\therefore q = -\frac{I_0}{\omega}[\cos \pi - \cos 0°]$$

$$\text{or } q = -\frac{I_0}{\omega}[-1-1] = \frac{2I_0}{\omega} \quad \dots (3)$$

If I_{av} is average value of AC over first half cycle, then from definition, charge

$$q = I_{av} \times \frac{T}{2} \quad \dots (4)$$

Then from eq - (3) and (4)

$$I_{av} \times \frac{T}{2} = \frac{2I_0}{\omega} = 2I_0 \times \frac{T}{2\pi}$$

$$\text{Or } \mathbf{I_{av}} = \frac{\mathbf{2I_0}}{\pi} \quad \dots (5)$$

$$\text{or } \mathbf{I_{av} = 0.637I_0} \quad \dots (6)$$

Note - Remember mean value of AC over complete cycle is always zero.

$$\text{for } 0 \to \frac{T}{2} = 0.637I_0$$

$$\& \text{ for } \frac{T}{2} \to T = (-)\ 0.637I_0$$

- Mean value or average value of alternating emf is given by

$$\mathbf{E_{av}} = \frac{\mathbf{2E_0}}{\pi} = \mathbf{0.637E_0}$$

Remember average value of alternating emf over complete cycle is always zero.

62. Root mean square value of alternating current

"It is that value of steady current, which generates same amount of heat energy in a given resistor in the same time as produced by alternating current in same time in same resistor."

We know that

$$I = I_0 \sin \omega t \quad \dots (1)$$

We assume that in time **dt** current through resistor **R** remains constant, then small amount of heat produced is given by

$$dH = I^2 R dt \quad \dots (2)$$

Now in one complete cycle **0 → T**, the total amount of heat produced will be

$$H = \int_0^T I^2 R dt \quad \dots (3)$$

From eq-(1) we can write the heat equation as

$$H = \int_0^T (I_0 \sin \omega t)^2 R dt$$

$$H = I_0^2 R \int_0^T \left(\frac{1-\cos 2\omega t}{2}\right) dt$$

$$H = \frac{I_0^2 R}{2}\left[\int_0^T 1dt - \int_0^T \cos 2\omega t dt\right]$$

$$H = \frac{I_0^2 R}{2}\left[t - \frac{\sin 2\omega t}{2\omega}\right]_0^T$$

$$H = \frac{I_0^2 R}{2}\left[(T-0) - \left(\frac{\sin 2\omega T}{2\omega} - \sin 0^\circ\right)\right]$$

$$\text{Now } \omega T = 2\pi$$

$$\therefore H = \frac{I_0^2 R}{2}\left[T - \frac{\sin 2\times 2\pi}{2\omega}\right]$$

$$\because \sin 4\pi = 0$$

$$\therefore \mathbf{H} = \frac{\mathbf{I_0^2 RT}}{\mathbf{2}} \quad \dots (4)$$

If I_{rms} is the root mean square value of alternating current over one complete cycle then from equation (4)

$$H = I_{rms}^2 RT = \frac{I_0^2 RT}{2}$$

$$\therefore \mathbf{I_{rms}} = \frac{\mathbf{I_0}}{\sqrt{\mathbf{2}}} = \mathbf{0.707I_0} \quad \dots (5)$$

Note - In all AC equations such as

$$\mathbf{I = I_0 \sin \omega t},$$

I represents instantaneous value of AC and $\mathbf{I_0}$ represents peak value.

- AC ammeter measures rms value.
- Alternating emf rating is also rms value.

Note: In numericals the value of emf for alternating source is always rms value.

- Root mean square value of alternating emf is given by

$$E_{rms} = \frac{E_0}{\sqrt{2}}$$

63. AC circuit containing resistor only (Phasor treatment)

"*Phasor is a rotating vector which represents a quantity varying sinusoidally with time.*"

Phasor is shown by an arrow & its length represents the peak value of that quantity. They rotate anticlockwise and their angular velocity is equal to angular velocity of that quantity. Projection of phasor on any axis gives instantaneous value of that quantity. When projection is along vertical axis, the value comes in sine form and if taken on horizontal axis it comes in cosine form. For AC circuit containing resistor only, the source emf is-

$$E = E_0 \sin \omega t \qquad \dots (1)$$

If **I** is instantaneous value of current then-

$$I = \frac{E}{R} = \frac{E_0}{R} \sin \omega t$$

$$\text{or } I = I_0 \sin \omega t \qquad \dots (2)$$

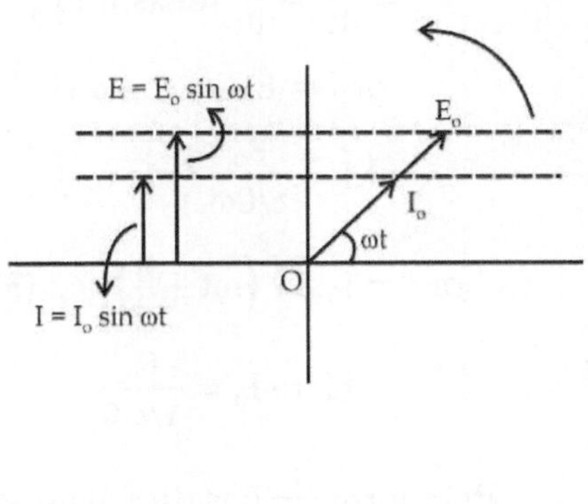

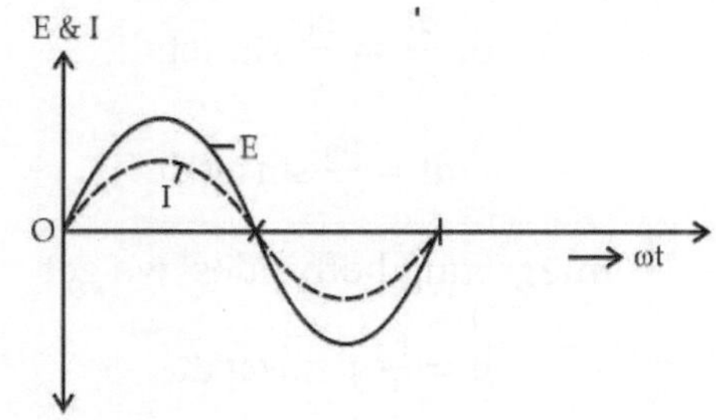

Note: AC Circuit containing resistor only, **E** and **I** are in phase. The magnitude of peak value of current is lesser than peak value of emf because $\mathbf{I_0} = \frac{\mathbf{E_0}}{\mathbf{R}}$

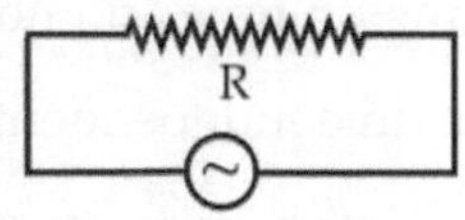

64. AC circuit containing inductor only (Phasor treatment)

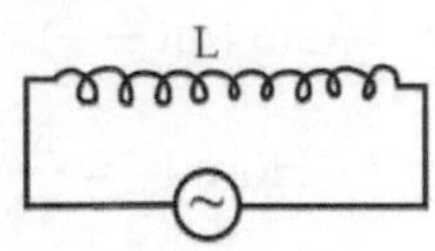

The alternating emf supplied in the circuit by source is given by

$$E = E_0 \sin \omega t \quad \dots (1)$$

In AC circuit current changes with time due to which emf is induced in inductor.

If $\frac{dI}{dt}$ is rate of change of current then induced emf will be

$$E = -L\frac{dI}{dt} \qquad \dots (2)$$

(−) sign is there because the induced emf opposes the change of current.

Now to maintain the flow of current in this circuit, applied voltage must be equal and opposite to the induced voltage i.e.,

$$E = (-)\left[-L\frac{dI}{dt}\right] = E_0 \sin \omega t$$

$$\text{or } \frac{dI}{dt} = \frac{E_0}{L} \sin \omega t$$

$$\therefore dI = \frac{E_0}{L} \sin \omega t\, dt$$

Integrating both sides, we get

$$I = \frac{E_0}{L} \int \sin \omega t\, dt$$

$$I = \frac{E_0}{L}\left(-\frac{\cos \omega t}{\omega}\right) + \text{constant} \quad \ldots (3)$$

In equation- (3) the integration constant has dimensions of current and it is time independent. But in AC circuit, current is depending on time and no component which is time independent exists, thus value of integration constant is zero.

Then from equation (3)

$$I = (-)\frac{E_0}{\omega L} \cos \omega t = -\frac{E_0}{\omega L} \sin\left[\frac{\pi}{2} - \omega t\right]$$

$$\text{Or } I = \frac{E_0}{\omega L} \sin\left(\omega t - \frac{\pi}{2}\right) \quad \ldots\ldots (4)$$

$$\text{Here } I_0 = \frac{E_0}{\omega L}$$

$$\therefore \mathbf{I = I_0 \sin\left(\omega t - \frac{\pi}{2}\right)} \quad \ldots\ldots (5)$$

Thus from eq-(5) we can see that current lags behind the alternating voltage **E** by a phase angle of **90°**. In equation- (4) term **ωL** represents inductive reactance X_L, it is the resistance offered by the inductor in the circuit. It has dimensions of resistance.

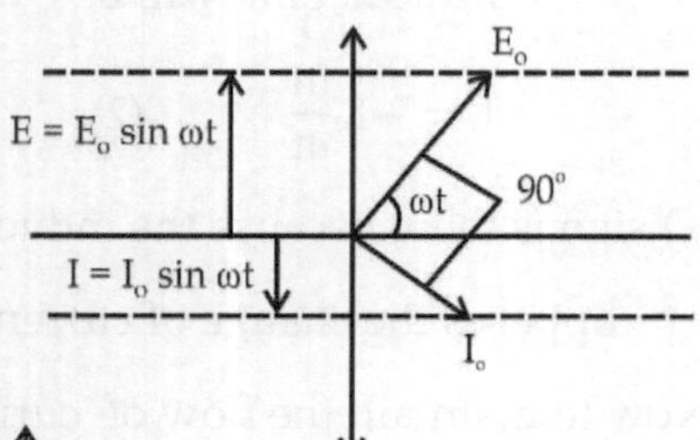

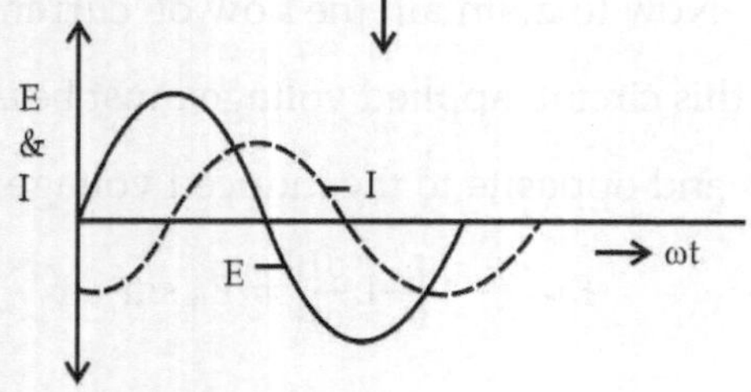

65. AC circuit containing capacitor only (Phasor treatment)

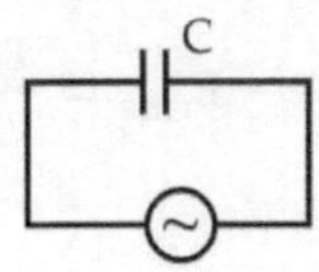

Let an AC source be connected to a capacitor only as shown in the figure. Suppose the emf supplied to the circuit is

$$E = E_0 \sin \omega t \quad \ldots (1)$$

When current is flowing in the circuit, for the first half cycle capacitor, it is getting charged and for the next half cycle it is getting discharged. If at an instant **t, q** charge is present on the capacitor then potential difference across the plates of the capacitor will be $V = q/C$

At every instant, this p.d. **V** must be equal to the emf applied i.e.,

$$V = \frac{q}{C} = E = E_0 \sin \omega t$$

$$V = \frac{q}{C} = E = E_0 \sin \omega t$$

$$\text{or } q = CE_0 \sin \omega t \quad \ldots (2)$$

If **I** is the instantaneous value of current in the circuit at instant **t** then

$$I = \frac{dq}{dt} = \frac{d}{dt}[CE_0 \sin \omega t]$$

$$\text{or } I = E_0 C\omega \cos \omega t$$

$$\text{or } I = \frac{E_0}{1/C\omega}(\cos \omega t)$$

$$\text{or } I = I_0 \sin\left(\omega t + \frac{\pi}{2}\right) \quad \ldots (3)$$

$$\text{Here } I_0 = \frac{E_0}{1/\omega C}$$

Here term $\frac{1}{\omega C}$ has dimensions of resistance, it is the resistance offered by capacitor and it is called capacitative reactance $\mathbf{X_C}$ and $\mathbf{X_C = \frac{1}{\omega C}}$

From equation (1) & (3) we can say that in AC circuit containing capacitor only, current leads voltage by **90°**.

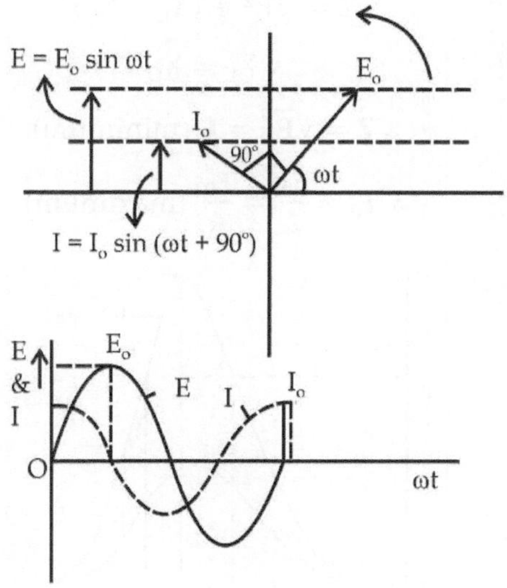

66. LCR series circuit (Phasor treatment)

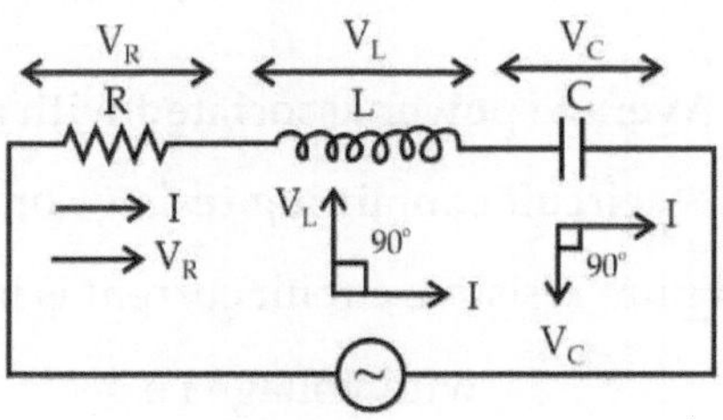

Let a pure resistor, inductor and capacitor be connected in series to a source of alternating emf. Current is same through all of them. It is represented by -

$$I = I_0 \sin \omega t \quad \ldots (1)$$

Since we know that current and voltage are in same phase in resistor.

Then phasor $\vec{V}_R = \vec{I}_0 R \quad \ldots (2)$

We represent current phasor $\vec{\mathbf{I}}_0$ along x-axis & voltage phasor for resistor $\vec{\mathbf{V}}_R$ also along x-axis. Now maximum voltage across **L** is $\vec{V}_L = \vec{I}_0 X_L$.

Since we know that voltage across inductor leads current by **90°** we represent $\vec{\mathbf{V}}_L$ along +y-axis.

The maximum voltage across C will be

$$\vec{V}_c = \vec{I}_0 X_c$$

Since voltage across capacitor lags behind current by **90°** we show $\vec{\mathbf{V}}_c$ along (-) y-axis.

If we assume $\vec{V}_L > \vec{V}_C$ & since the phase difference between them is **180°** then the net voltage is

$\vec{V}_L - \vec{V}_C$ along (+) y-axis.

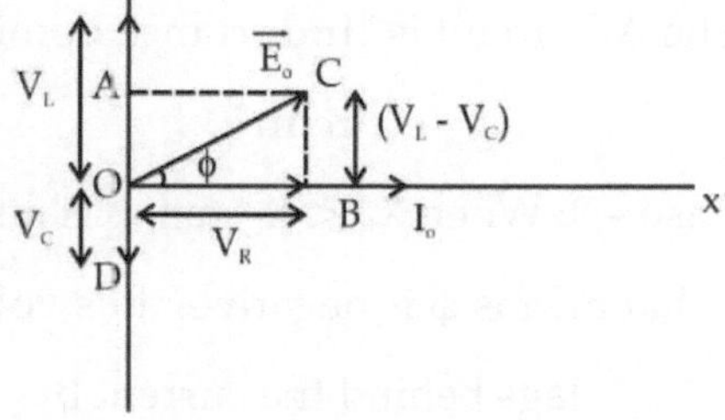

$\vec{E}_0$ is the vector sum of $\vec{V}_R, \vec{V}_L$ and $\vec{V}_c$, which makes angle $\boldsymbol{\phi}$ with the current phasor $\vec{\mathbf{I}}_0$.

$$\text{Now } E_0 = \sqrt{V_R^2 + (V_L - V_C)^2}$$

$$\text{or } E_0 = \sqrt{(I_0R)^2 + (I_0X_L - I_0X_c)^2}$$

$$\text{or } E_0 = I_0\sqrt{R^2 + (X_L - X_C)^2}$$

$$\text{or } \frac{E_0}{I_0} = \sqrt{R^2 + (X_L - X_c)^2} = Z \quad \ldots (3)$$

Here **Z** is "**Impedence**" of the circuit which is total effective resistance of LCR series circuit.

From phasor diagram we can see that voltage leads current by a phase angle $\boldsymbol{\phi}$

$$\& \tan\phi = \frac{V_L - V_c}{V_R} = \frac{I_0X_L - I_0X_C}{I_0R}$$

$$\text{Or } \tan\phi = \frac{X_L - X_C}{R} \quad \ldots (4)$$

& emf can be represented as -

$$E = E_0 \sin(\omega t + \phi) \quad \dots (5)$$

Case - I : When $X_L = X_C, \tan\phi = 0$

$\therefore \phi = 0°$, this condition is called **resonance** and in this condition

$$V_L = V_C \;\&\; Z = R$$

also source emf $= V_R$ and current is in phase with the voltage.

- Circuit is non-inductive.

Case - 2 : When $X_L > X_C, \tan\phi$ is positive that means ϕ is positive and voltage leads current by a phase angle ϕ.

The AC circuit is "**Inductance dominated circuit**".

Case - 3: When $X_L < X_C, \tan\phi$ is negative that means ϕ is negative, thus voltage lags behind the current by a phase angle ϕ.

The AC circuit is "**capacitance dominated circuit**".

67. Resonance in LCR series circuit

In LCR series circuit, at a particular frequency, the current becomes maximum, impedance becomes minimum, potential difference across inductor becomes equal to the potential difference across capacitor and current is in phase with the voltage, it is called **series resonance circuit.**

At resonance $X_L = X_C$

$$\boldsymbol{\omega_r L = \frac{1}{\omega_r C}} \quad \dots (1)$$

ω_r = Angular frequency at resonance

$$\& \; \omega_r = 2\pi\upsilon_r$$

$$\therefore 2\pi\upsilon_r \times L = \frac{1}{2\pi\upsilon_r \times C}$$

$$\text{or } \boldsymbol{\upsilon_r = \frac{1}{2\pi\sqrt{LC}}} \quad \dots (2)$$

$$\because Z = \sqrt{R^2 + (X_L - X_C)^2}$$

$$\because X_L = X_C$$

$$\therefore \mathbf{Z = \sqrt{R^2} = R} \text{ (minimum)}$$

$$\therefore \mathbf{I_0 = \frac{E_0}{Z} = \frac{E_0}{R}} \text{ (maximum)}$$

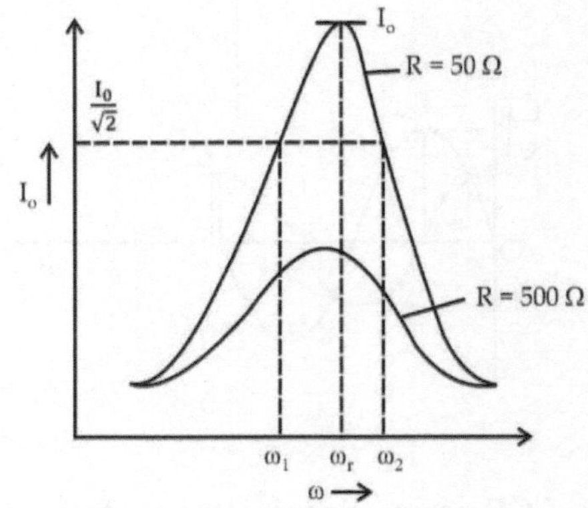

68. Average power associated with an AC circuit containing resistor only

In pure resistive circuit current is in phase with voltage i.e.,

$$E = E_0 \sin\omega t \quad \text{and} \quad I = I_0 \sin\omega t$$

thus instantaneous power will be

$$P = E \times I = (E_0 \sin\omega t)(I_0 \sin\omega t)$$

$$\text{or } EI = E_0 I_0 \sin^2 \omega t \quad \dots (1)$$

If instantaneous power remains constant for a small time **dt** then we can say that

$$\frac{dW}{dt} = E_0 I_0 \sin^2 \omega t$$

$$\text{Or } dW = E_0 I_0 \sin^2 \omega t \, dt \quad \dots (2)$$

And total work done over one full cycle will be

$$W = \int_0^T E_0 I_0 \sin^2 \omega t \, dt$$

$$W = E_0 I_0 \int_0^T \left(\frac{1-\cos 2\omega t}{2}\right) dt$$

$$\text{or } W = \frac{E_0 I_0}{2}\int_0^T dt - \frac{E_0 I_0}{2}\int_0^T \cos 2\omega t dt$$

$$\text{Here } \int_0^T \cos 2\omega t\, dt = 0$$

$$\text{Or } W = \frac{E_0 I_0}{2} T - 0$$

∴ Average power supplied to **R** over a complete cycle

$$P = \frac{W}{T} = \frac{E_0 I_0}{2}$$

$$\therefore P = \frac{E_0}{\sqrt{2}} \times \frac{I_0}{\sqrt{2}}$$

Or $\mathbf{P = E_{rms} I_{rms}}$ ... (3)

69. Average power associated with an AC circuit containing Inductor only

In inductor containing **AC** circuit, current lags behind voltage by $\boldsymbol{\pi/2}$ thus instantaneous emf and current can be written as -

$$E = E_0 \sin \omega t$$

$$I = I_0 \sin\left(\omega t - \frac{\pi}{2}\right)$$

$$\text{or } I = -I_0 \cos \omega t$$

Instantaneous power can be written as

$$\frac{dW}{dt} = E \times I$$

$$\text{or } dW = EIdt \ldots (1)$$

Total work done over a complete cycle i.e. from $0 \rightarrow T$ will be

$$W = \int_0^T EIdt \ldots (2)$$

Substituting **E** and **I** values, we get

$$W = -\int_0^T E_0 \sin \omega t \cdot I_0 \cos \omega t\, dt$$

$$\text{Or } W = -E_0 I_0 \int_0^T \sin \omega t \cos \omega t\, dt$$

$$W = -\frac{E_0 I_0}{2}\int_0^T 2 \sin \omega t \cos \omega t\, dt$$

$$W = -\frac{E_0 I_0}{2}\int_0^T \sin 2\omega t\, dt$$

$$W = -\frac{E_0 I_0}{2}\left[-\frac{\cos 2\omega t}{2\omega}\right]_0^T = 0$$

This means average power over a complete cycle of AC circuit through an inductor is zero.

70. Average power associated with an AC circuit containing capacitor only

In capacitor containing **A. C.** circuit, current leads voltage by $\boldsymbol{\pi/2}$, thus instantaneous emf and current can be written as -

$$E = E_0 \sin \omega t \;\&$$

$$I = I_0 \sin\left(\omega t + \frac{\pi}{2}\right)$$

$$\text{or } I = I_0 \cos \omega t$$

Instantaneous power can be written as

$$\frac{dW}{dt} = E \times I$$

$$\text{or } dW = EIdt \quad \ldots (1)$$

Total work done over a complete cycle i.e. from $0 \rightarrow T$ will be

$$W = \int_0^T EIdt \quad \ldots (2)$$

substituting **E** and **I** values we get

$$w = \int_0^T E_0 \sin \omega t \cdot I_0 \cos \omega t\, dt$$

$$w = \frac{E_0 I_0}{2}\int_0^T 2 \sin \omega t \cos \omega t\, dt$$

$$w = \frac{E_0 I_0}{2}\int_0^T \sin 2\omega t\, dt$$

$$\text{or } w = \frac{E_0 I_0}{2}\left[-\frac{\cos 2\omega t}{2\omega}\right]_0^T = \text{zero}$$

This means average power over a complete cycle of AC circuit containing capacitor only is zero.

71. Average power associated with LCR series circuit

Assuming a **LCR**-series circuit in which $V_L > V_c$ then instantaneous emf and current are given by

$$E = E_0 \sin \omega t \text{ and}$$

$$I = I_0 \sin (\omega t - \phi)$$

& instantaneous power will be given by

$$\frac{dW}{dt} = EI \quad \dots (1)$$

$$\text{or } \frac{dW}{dt} = E_0 \sin \omega t \times I_0 \sin (\omega t - \phi)$$

$$\text{or } \frac{dW}{dt} = E_0 I_0 \sin \omega t(\sin \omega t \cos\phi - \cos \omega t \sin\phi)$$

$$\text{or } = E_0 I_0 \sin^2 \omega t \cos \phi - E_0 I_0 \sin \omega t \cos \omega t \sin \phi$$

$$= E_0 I_0 \sin^2 \omega t \cos \phi - \frac{E_0 I_0}{2} \sin 2\omega t \sin \phi$$

We assume that instantaneous power remains constant for a small time **dt** then small amount of work done will be

$$dW = \left[E_0 I_0 \sin^2 \omega t \cos \phi - \frac{E_0 I_0}{2} \sin 2\omega t \sin \phi\right] dt$$

or Total work done over a complete cycle will be i.e., from $0 \to T$

$$W = \int_0^T E_0 I_0 \sin^2 \omega t \cos \phi dt - \int_0^T \frac{E_0 I_0}{2} \sin 2\omega t \sin \phi dt$$

$$\text{Or } W = E_0 I_0 \cos \phi \int_0^T \sin^2 \omega t \, dt - \frac{E_0 I_0}{2} \sin \phi \int_0^T \sin 2\omega t \, dt$$

$$\because \int_0^T \sin^2 \omega t \, dt = \frac{T}{2} \text{ and } \int_0^T \sin 2\omega t dt = 0$$

$$\therefore W = E_0 I_0 \cos \phi \times \frac{T}{2} \quad \dots (2)$$

$$\text{or } \frac{W}{T} = \frac{E_0 I_0 \cos \phi}{2}$$

$$\text{Or } P = \frac{E_0}{\sqrt{2}} \frac{I_0}{\sqrt{2}} \cos \phi \quad \dots (3)$$

$$\text{Or } \mathbf{P = E_{rms} I_{rms} \cos \phi} \quad \dots\dots (4)$$

72. Power factor of an AC circuit

Power factor of an AC circuit is the ratio of true power to apparent power of the circuit.

$$\text{Apparent power} = E_{rms} I_{rms}$$

$$\text{\& True power } P = E_{rms} I_{rms} \cos \phi$$

$$\therefore \text{Power factor} = \frac{P}{E_{rms} I_{rms}} = \cos \phi$$

From impedance triangle

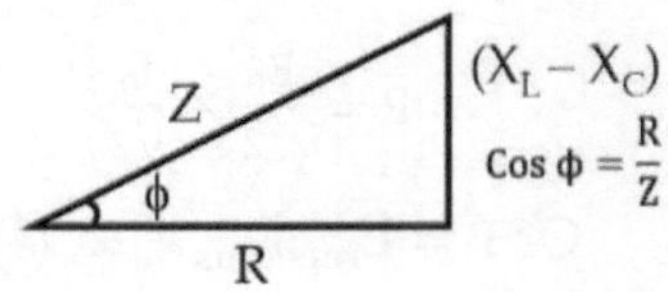

$$\cos \phi = \frac{R}{\sqrt{R^2 + (x_L - x_c)^2}}$$

For pure resistive circuit $\phi = 0°$

$$\therefore \cos \phi = 1$$

For pure inductive or capacitative circuit

$$\phi = 90°$$

$$\therefore \cos \phi = 0$$

Wattless Current: In pure inductive or capacitative circuit, current consumes no power and it is called **Wattless Current.**

73. AC Generator

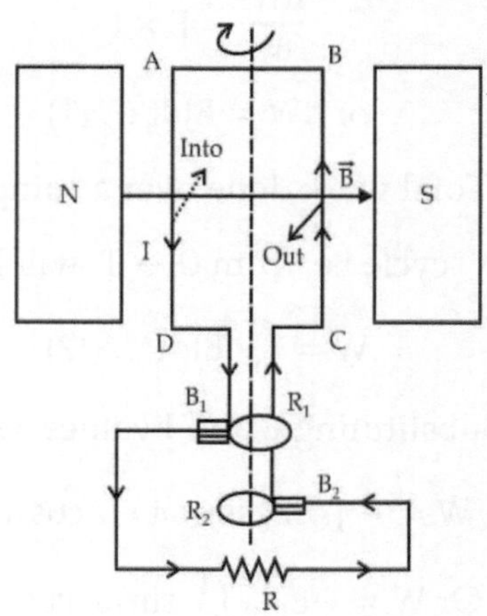

[Current induced for the first half rotation in coil **ABCD** is shown using flemming's right hand rule]

An AC generator is a device which produces alternating current. i.e., electrical energy when mechanical energy is spent on it.

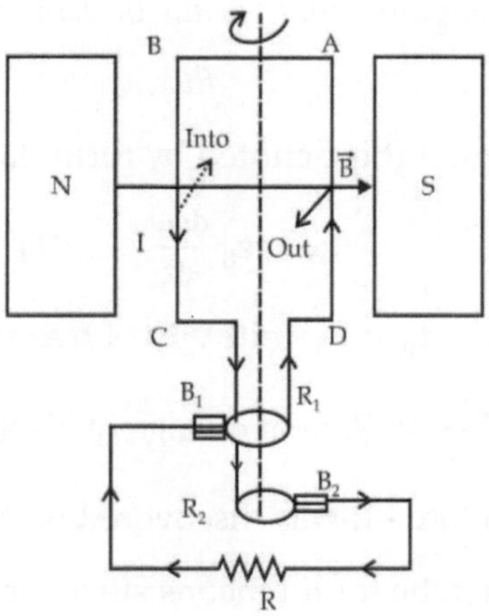

[Direction of induced current for second half cycle of rotation using flemming's]

Principle: An AC generator which is also called Dynamo is based on the phenomenon of electromagnetic Induction. When a coil is rotated in external magnetic field, the flux linked with the coil changes & thus current & emf is induced in the coil. If **N** is total number of turns and **A** is the area enclosed by each turn in the coil then

$$\phi = N(\vec{B} \cdot \vec{A}) = NBA\cos\theta$$

Here **B** is external magnetic field

If **ω** is angular velocity of the coil then

$$\phi = NBA\cos\omega t \quad \dots (1)$$

and at any instant **t** if **ε** is induced emf then

$$\varepsilon = (-)\frac{d\phi}{dt} = (-)\frac{d}{dt}(NBA\cos\omega t)$$

$$\text{or } \varepsilon = -NBA\frac{d}{dt}(\cos\omega t)$$

$$\varepsilon = -NBA(-\sin\omega t)\omega$$

$$\therefore \boldsymbol{\varepsilon = NBA\omega\sin\omega t} \quad \dots (2)$$

When $\sin\omega t = 1$

$$E = E_0 = NBA\omega \quad \dots (3) \text{ (Maximum value)}$$

The current will be given by

$$I = \frac{E}{R} = \frac{E_0\sin\omega t}{R}$$

$$\text{or } I = I_0\sin\omega t \quad \dots (4)$$

$$\text{Here } I_0 = \frac{E_0}{R}$$

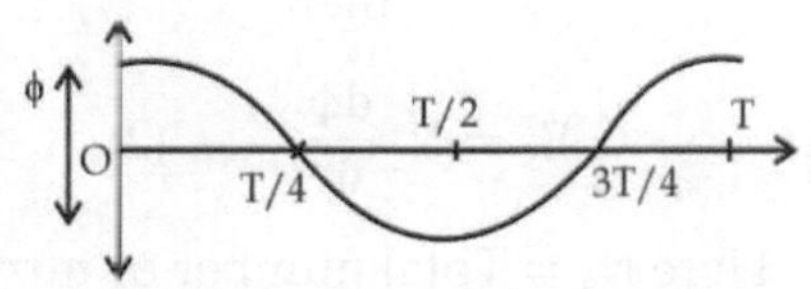

Fig. Variation of flux with time

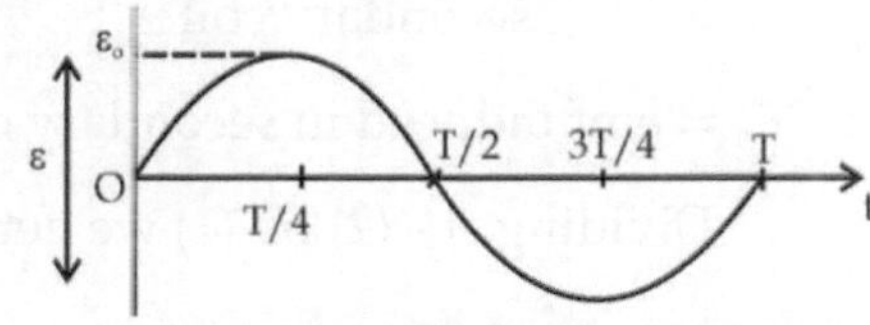

Fig. Variation of induced emf with time

74. Transformer

It is a device which is used for changing the AC voltage.

Principle: *It's working is based on principle of mutual induction according to which when the magnetic flux linked with a coil changes, an emf is induced in the nearby coil.*

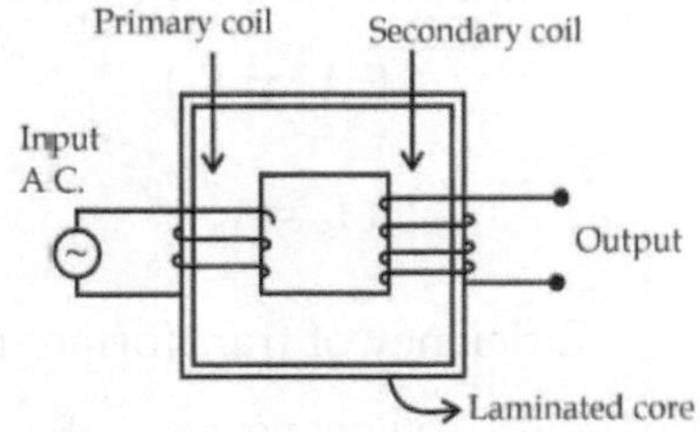

In transfromer two coils primary and secondary are wound over a laminated soft iron core. We assume that coils are pure inductors. When AC is fed in primary coil, the induced emf magnitude will be-

$$E_p = N_P \frac{d\phi}{dt} \quad \dots (1)$$

Here N_P = Total no.of turns in primary coil

E_p = emf induced in primary coil

We also assume that there is no flux loss and the same flux change is associated with secondary coil,

then

$$E_s = N_s \frac{d\phi}{dt} \quad \dots (2)$$

Here N_s = Total number of turns in secondary coil

E_s = emf induced in secondary coil

Dividing eq- (2) by (1) we get

$$\frac{E_s}{E_p} = \frac{N_s}{N_p}$$

$$\text{or } E_s = \frac{N_s}{N_p} \times E_p \quad \dots (3)$$

When $N_s > N_p$ then $E_s > E_p$, the transformer is called step-up transformer.

If $N_p > N_s$ then $E_p > E_s$ and the transformer is step down transformer.

$\frac{N_s}{N_p}$ is called transformation ratio.

If we assume no energy loss then

$$E_p I_p = E_s I_s$$

$$\therefore I_s = I_p \frac{E_p}{E_s}$$

Efficiency of transformer (η)

$$\boldsymbol{\eta = \frac{\text{output power}}{\text{Input power}} = \frac{E_s I_s}{E_p I_p}}$$

Chapter - 8

Electromagnetic Waves

75. Displacement current

Definition - *Displacement current comes into play in a region in which electric field is changing with time due to change in electric flux.*

It is denoted by formula

$$\boldsymbol{I_D = \varepsilon_0 \frac{d\phi_E}{dt}} \quad \dots (1)$$

ε_0 = permittivity of free space

$\frac{d\phi_E}{dt}$ = Rate of change of electric flux.

Note - It was discovered by Maxwell when he found inconsistency in Ampere's circuital law for conduction current during charging of a capacitor.

Modified Ampere's circuital law is also called "**Maxwell Amperes circuital law**".

$$\oint \vec{B} \cdot d\vec{l} = \mu_0 (I + I_D)$$

$$\boldsymbol{\oint \vec{B} \cdot d\vec{l} = \mu_0 \left[I + \varepsilon_0 \frac{d\phi_E}{dt}\right]} \quad \dots (2)$$

Note: Please remember, displacement current exists only when capacitor is charging.

When capacitor is fully charged, displacement current vanishes.

Displacement current can also be expressed as follows

$$I_D = \varepsilon_0 \frac{d}{dt} \int \vec{E} \cdot d\vec{S} \quad \dots (3)$$

$$[\because \phi_E = \int \vec{E} \cdot d\vec{S}]$$

$$\text{Or } I_D = \varepsilon_0 A \frac{dE}{dt}$$

$$\text{Or } I_D = \varepsilon_0 A \frac{d}{dt}\left(\frac{V}{d}\right) \quad \left[\because E = \frac{V}{d}\right]$$

$$\text{Or } I_D = \frac{\varepsilon_0 A}{d}\left(\frac{dV}{dt}\right)$$

$$\therefore \mathbf{I_D = C\frac{dV}{dt}} \quad .....(4)$$

Here, E = Electric field between the plates of the capacitor

V = Potential difference

d = Separation between the plates

C = Capacitance of the capacitor

A = Area of plate

76. Electromagnetic waves and expressions of electric & magnetic fields

"*Electromagnetic waves are those waves in which electric and magnetic field coexist & they are at right angles to each other as well as direction of propagation of wave, varying sinusoidally.*"

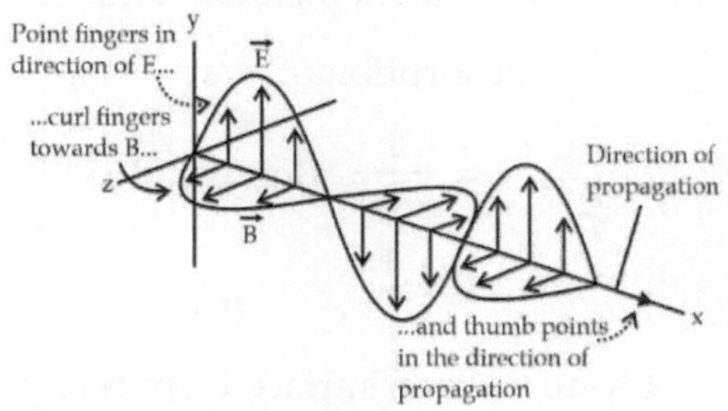

At any instant, the electric and magnetic fields varying sinusoidally with **x** at **t** which can be represented by equations

$$E = E_y = E_0 \sin(\omega t - kx) \quad ...(1)$$

$$B = B_z = B_0 \sin(\omega t - kx) \quad ...(2)$$

ω = Angular frequency

k = propagation vector

[Here $\omega = ck$]

c = speed of light

The speed of electromagnetic wave in free space is given by

$$c = \frac{E_0}{B_0} \quad \text{Also } \mathbf{c = \frac{1}{\sqrt{\mu_0 \varepsilon_0}}}$$

The velocity of electromagnetic waves in any medium is given by $v = \frac{1}{\sqrt{\mu\varepsilon}}$

Here $\mu = \mu_0 \mu_r$

And $\varepsilon = \varepsilon_0 \varepsilon_r$

$$\therefore \mathbf{v = \frac{1}{\sqrt{\mu_0 \mu_r \varepsilon_0 \varepsilon_r}} = \frac{c}{\sqrt{\mu_r \varepsilon_r}}}$$

77. Energy density of electromagnetic waves

Electromagnetic waves carry energy when they travel through space & this energy is equally shared between electric field and magnetic field. It is energy per unit volume & denoted by $\mathbf{u_E}$ for the energy density of electric field and $\mathbf{u_B}$ for energy density of magnetic field & their formulas are

$$u_E = \frac{1}{2}\varepsilon_0 E_{rms}^2 \quad ...(1) \quad \left[E_{rms} = \frac{E_0}{\sqrt{2}}\right]$$

$$u_B = \frac{1}{2}\frac{B_{rms}^2}{\mu_0} \quad ...(2) \quad \left[B_{rms} = \frac{B_0}{\sqrt{2}}\right]$$

Since total energy of electromagnetic waves is equally divided into electric and magnetic field, we can say that

Average energy density

$$\mathbf{u_{av} = \frac{1}{2}\varepsilon_0 E_{rms}^2 + \frac{1}{2}\frac{B_{rms}^2}{\mu_0}} \quad ...(3)$$

$$\because c = \frac{E_{rms}}{B_{rms}} \qquad \therefore B_{rms} = \frac{E_{rms}}{c}$$

Then equation (3) can be written as

$$u_{av} = \frac{1}{2}\varepsilon_0 E_{rms}^2 + \frac{1}{2\mu_0}\frac{E_{rms}^2}{c^2}$$

$$\& \ c = \frac{1}{\sqrt{\mu_0 \varepsilon_0}} \text{ or } c^2 = \frac{1}{\mu_0 \varepsilon_0}$$

$$\therefore u_{av} = \frac{1}{2}\varepsilon_0 E_{rms}^2 + \frac{1}{2\mu_0}\frac{E_{rms}^2}{1} \times \mu_0\varepsilon_0$$

$$\therefore u_{av} = \frac{1}{2}\varepsilon_0 E_{rms}^2 + \frac{1}{2}\varepsilon_0 E_{rms}^2$$

$$\text{or } \mathbf{u_{av} = \varepsilon_0 E_{rms}^2} \quad \ldots (4)$$

Similarly we can prove that total energy density

$$\mathbf{u_{av}} = \frac{\mathbf{B_{rms}^2}}{\boldsymbol{\mu_0}} \quad \ldots (5)$$

78. Intensity (I) and momentum (P) carried by electromagnetic waves

Intensity of electromagnetic waves is the energy falling per unit area per unit time on a given surface.

Denoted by symbol **I**.

$$I = \frac{E}{A \times t} \quad \ldots (1)$$

If we multiply and divide it by distance then -

$$I = \frac{E}{A \times t} \times \frac{d}{d}$$

$$\text{Then } I = \frac{E}{V} \times v$$

$\because A \times d = V$ volume & $\frac{d}{t} = v$ speed

$\frac{E}{V}$ is energy density &

v speed of E.M. waves i.e. c

$$\therefore I = u_{av} \times c \quad \ldots (2)$$

$$I = \varepsilon_0 E_{rms}^2 c = c\frac{B_{rms}^2}{\mu_0} \quad \ldots (3)$$

Momentum carried by EM waves will be

$$p = \frac{\text{Energy of the wave (E)}}{\text{Speed of the wave (c)}}$$

$$\mathbf{p = \frac{E}{c}} \quad \ldots (4)$$

79. Radiation Pressure

"It is the pressure exerted by electromagnetic waves on a given area."

Radiation pressure is different for different surfaces. Since surface can completely absorb the radiation or reflect them thus we have two cases -

Case-I: When surface is absorbing radiation completely then force exerted on the surface will be

$$F = \frac{-\Delta p}{\Delta t} = -\left[\frac{0-p}{\Delta t}\right]$$

$$\text{or } F = \frac{p}{\Delta t} \quad \ldots (1)$$

If A is the area of absorbing surface then the radiation pressure will be-

$$P = \frac{F}{A} = \frac{p}{A\Delta t} = \frac{E}{cA\Delta t} \quad \ldots (2) \quad \because p = \frac{E}{c}$$

Here $\mathbf{\Delta t}$ is time taken for complete absorption of wave.

$$\because I = \frac{E}{A \times \Delta t} \text{ (Intensity)}$$

$$\therefore \mathbf{P = \frac{I}{c}} \quad \ldots (3)$$

Case-II: When surface is reflecting radiation completely then change in momentum will be

$$(\Delta p) = 2p \quad \ldots. (1)$$

$$\& \text{ Force (F)} = \frac{\Delta p}{\Delta t} = \frac{2p}{\Delta t}$$

& Radiation pressure

$$P = \frac{F}{A} = \frac{2p}{A\Delta t} = \frac{2E}{cA\Delta t} \quad \ldots (2)$$

$$\because I = \frac{E}{A \times \Delta t} \text{ (Intensity)}$$

$$\therefore \mathbf{P = \frac{2I}{c}} \quad \ldots(3)$$

80. Prove that displacement current is $\varepsilon_0 \frac{d\phi}{dt}$ and it is equal to conduction current (I_C).

When the capacitor is getting charged then the electric flux between the plates of capacitor changes, which is given by

$$\frac{d\phi}{dt} = \frac{d}{dt}(EA) \quad \dots (1)$$

Here E = Electric field between the plates & A = Area of the plate

$$\because E = \frac{\sigma}{\varepsilon_0} = \frac{Q}{A\varepsilon_0}$$

then from equation - (1)

$$\therefore \frac{d\phi}{dt} = \frac{d}{dt}\left(\frac{Q}{A\varepsilon_0} \times A\right) \quad \dots (2)$$

$$\text{or } \frac{d\phi}{dt} = \frac{1}{\varepsilon_0}\frac{dQ}{dt} = \frac{1}{\varepsilon_0} I_D$$

$$\text{or } \mathbf{I_D = \varepsilon_0 \frac{d\phi}{dt}} \quad \dots (3)$$

- **Proof for $I_C = I_D$**

Conduction current (I_C) is due to flow of electrons in the circuit and it is

$$I_C = \frac{dQ}{dt} \quad \dots (1)$$

Displacement current is due to change in electric flux between the plates of capacitor and is given by

$$I_D = \varepsilon_0 \frac{d\phi}{dt} \quad \dots (2)$$

$$\text{Since } \phi = \frac{Q}{\varepsilon_0}$$

∴ From equation -(2)

$$I_D = \varepsilon_0 \frac{d}{dt}\left(\frac{Q}{\varepsilon_0}\right)$$

$$\text{or } I_D = \frac{\varepsilon_0}{\varepsilon_0}\frac{dQ}{dt} = I_C$$

$$\therefore I_D = I_C \quad \dots (3)$$

Please note: If in the numericals, conduction current is given and displacement current is asked then it is equal to conduction current when capacitor is charging.

- Once capacitor is fully charged, displacement current becomes zero.
- Maxwell used four equations to explain the nature of EM waves

(1) Gauss's Law of electrostatics

$$\oint \vec{E} \cdot d\vec{S} = Q/\varepsilon_0$$

(2) Gauss theorem for magnetism

$$\oint \vec{B} \cdot d\vec{S} = 0$$

(3) Faraday's law of Electromagnetic Induction $\oint \vec{E} \cdot d\vec{\ell} = (-)\frac{d\phi_B}{dt}$

(4) MaxWell-Amperes circuital law

$$\oint \vec{B} \cdot \vec{d\ell} = \mu_0\left(I_C + \varepsilon_0 \frac{d\phi}{dt}\right)$$

Chapter - 9

Ray Optics

81. Relation between f and R

(a) Concave mirror-

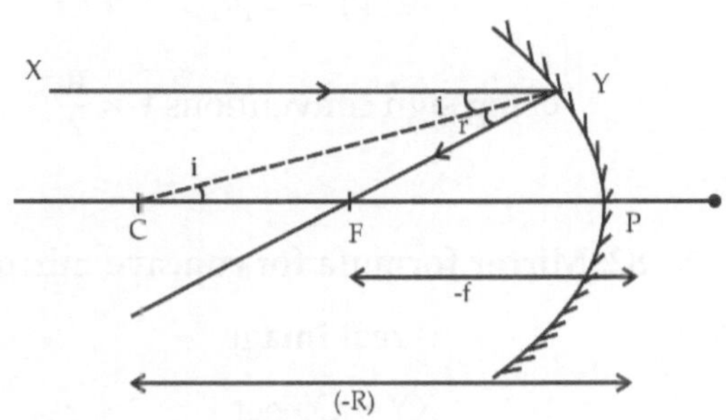

Here ∠XYC = i (angle of incidence)

∠CYF = r (angle of reflection)

Now ∠CYF = ∠XYC (Law of reflection)

And ∠XYC = ∠YCF = i (alternate angles)

In △ CYF as ∠i = ∠r ∴ CF = FY

But FY = FP [∵ aperture is small]

∴ CF = FP i.e. F is the centre of PC

$\therefore\ PF = \frac{1}{2}PC$, using sign conventions

$\because\ PF = -f$ and $PC = -R$

$\therefore\ -f = -\frac{R}{2}$ or

$$\mathbf{f = \frac{R}{2}}$$

(b) Convex Mirror-

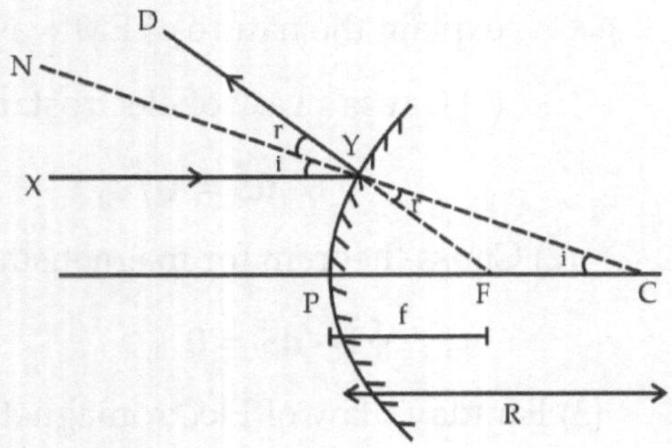

$\angle XYN = i$

$\angle DYN = r$

$\angle FYC = \angle DYN = r$ (opposite angles)

$\angle YCF = \angle NYX = i$ (corresponding angles)

In Δ CYF, as $i = r$

$\therefore\ CF = FY$ also $FY = FP$

($\because$ aperture is small)

$\therefore\ CF = FP$ or F is the middle point of PC

$\therefore\ PF = \frac{1}{2}PC$

using sign conventions $\mathbf{f = \frac{R}{2}}$

82. Mirror formula for concave mirror real image

XY = Object

X'Y' = Image

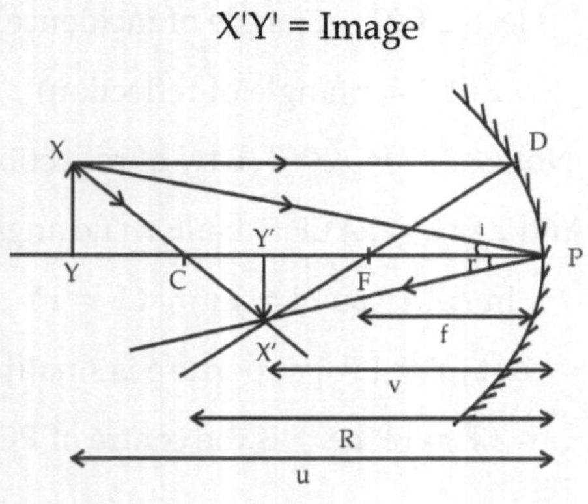

From the figure given above

Δ_s XYC and X'Y'C are similar

$$\therefore\ \frac{XY}{X'Y'} = \frac{CY}{CY'} \quad \dots (1)$$

again Δ_s XYP and X'Y'P are similar

$$\therefore\ \frac{XY}{X'Y'} = \frac{PY}{PY'} \quad \dots (2)$$

From eq- (1) and (2)

$$\frac{CY}{CY'} = \frac{PY}{PY'} \quad \dots (3)$$

From figure-

$$CY = PY - PC$$

$$CY' = PC - PY'$$

$\therefore$ From eq. (3)

$$\frac{PY - PC}{PC - PY'} = \frac{PY}{PY'} \quad \dots (4)$$

According to new cartesian sign conventions

$$PY = -u, PC = -R, PY = -v$$

$$\therefore\ \frac{-u+R}{-R+v} = \frac{-u}{-v}$$

$$\text{or } uv - vR = +uR - uv$$

$$2uv = uR + vR$$

Dividing both sides by **uvR**, we get

$$\frac{1}{v} + \frac{1}{u} = \frac{2}{R} \because f = \frac{R}{2}$$

$$\therefore\ \mathbf{\frac{1}{v} + \frac{1}{u} = \frac{1}{f}}$$

Note: Mirror formula for concave mirror virtual image can be derived using similar approach and the following ray diagram.

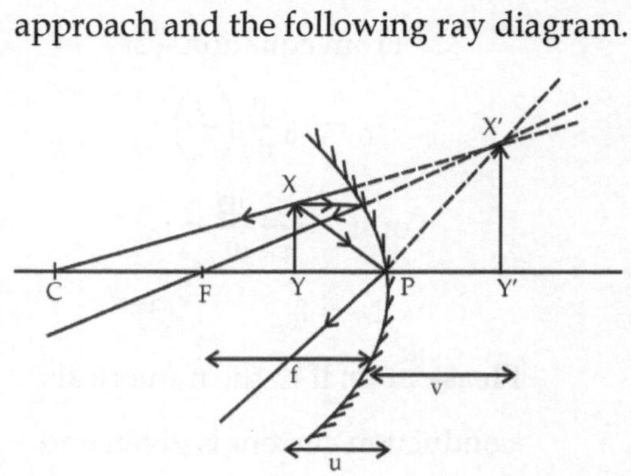

83. Mirror formula for convex mirror.

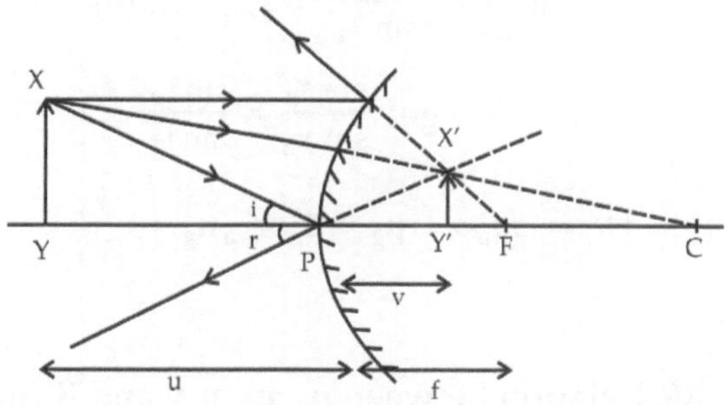

XY = Object

X'Y' = Image

From the figure, Δ_S XYC and X'Y'C are similar

$$\therefore \frac{XY}{X'Y'} = \frac{CY}{CY'} \quad \ldots (1)$$

Again Δ_S **XYP** and **X'Y'P** are similar

$$\therefore \frac{XY}{X'Y'} = \frac{PY}{PY'} \quad \ldots (2)$$

From eq. (1) and (2)

$$\frac{CY}{CY'} = \frac{PY}{PY'} \quad \ldots (3)$$

From figure -

$$CY = PY + PC; \;\; CY' = PC - PY'$$

$$\text{Or } \frac{PY+PC}{PC-PY'} = \frac{PY}{PY'} \quad \ldots (4)$$

According to new cartesian sign conventions, $PY = -u, PY' = +v, PC = +R$

$\therefore$ From equation (4)

$$\frac{-u+R}{R-v} = \frac{-u}{v}$$

$$\text{or } -uv + vR = -uR + uv$$

$$\text{or } vR + uR = 2uv$$

Dividing both sides by uvR, we get -

$$\frac{1}{u} + \frac{1}{v} = \frac{2}{R} = \frac{1}{f}$$

$$\therefore \frac{1}{f} = \frac{1}{v} + \frac{1}{u}$$

84. Linear magnification of a spherical mirror

Linear magnification is the ratio of the size of the image formed by the mirror to the size of the object.

$$m = \frac{X'Y'}{XY} = \frac{h_2 \rightarrow \text{Image height}}{h_1 \rightarrow \text{Object height}}$$

h_1 is always positive.

m is positive if h_2 is positive.

Case -I - Concave mirror real image-

$$m = \frac{h_2}{n_1} = -\frac{v}{u}$$

Case-II - Concave mirror virtual image -

$$m = \frac{h_2}{h_1} = -\frac{v}{u}$$

Case-III - Convex mirror -

$$m = \frac{h_2}{h_1} = -\frac{v}{u}$$

For all cases same.

Note- Please use sign conventions again with numerical values while solving problems.

Other formulae for magnification

$$\because \frac{1}{u} + \frac{1}{v} = \frac{1}{f}$$

Multiplying both sides by v, we get

$$\frac{v}{u} + \frac{v}{v} = \frac{v}{f}$$

$$\text{Or } \frac{v}{u} + 1 = \frac{v}{f}$$

$$\text{or } \frac{v}{u} = \frac{v}{f} - 1 = \frac{v-f}{f}$$

$$\because m = -\frac{v}{u}$$

$$\therefore \mathbf{m = \frac{f-u}{f}} \quad \ldots (1)$$

Again from mirror formula

$$\frac{1}{u} + \frac{1}{v} = \frac{1}{f}$$

multiplying both sides by u

$$\frac{u}{u}+\frac{u}{v}=\frac{u}{f}$$

$$\text{or } 1+\frac{u}{v}=\frac{u}{f} \text{ or } \frac{u}{v}=\frac{u}{f}-1$$

$$\text{or } \frac{u}{v}=\frac{u-f}{f} \text{ or } \frac{v}{u}=\frac{f}{u-f}$$

$$\because m=-\frac{v}{u}$$

$$\therefore \mathbf{m}=\frac{\mathbf{f}}{\mathbf{f-u}} \quad \ldots (2)$$

85. Refraction of light through a rectangular glass slab and principle of reversibility

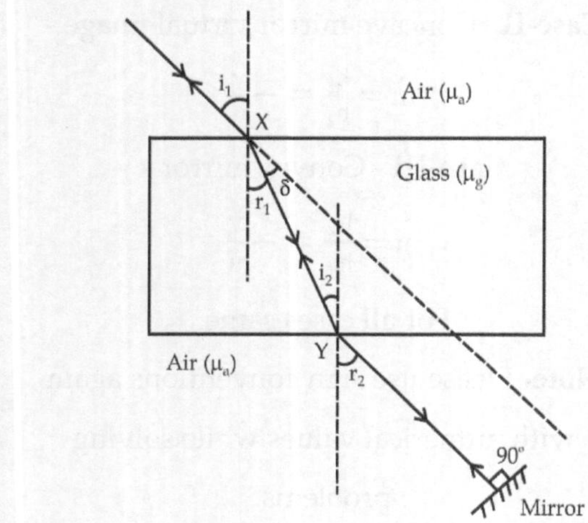

Applying Snell's Law at **L**

$$\mu_a \sin i_1 = \mu_g \sin r_1$$

$$\text{Or } \frac{\sin i_1}{\sin r_1}=\frac{\mu_g}{\mu_a}={}_a\mu_g \quad \ldots (1)$$

Applying Snell's Law at **M**

$$\mu_g \times \sin i_2 = \mu_a \sin r_2$$

$$\frac{\mu_a}{\mu_g}=\frac{\sin i_2}{\sin r_2}={}_g\mu_a \quad \ldots (2)$$

According to principle of reversibility of light, when the path of a ray of light, after any number of refractions is reversed, the ray retraces the same path.

For reverse ray applying snell's law at **M**

$$\mu_a \sin r_2 = \mu_g \sin i_2$$

$$\text{or } \frac{\mu_g}{\mu_a}=\frac{\sin r_2}{\sin i_2}={}_a\mu_g \quad \ldots (3)$$

$$\text{We get } \frac{\sin i_2}{\sin r_2}\times\frac{\sin r_2}{\sin i_2}$$

$$= {}_g\mu_a \times {}_a\mu_g = 1 \text{ or } {}_a\mu_g = \frac{1}{{}_g\mu_a}$$

86. Relation between μ_1, μ_2, u, v and R for refraction from rarer to denser medium at convex spherical surface (real image)

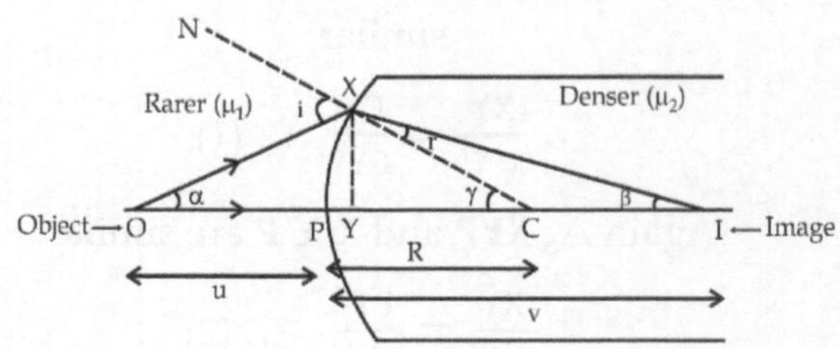

From figure $\angle XOY = \alpha$,

$\angle XIY = \beta$ & $\angle XCY = \gamma$

Also from figure $\gamma = r + \beta$

or $r = \gamma - \beta$ ... (1)

and $i = \alpha + \gamma$ ... (2)

According to Snell's law

$$\mu_2 \sin r = \mu_1 \sin i$$

$$\text{Or } \frac{\mu_2}{\mu_1}=\frac{\sin i}{\sin r}=\frac{i}{r} \; (\because \text{Angles are small})$$

$$\therefore \mu_1 i = \mu_2 r$$

substituting equations (1) & (2) in the above equation we get

$$\mu_1(\alpha+\gamma)=\mu_2(\gamma-\beta)$$

As angles α, β and γ are small using

$$\theta=\frac{\text{arc}}{\text{radius}}, \text{ we get}$$

$$\mu_1\left[\frac{XY}{YO}+\frac{XY}{YC}\right]=\mu_2\left[\frac{XY}{YC}-\frac{XY}{YI}\right] \quad \ldots (3)$$

As aperture of spherical surface is small, **Y** is close to **P**

$\therefore$ YO $\approx$ PO, YI $\approx$ PI and YC $\approx$ PC

Substituting in eq. (3) we get

$$\mu_1\left[\frac{1}{PO}+\frac{1}{PC}\right]=\mu_2\left[\frac{1}{PC}-\frac{1}{PI}\right]$$

$$\text{or } \frac{\mu_1}{PO}+\frac{\mu_1}{PC}=\frac{\mu_2}{PC}-\frac{\mu_2}{PI}$$

$$\because PO=-u,\ PI=+v,\ PC=+R$$

$$\therefore \frac{\mu_1}{-u}+\frac{\mu_2}{v}=\frac{\mu_2-\mu_1}{R} \quad \dots (4)$$

Note: The formula for refraction from rarer to denser medium at convex spherical surface (Virtual image) can be derived using similar approach and the following the ray diagram.

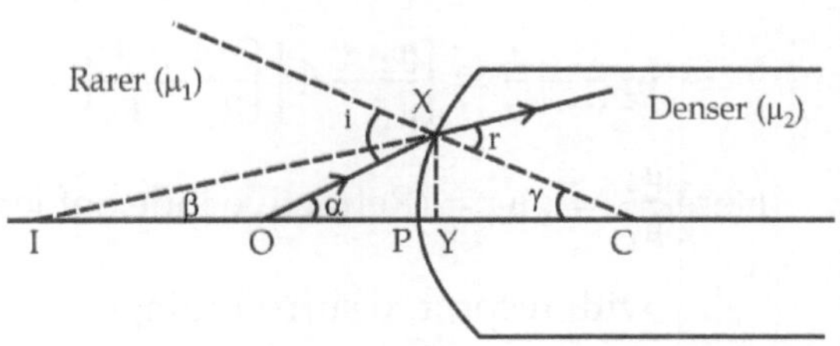

Note: The formula for Refraction from rarer to denser medium at concave spherical surface can be derived using similar approach and the following the ray diagram.

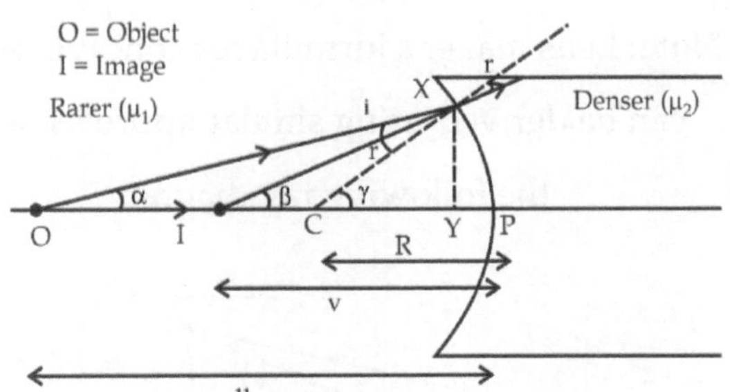

87. Relation between μ_1, μ_2, u, v and R for refraction from denser to rarer medium at a convex refracting surface

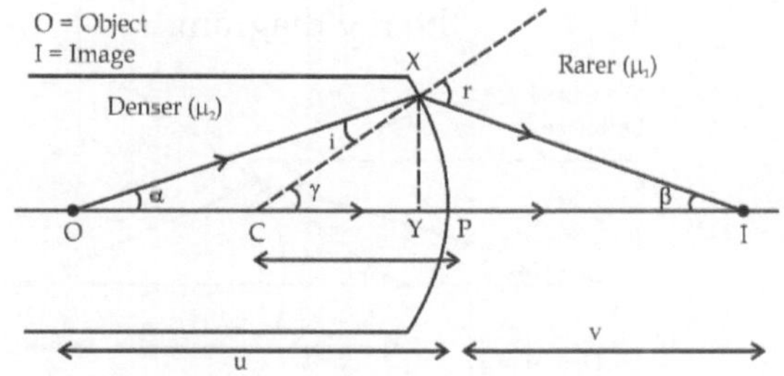

Using Snell's Law

$$\mu_2 \sin i=\mu_1 \sin r$$

Since **i** and **r** are small

$$\mu_2 i=\mu_1 r \quad \dots (1)$$

In $\triangle$ OXC,

$$i=\gamma-\alpha \quad \dots (2)$$

$$\&\ \triangle IXC,\ r=\gamma+\beta \quad \dots (3)$$

substituting **i** and **r** values in equation (1) we get

$$\mu_2(\gamma-\alpha)=\mu_1(\gamma+\beta) \quad \dots (4)$$

since angles are small,

using $\theta=\frac{\text{arc}}{\text{radius}}$, we get

$$\mu_2\left[\frac{XY}{YC}-\frac{XY}{YO}\right]=\mu_1\left[\frac{XY}{YC}+\frac{XY}{YI}\right]$$

$$\text{or } \mu_2\left[\frac{1}{YC}-\frac{1}{YO}\right]=\mu_1\left[\frac{1}{YC}+\frac{1}{YI}\right]$$

Since aperture is small, **Y** is close to **P**

Then YC $\approx$ PC, YO $\approx$ PO & YI $\approx$ PI

$$\therefore \mu_2\left[\frac{1}{PC}-\frac{1}{PO}\right]=\mu_1\left[\frac{1}{PC}+\frac{1}{PI}\right]$$

$$\text{or } \frac{(-)\mu_2}{PO}-\frac{\mu_1}{PI}=-\frac{\mu_2}{PC}+\frac{\mu_1}{PC}$$

$$PO=-u, PI=v, PC=-R$$

$$\therefore \frac{(-)\mu_2}{-u}-\frac{\mu_1}{v}=\frac{-(\mu_2-\mu_1)}{-R}$$

$$\text{or } -\frac{\mu_2}{u}+\frac{\mu_1}{v}=\frac{\mu_1-\mu_2}{R} \quad \dots (5)$$

Note: The formula for refraction from denser

to rarer medium at concave spherical surface (Virtual image) can be derived using similar approach and the following the ray diagram.

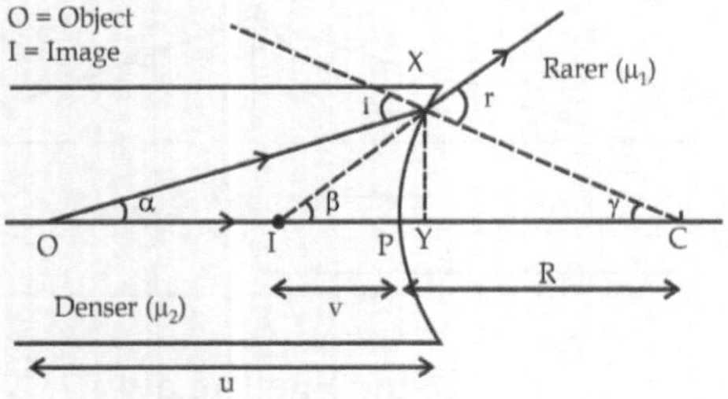

88. Lens Maker's formula for convex lens

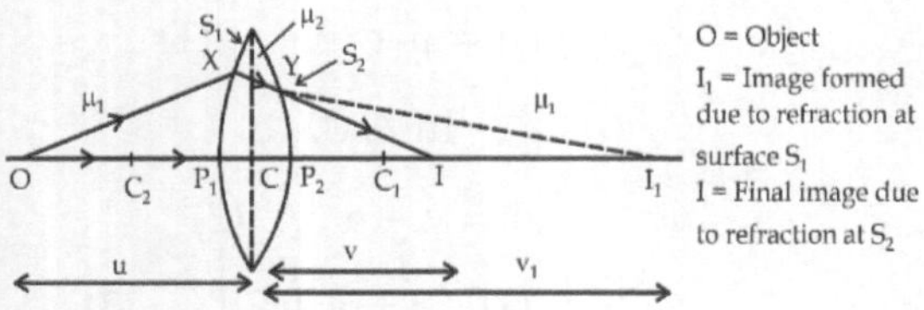

Lens is thin, we take $CI_1 \approx P_1I_1 = v_1$ and

$CC_1 \approx P_1C_1 = R_1$

and $CO \approx P_1O = u$

Refraction at first spherical surface.

Light goes from rarer ($\boldsymbol{\mu_1}$) to denser ($\boldsymbol{\mu_2}$) medium, if second surface $\mathbf{S_2}$ is absent and denser medium is continued, we would get image of the object at $\mathbf{I_1}$,

Now formula for refraction from rarer to denser medium at spherical surface is given by

$$\frac{\mu_1}{-u} + \frac{\mu_2}{v_1} = \frac{\mu_2 - \mu_1}{R_1} \quad \dots (1)$$

(Here $\mathbf{v = v_1}$ from first spherical surface)

(**Note:** We take formula as such derived before) Since denser medium is not continuous, the refracted ray **XY** Again gets refracted and final image is formed at **I**.

Note- For refraction at second surface $\mathbf{I_1}$ acts like a virtual object, whose real image is formed at **I**.

Now $u = CI_1 \approx P_2I_1 = v_1$

& $CI \approx P_2I = v$ & R_2 is radius of curvature of second surface of lens, now light goes from denser to rarer

$$\therefore -\frac{\mu_2}{v_1} + \frac{\mu_1}{v} = \frac{\mu_1 - \mu_2}{R_2} = \frac{\mu_2 - \mu_1}{-R_2} \quad \dots (2)$$

Adding eq. (1) & (2) we get

$$\frac{\mu_1}{-u} + \frac{\mu_2}{v_1} - \frac{\mu_2}{v_1} + \frac{\mu_1}{v} = (\mu_2 - \mu_1)\left[\frac{1}{R_1} - \frac{1}{R_2}\right]$$

$$\text{or } \mu_1\left[\frac{1}{v} - \frac{1}{u}\right] = (\mu_2 - \mu_1)\left[\frac{1}{R_1} - \frac{1}{R_2}\right]$$

$$\text{or } \frac{1}{v} - \frac{1}{u} = \left[\frac{\mu_2}{\mu_1} - 1\right]\left[\frac{1}{R_1} - \frac{1}{R_2}\right]$$

Here $\frac{\mu_2}{\mu_1} = {}_1\mu_2 =$ (Refractive index of lens with respect to surroundings)

Also when object is at infinity, image is formed at principal focus of the lens

$\because$ When $u = \infty, v = f$

$$\therefore \frac{\mathbf{1}}{\mathbf{f}} = (\boldsymbol{\mu} - \mathbf{1})\left[\frac{\mathbf{1}}{\mathbf{R_1}} - \frac{\mathbf{1}}{\mathbf{R_2}}\right]$$

This is lens makers formula.

Note: Lens maker's formula for concave lens can be derived using similar approach and the following ray diagram.

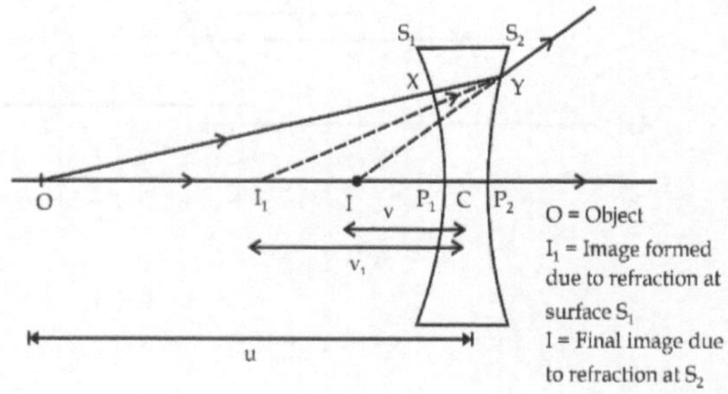

Note: In the above derivation we directly took formula of refraction from denser to rarer & rarer to denser without using sign convention again, but when we solve numericals, we need to use sign convention with numerical values.

89. Lens formula

Convex Lens - Real Image

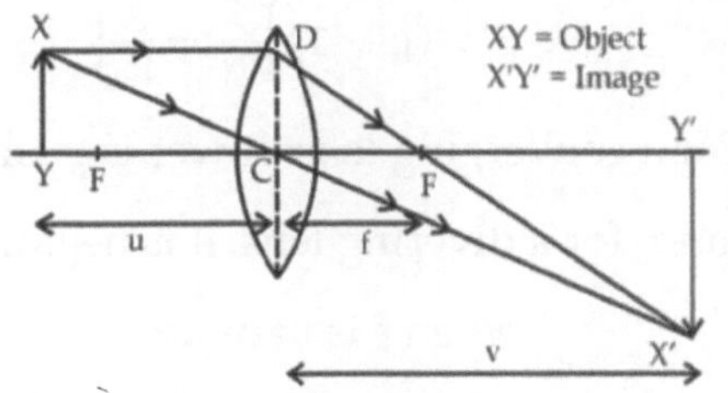

Here, C = Optical Centre

XY = Object

X' Y' = Image

CF = Focal Length (f)

CY' = Image distance (v)

CY = Object distance (u)

From the figure -

Δ_s **X'Y'C** and **XYC** are similar

$$\therefore \frac{X'Y'}{XY} = \frac{CY'}{CY} \quad \ldots (1)$$

Also $\mathbf{\Delta_S X'Y'F}$ and **CDF** are similar

$$\therefore \frac{X'Y'}{CD} = \frac{FY'}{CF}$$

Since **CD = XY** $\therefore \frac{X'Y'}{XY} = \frac{FY'}{CF} \quad \ldots (2)$

From equations (1) and (2)

$$\frac{CY'}{CY} = \frac{FY'}{CF} = \frac{CY' - CF}{CF} \quad \ldots (3)$$

$$[\because FY' = CY' - CF]$$

As per new cartesian sign conventions

$$CY = -u, CY' = +v, CF = +f$$

Hence from equation (3)

$$\therefore \frac{v}{-u} = \frac{v-f}{f}$$

$$\text{or } vf = -uv + uf$$

$$\text{or } uv = uf - vf$$

Dividing both sides **uvf**

$$\frac{uv}{uvf} = \frac{uf}{uvf} - \frac{vf}{uvf}$$

$$\text{Or } \frac{1}{f} = \frac{1}{v} - \frac{1}{u}$$

This is Lens formula.

Note: Lens formula for Convex Lens (Virtual image) can be derived using similar approach and the following ray diagram.

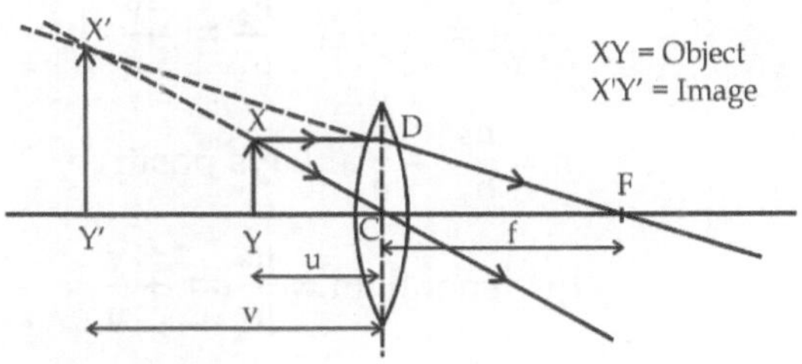

Note: Lens formula for Concave Lens can be derived using similar approach and the following ray diagram.

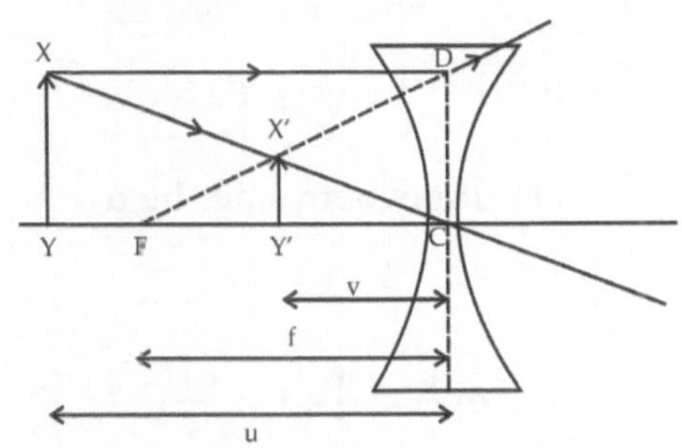

90. Linear magnification produced by a Lens

Linear magnification produced by a lens is the ratio of the size of image ($\mathbf{h_2}$) to the size of object $\mathbf{h_1}$. Denoted by **m**.

$$m = \frac{X'Y'}{XY} = \frac{h_2}{h_1}$$

For convex lens real image

$$\frac{X'Y'}{XY} = \frac{CY'}{CY} \therefore m = \frac{-h_2}{h_1} = \frac{+v}{-u}$$

$$\text{or } m = -\frac{h_2}{h_1} = -\frac{v}{u}$$

It's always negative for real image

For convex lens virtual image

$$\frac{X'Y'}{XY} = \frac{CY'}{CY} \text{ i.e. } \frac{h_2}{h_1} = \frac{-v}{-u}$$

$$\text{or } \mathbf{m} = \frac{\mathbf{h_2}}{\mathbf{h_1}} = \frac{v}{u}$$

(It's positive for virtual image.)

For concave lens [always virtual]

$$\frac{X'Y'}{XY} = \frac{CY'}{CY} \text{ i.e. } \frac{h_2}{h_1} = \frac{-v}{-u}$$

$$\therefore m = \frac{h_2}{h_1} = \frac{v}{u} \text{ always positive.}$$

$$\text{In General, } m = \frac{h_2}{h_1} = +\frac{v}{u}$$

- Please use sign with numerical values to solve the numericals.
- Magnification in terms of u and f-

From lens formula

$$\frac{1}{v} - \frac{1}{u} = \frac{1}{f}$$

multiply both sides by **u**

$$\frac{u}{v} - 1 = \frac{u}{f}$$

$$\text{or } \frac{u}{v} = \frac{u}{f} + 1 = \frac{u+f}{f}$$

$$\because m = \frac{v}{u}$$

$$\therefore \mathbf{m} = \frac{\mathbf{f}}{\mathbf{u+f}}$$

- Magnification in terms of **v** and **f**.

From lens formula $\frac{1}{v} - \frac{1}{u} = \frac{1}{f}$

multiply both sides by **v**

$$\frac{v}{v} - \frac{v}{u} = \frac{v}{f}$$

$$\text{or } 1 - \frac{v}{u} = \frac{v}{f} \quad \text{or} \quad \frac{v}{u} = 1 - \frac{v}{f}$$

$$\text{or} \quad \mathbf{m} = \frac{\mathbf{v}}{\mathbf{u}} = \frac{\mathbf{f-v}}{\mathbf{f}}$$

91. Power of a Lens

It is the ability of the lens to converge/diverge a beam of light falling on the lens.

$$\mathbf{P = \frac{1}{f}}$$

$$\because \frac{1}{f} = (\mu - 1)\left[\frac{1}{R_1} - \frac{1}{R_2}\right]$$

$$\therefore P = (\mu - 1)\left[\frac{1}{R_1} - \frac{1}{R_2}\right]$$

For a converging lens power is positive and for a divering lens, it is negative.

When **f** is in metre

$$\mathbf{P = \frac{1}{f_{(m)}} \text{ Dioptre}}$$

When **f** is in centimeter

$$\mathbf{P = \frac{100}{f_{cm}} \text{ Dioptre}}$$

92. Combination of thin lenses in contact

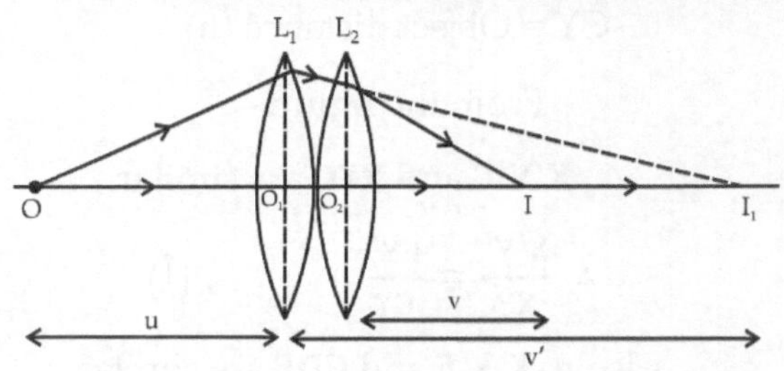

O = Object

I_1 = Image produced by L_1

I = Image produced by L_2

OO_1 = u. Lens L_1 produces image $\mathbf{I_1}$ at a distance **v′** from $\mathbf{L_1}$.

$$\therefore \frac{1}{v'} - \frac{1}{u} = \frac{1}{f_1} \quad \ldots (1)$$

$\mathbf{I_1}$ behaves as object for lens $\mathbf{L_2}$ which forms final image **I** at distance $\mathbf{O_2I}$. Please note the image formed by lens $\mathbf{L_1}$ acts like a virtual object for $\mathbf{L_2}$ i.e., $\mathbf{O_2I_1 = u}$ **which is v′** for $\mathbf{L_2}$. Here we assume the lenses are thin.

$$O_1I_1 = O_2I_1$$

$\therefore$ For L_2 using lens formula

$$\frac{1}{v} - \frac{1}{v'} = \frac{1}{f_2} \quad \dots (2)$$

Adding equations (1) and (2), we get

$$\frac{1}{v'} - \frac{1}{u} + \frac{1}{v} - \frac{1}{v'} = \frac{1}{f_1} + \frac{1}{f_2}$$

$$\text{or } \frac{1}{v} - \frac{1}{u} = \frac{1}{f_1} + \frac{1}{f_2} \quad \dots (3)$$

When both the lenses are replaced by a single lens of focal length **F**, which forms image **I** at distance **v**, of an object at distance **u** from the lens then -

$$\frac{1}{v} - \frac{1}{u} = \frac{1}{f} \quad \dots (4)$$

Comparing (3) and (4) we get

$$\frac{1}{F} = \frac{1}{f_1} + \frac{1}{f_2} \quad \dots (5)$$

Note- We need to put the numerical values with sign.

93. Refraction through a prism & calculation of angle of deviation

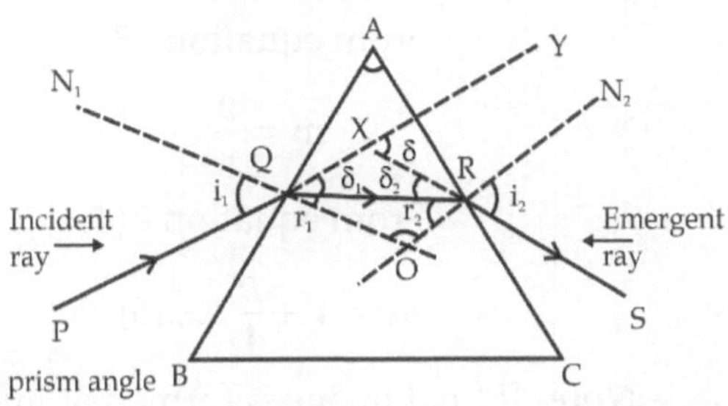

From figure $\delta_1 = i_1 - r_1$

$\& \; \delta_2 = i_2 - r_2$

Total deviation produced

$$\delta = \delta_1 + \delta_2$$

$$\therefore \delta = (i_1 - r_1) + (i_2 - r_2)$$

$$\text{Or } \delta = (i_1 + i_2) - (r_1 + r_2) \quad \dots (1)$$

sum of angles of $\triangle$ **AQR** is **180°** i.e.

$$X + (90° - r_1) + (90° - r_2) = 180°$$

$$\therefore X = r_1 + r_2 \quad \dots (2)$$

From eq. (1) and (2)

$$\boldsymbol{\delta = (i_1 + i_2) - A} \quad \dots (3)$$

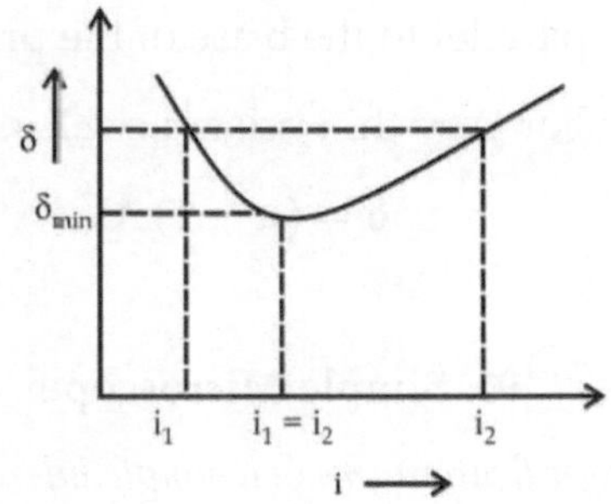

δ_{min} = angle of minimum deviation

Figure: Graph between angle of deviation and angle of incidence

94. Angle of minimum deviation and prism formula

From above graph we found that at minimum deviation, there is only one angle of incidence.

$\therefore$ When $\delta = \delta_{min}$; $i_1 = i_2 = i$

And according to Snell's Law

$$\mu_1 \sin i_1 = \mu_2 \sin r_1 \quad \dots (1)$$

$$\text{And } \mu_2 \sin r_2 = \mu_1 \sin i_2 \quad \dots (2)$$

$$\because i_1 = i_2 = i$$

Then from equation (1) & (2)

$$\therefore r_1 = r_2 = r$$

Now we know that $X = r_1 + r_2$

$$\text{or } X = r + r = 2r \text{ or } r = \frac{X}{2}$$

$$\text{And } \delta_{min} = (i_1 + i_2) - (r_1 + r_2)$$

$$\text{Or } \delta_{min} = 2i - X \quad \dots (1)$$

$$\text{or } i = \frac{X + \delta_{min}}{2} \quad \dots (2)$$

$$\because \mu = \frac{\sin i}{\sin r}$$

$$\boldsymbol{\because \mu = \frac{\sin (A + \delta_{min})/2}{\sin A/2}} \quad \dots (3)$$

This is prism formula.

Please remember that at minimum deviation, the refracted ray in the prism is parallel to the b ase of the prism.

For very thin prism i.e. $\angle X < 10°$

$$\boldsymbol{\delta = (\mu - 1) A}$$

95. Simple Microscope

Magnification (m) of a simple microscope is the ratio of the angle subtended by the image to the angle subtended by the object on the eye when they are at same distance from the eye.

Suppose $\boldsymbol{\alpha}$ is the angle subtended by the object when object is placed at near point (D) of the eye.

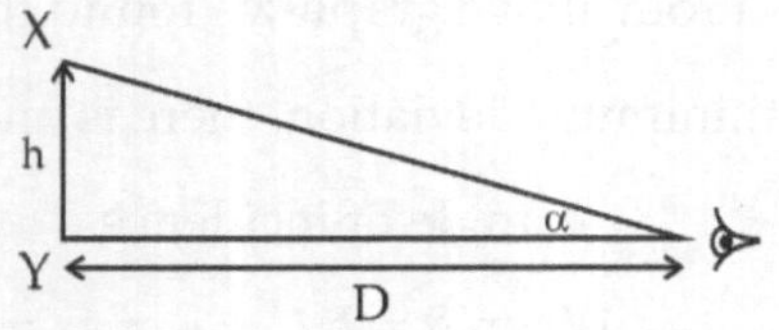

Then for small objects, we can write -

$$\alpha \approx \tan \alpha = \frac{h}{D} \quad ....(1)$$

Now when we use simple microscope [convex lens of small focal length], the image formation takes place as shown in the figure-

XY = Object

X'Y' = Image

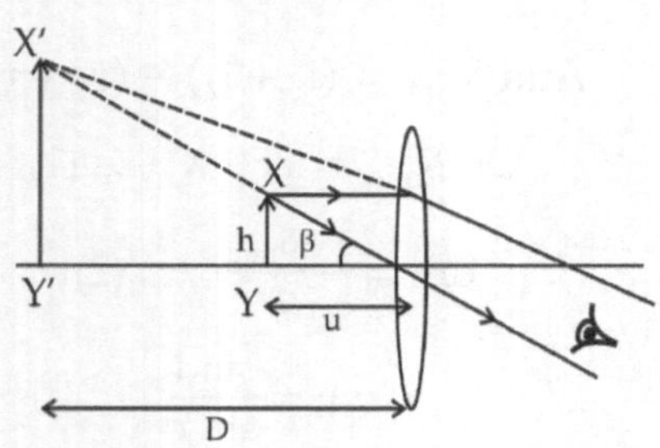

In this case, the angle subtended on the eye is $\boldsymbol{\beta}$, hence when image is formed at least distance of distinct vision (D).

From the figure -

$$\beta \approx \tan \beta = \frac{h}{u} \quad ....(2)$$

Here h = height of object

& u = object distance

Hence using equation (1) & (2)

magnification can be written as

$$m = \frac{\tan \beta}{\tan \alpha} \approx \frac{\beta}{\alpha} = \frac{h}{u} \times \frac{D}{h}$$

$$\text{or} \quad m = \frac{D}{u} \quad ....(3)$$

Now we can use the lens formula -

$$\frac{1}{f} = \frac{1}{v} - \frac{1}{u}$$

Here $v = -D$ & u – negative & f is positive

$$\therefore \frac{1}{f} = \frac{1}{-D} - \frac{1}{-u}$$

multiplying both sides by D we get

$$\frac{D}{f} = \frac{D}{-D} + \frac{D}{u}$$

$$\text{or} \frac{D}{f} = -1 + \frac{D}{u}$$

$$\text{or} \quad \frac{D}{u} = 1 + \frac{D}{f} \quad ....(4)$$

From equation - 3

$$m = \frac{D}{u}$$

∴ From equation - (4)

$$\mathbf{m = 1 + \frac{D}{f}} \quad ....(5)$$

Note- If final image is formed at infinity, then as shown in the figure, the size of the image will appear in proportional to the angle

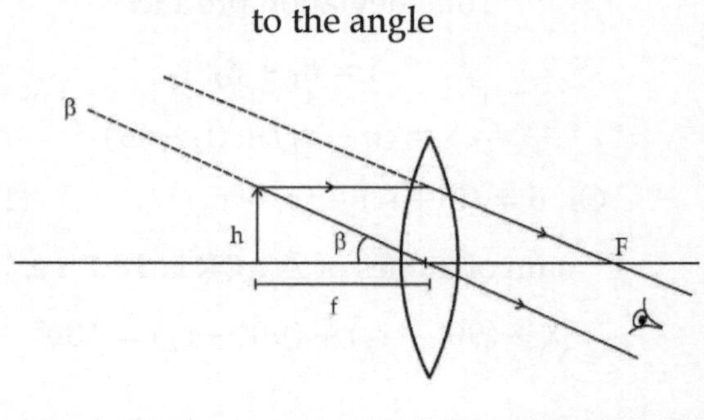

This will happen when object is placed at **F**.

$$\text{Now, } \beta \approx \tan\beta = \frac{h}{f} \quad \dots (6)$$

And from equation -(1)

$$\alpha = \tan\alpha = \frac{h}{D}$$

Thus magnification in this case will be

$$m = \frac{\beta}{\alpha} = \frac{h}{f} \times \frac{D}{h} = \frac{D}{f}$$

$$\therefore \mathbf{m} = \frac{\mathbf{D}}{\mathbf{f}} \quad \dots (7)$$

Equation -(7) represents, magnification formula, when final image is formed at infinity. Please note in questions normal adjustment means the relaxed eye or when final image is formed at infinity.

96. Compound Microscope

"*It is a device which is used to produce large magnification to see clearly very small objects*".

It consists of two convex lenses placed in a tube coaxially. One lens is kept close to object has small aperture, less focal length and high power. Another lens close to eye is called eye piece which further magnifies the real image formed by objective lens.

There are two cases of image formation using compound microscope-

Case-I - When final image is formed at least distance of distinct vision-(D)

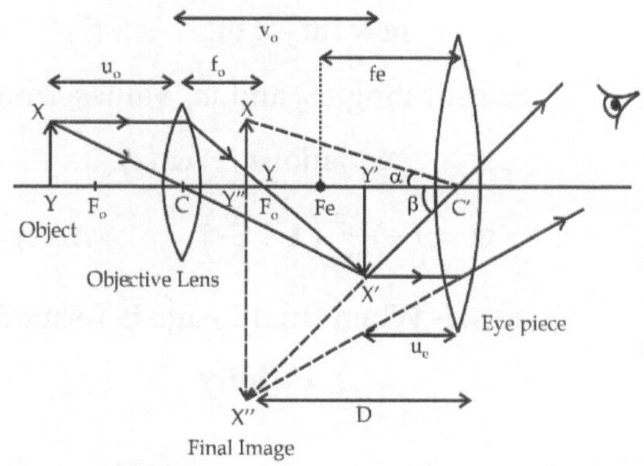

The object is placed very close of $\mathbf{F_0}$ so that a real and magnified image can be obtained. Arrangement between lenses is made in such a manner so that the real image $\mathbf{X'Y'}$ formed by objective lens lies between optical centre and focal point of eye piece. i.e. between $\mathbf{C'}$ & $\mathbf{F_e}$. $\mathbf{X'Y'}$ acts like an object for eyepiece & it forms virtual image of $\mathbf{X'Y'}$ at least distance of distinct vision D, $\mathbf{X''Y''}$.

Now magnification of objective lens

$$m_0 = \frac{+v_0}{-u_0} = -\frac{v_0}{u_0} \quad \dots (1)$$

And for eye piece which is used as simple microscope lens

$$m_e = 1 + \frac{D}{F_e} \quad \dots (2)$$

Now

$$\mathbf{m} = \frac{\textbf{Angle substended by final image}}{\textbf{Angle subtended by object when placed at same position as final image}}$$

$$\therefore \text{From figure } m = \frac{\beta}{\alpha} = \frac{\tan\beta}{\tan\alpha}$$

($\because$ α & β) are small

$$m = \frac{X''Y''/C'Y''}{XY/C'Y''} = \frac{X''Y''}{XY}$$

If we multiply and divide by $X'Y'$ we get

$$m = \frac{X''Y''}{X'Y'} \times \frac{X'Y'}{XY} = m_e \times m_0$$

$\therefore$ $\mathbf{m = m_0 \times m_e}$ ... (3)

Substituting m_0 and m_e values from equations (1) & (2)

$$\mathbf{m = (-)\frac{v_0}{u_0}\left(1+\frac{D}{F_e}\right)} \quad ... (4)$$

Case-II- When final image is formed at infinity

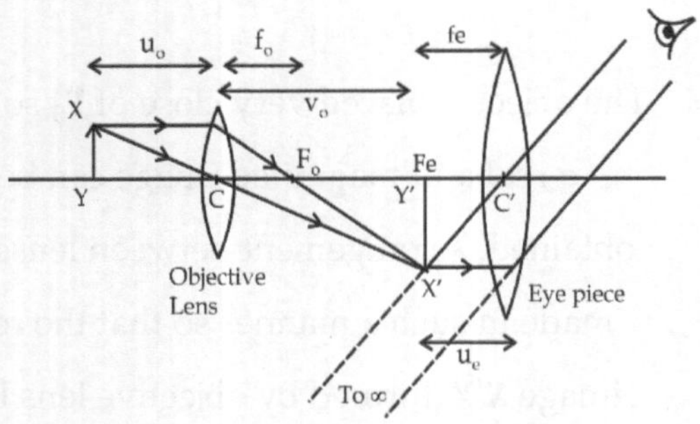

In this case eyes are least strained and objective lens forms real image of object **XY** in the focal plane of eye piece & final image is formed by eye piece at infinity.

$$\because m_0 = (-)\frac{v_0}{u_0} \quad ... (1)$$

$$\text{And } m_e = \frac{D}{f_e} \quad ... (2)$$

(Derived in simple microscope derivation)

$$\therefore \mathbf{m} = (-)\frac{\mathbf{v_0}}{\mathbf{u_0}}\frac{\mathbf{D}}{\mathbf{f_e}} \quad ... (3)$$

Please note:

- When final image is formed at infinity magnification is minimum and when final image is formed at least distance of distinct vision, magnification is maximum.
- If $\boldsymbol{\ell}$ is the tube length and real image formed by objective is very close to eye piece then-

$$v_0 = \ell$$

$$\text{Then } \mathbf{m} = (-)\frac{\ell}{f_0}\left(\frac{\mathbf{D}}{\mathbf{f_e}}\right) \quad ... (4)$$

(When final image is formed at infinity)

$$\text{And m} = (-)\frac{\ell}{f_0}\left(1+\frac{\mathbf{D}}{\mathbf{f_e}}\right) \quad ... (5)$$

(When final image is formed at least distance of distinct vision)

- We need eyepiece of larger aperture which enables us to see highly magnified image.
- When final image is formed at least distance of distinct vision, eyes are most strained and magnification is maximum.
- In the question, if description of compound microscope in normal adjustment is asked then we take the case when final image is formed at ∞.

97. Astronomical Telescope

"*A device which is used to see distant objects like stars, planets or any distant object on ground is called telescope*".

It is composed of two convex lenses, objective lens is of large aperture so that it can capture maximum amount of light coming from distant star so that we get a brighter image of the object. Since object (star, planets) is at infinite distance, the image is formed in the focal plane of objective lens.

Here also two cases are there.

Case-I - When final image is formed at least distance of distinct vision

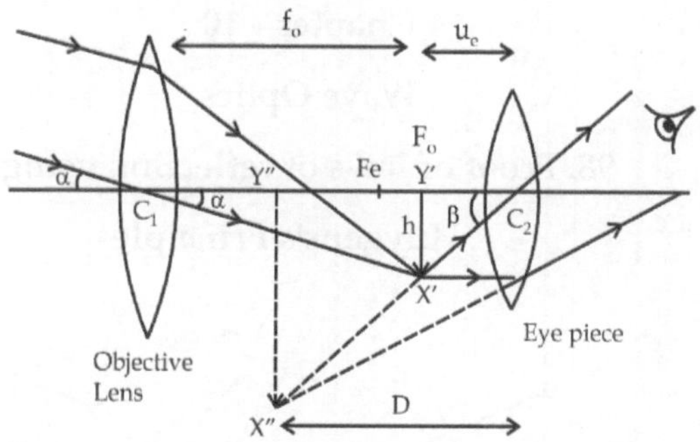

Here angular magnification will be from figure

$$m = (-)\frac{\beta}{\alpha} \quad \dots (1)$$

(Because α and β are on the opposite sides of the principle axis)

From figure

$$\alpha \approx \tan \alpha$$

$$\text{or } \alpha = \frac{X'Y'}{C_1Y'} = \frac{h}{f_0} \quad \dots (2)$$

[Here h is the height of image X'Y']

$$\& \beta \approx \tan \beta = \frac{X'Y'}{C_2Y'} = \frac{h}{u_e} \quad \dots (3)$$

$$\therefore m = (-)\frac{h}{u_e} \times \frac{f_0}{h} = (-)\frac{f_0}{u_e} \dots (4)$$

Please note that if focal length of objective lens is more, a magnified image is formed.

For eyepiece we can write -

$$\frac{1}{f} = \frac{1}{v} - \frac{1}{u} \text{ or } \frac{1}{f_e} = \frac{1}{-D} - \frac{1}{-u_e}$$

$$\text{Or } \frac{1}{u_e} = \frac{1}{D} + \frac{1}{f_e} \quad \dots (5)$$

If we substitute the above value of $\frac{1}{u_e}$ in equation (4) we get

$$m = -f_0\left[\frac{1}{D} + \frac{1}{f_e}\right]$$

$$\text{Or } \mathbf{m} = (-)\frac{f_0}{f_e}\left[\frac{f_e}{D} + 1\right] \quad \dots (6)$$

From equation -(6) we can say that when f_0 is more, magnification is more i.e. f_e should be less & f_0 should be more for good magnification.

Here tube length $\mathbf{L = f_0 + u_e}$

Case-II- When final Image is formed at infinity

In this case the distance between objective lens and eye piece is adjusted in such a manner that their focal points coincide i.e., the image formed by objective lens lies in the focal plane of both the lenses.

The real image formed by objective lens acts like an object for eyepiece which produces its final image at infinity.

The angle subtended by object on eye is **α** and by final image is **β**.

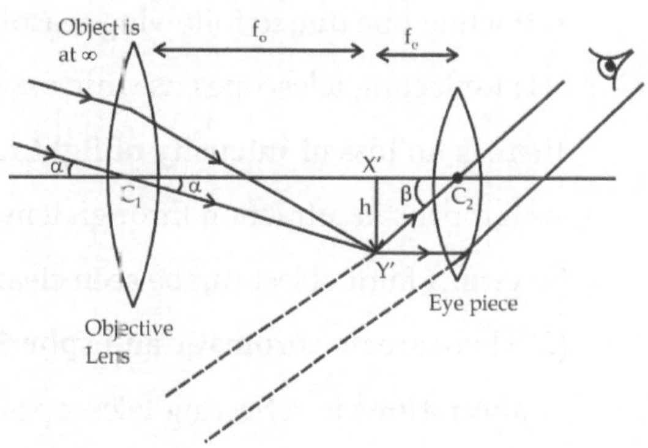

In this case eyes are least strained and magnification produced by lens is minimum.

From figure

$$\alpha = \tan \alpha = \frac{X'Y'}{C_1X'}$$

$$\therefore \alpha = \frac{h}{f_0} \quad \dots (1)$$

$$\& \beta = \tan \beta = \frac{X'Y'}{C_2X'} = \frac{h}{f_e} \quad \dots (2)$$

$\& m = (-)\frac{\beta}{\alpha}$ (Because α & β are on the opposite sides of principle axis) substituting α & β values in magnification formula, we get

$$m = (-)\frac{h}{f_e} \times \frac{f_0}{h} = (-)\frac{f_0}{f_e}$$

$$\mathbf{m = (-)\frac{f_0}{f_e}} \quad \dots (3)$$

Please remember, in questions if normal adjustment is asked we consider final image is formed at infinity.

In this case tube length is

$$\mathbf{L = f_0 + f_e} \quad \dots (4)$$

Note: - The optical instruments studied so far are using lenses.

We have reflecting telescopes also which use large size concave mirror.

Reflecting telescope is better than the refracting one due to following reasons -

(1) Reflecting telescopes use mirrors so there is **no loss of intensity of light** rays as happens in refraction through lenses. So even a faint object can be seen clearly.

(2) There are **no chromatic and spherical aberrations** in reflecting telescopes.

(3) Making large size mirrors is easier than making lenses of equivalent size. Lenses are costlier than mirrors.

We have in our syllabus Cassegrain telescope.

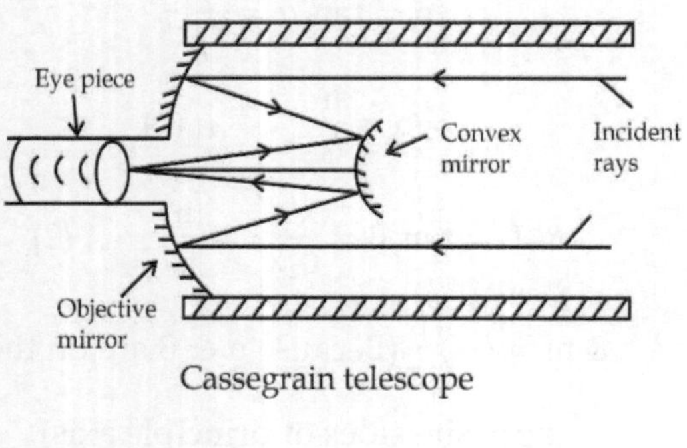

Cassegrain telescope

Chapter - 10

Wave Optics

98. Proof of laws of reflection using Huygens's Principle

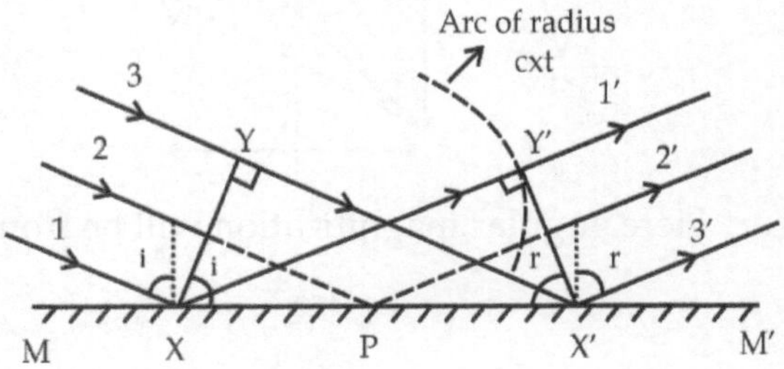

Here **MM′** is the mirror surface, **XY** = Incident wave front, **∠YXX′** = **∠i**, **1, 2, 3** are incident rays which are perpendicular to the wavefront **XY**.

Now as per Huygen's principle every point on **XY** acts like a source of light which emits secondary wavelets.

We assume one secondary wavelet from point **Y** hits the mirror at point **X′** in time **t**. Since speed of light in air is taken as **c** then **YX′ = c × t**. Now the reflected secondary wavelet from point **X** also travels the same distance **ct** in time **t**.

So we draw an arc of radius **c × t** keeping **X** at the centre. Now we draw a tangent from this arc, which joins point **X′** on the mirror and thus we get reflected Wavefront **X′Y′** and angle of reflection

$$\angle r = \angle Y'X'X$$

In triangles XX′Y & XX′Y′; XX′ is common and YX′ = XY′ = c × t

Also ∠Y = ∠Y′ = 90°

that means triangles XX′Y and XX′Y′ are congruent

$$\therefore \angle YXX' = \angle Y'X'X$$

i.e. $\angle i = \angle r$

This is **I** law of reflection.

Also incident wavefront **XY**, the mirror plane **MM′** and reflecting wavefront **X′Y′** are all perpendicular to the plane of paper, therefore incident ray, normal to the mirror and reflected ray all lie in the plane of paper which is **II** law of reflection.

99. Proof of Laws of refraction using Huygens's Principle

As shown in the figure, rays **1, 2 & 3** are incident on a part of interface of two medium shown as **AB**.

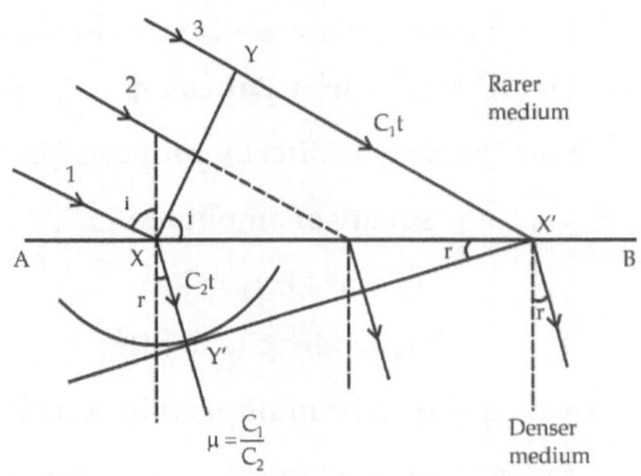

XY is incident wave front. According to Huygens's principle each point on this incident wavefront acts like a new source of light and emits secondary wavelets. Speed of light in rarer medium is $\mathbf{c_1}$ & in time **t** a secondary wavelet from point **Y** strikes the interface at point **X′** then.

$$YX' = c_1t \quad \dots (1)$$

Speed of light in denser medium is $\mathbf{c_2}$. In the same time **t** a secondary wavelet will cover distance $\mathbf{c_2t}$ in denser medium. We draw an arc of radius $\mathbf{c_2t}$ keeping point **X** at the centre of arc and draw a tangent from **X′** on this arc which meets at point **Y′**. Thus we get refracted wavefront **X′Y′**.

$$\text{Now } XY' = c_2t \quad \dots (2)$$

Since refractive index of denser medium with respect to rarer medium is the ratio of their speeds of light, we can say that

$$\mu = \frac{c_1}{c_2} \dots (3)$$

Now $\angle YXX' = \angle i$ & $\angle XX'Y' = \angle r$

In triangle XX′Y

$$\sin i = \frac{YX'}{XX'}$$

$$\text{or } \sin i = \frac{c_1t}{XX'} \dots (4) \quad (\because YX' = c_1t)$$

$$\text{In } \triangle XX'Y' \ \sin r = \frac{XY'}{XX'} = \frac{c_2t}{XX'}$$

$$(\because AB' = c_2t)$$

$$\therefore \frac{\sin i}{\sin r} = \frac{c_1t}{AA'} \times \frac{AA'}{c_2t}$$

$$\text{Or } \frac{\sin i}{\sin r} = \frac{c_1}{c_2}$$

$$\text{since } \frac{c_1}{c_2} = \mu$$

$$\therefore \mu = \frac{\sin i}{\sin r}$$

This is snell's law of refraction.

Also from the figure, The incident ray, normal to the interface **AB** and refracted ray all three lie in the same plane because incident wavefront, interface **AB** and refracted wavefront are perpendicular to the plane of the paper.

This is **II** law of refraction.

100. Interference of light: Young's double slit experiment and conditions for constructive and destructive interference

When light waves from two coherent sources reach at a same point, then they interfere with each other and resultant displacement at the point of interference is vector sum of individual wave displacement. That means

$$\vec{y} = \vec{y}_1 + \vec{y}_2 \quad ...... (1)$$

Due to interference of light, at some points, the secondary wavelets are meeting in phase and we get a bright spot on the screen which is called bright fringe and at some points on the screen, we get dark spots where the wavelets are out of phase with each other, these are called dark fringes. We need coherent sources of light to see an interference pattern.

Coherent sources: *Those sources of light which emit continuous light waves of same frequency and wavelength having a constant phase difference or no phase difference are called coherent sources of light.*

Conditions for constructive and destructive interference –

Let the waves from two coherent sources of light are presented as

$$y_1 = a \sin \omega t \quad \ldots (1)$$

$$\& \quad y_2 = b \sin(\omega t + \phi) \quad \ldots (2)$$

Here **a** and **b** are amplitudes of two wavelets and **ϕ** is the constant phase difference, then using superposition principle

$$y = y_1 + y_2$$

$$y = a \sin \omega t + b \underbrace{\sin(\omega t + \phi)}_{\text{Like } \sin(a+b)}$$

$$\because \sin(a + b) = \sin a \cos b + \cos a \sin b$$

$$\therefore y = a\sin\omega t + b\sin\omega t \cos\phi + b\cos\omega t \sin\phi \quad \ldots (3)$$

$$\text{Or } y = \sin \omega t(a + b\cos\phi) + b\sin\phi \cos\omega t$$

$$\text{Assume } a + b\cos\phi = R\cos\theta \quad \ldots (4)$$

$$\& \quad b\sin\phi = R\sin\theta \quad \ldots (5)$$

$$\text{then } y = R\cos\theta \sin \omega t + R\sin\theta \cos \omega t$$

$$\therefore \mathbf{y = R\sin(\omega t + \theta)} \quad \ldots (6)$$

Equation -(6) represents a simple harmonic wave having amplitude **R**.

Now squaring equations (4) and (5) and adding them, we get

$$R^2(\cos^2\theta + \sin^2\theta) = (a + b\cos\phi)^2 + (b\sin\phi)^2$$

$$\therefore R^2 \times 1 = a^2 + b^2\cos^2\phi + 2ab\cos\phi + b^2\sin^2\phi$$

$$\text{Or } \mathbf{R^2 = a^2 + b^2 + 2ab\cos\phi} \quad \ldots (7)$$

Now intensity is directly proportional to square of amplitude

$$I_1 \propto a^2 \text{ Or } \mathbf{I_1 = ka^2}$$

$$\& \mathbf{I_2 = kb^2} \ \& \mathbf{I_R = kR^2}$$

Then eq-(7) can be multiplied by **k** on both sides and we get intensity equation as

$$kR^2 = ka^2 + kb^2 + 2\sqrt{k}a\sqrt{k}b \cos\phi$$

$$\text{Or} \quad \mathbf{I_R = I_1 + I_2 + 2\sqrt{I_1 I_2}\cos\phi} \quad \ldots (8)$$

- Now **for constructive interference**

I_R is maximum i.e. $\cos\phi = \max = +1$

i.e. $\phi = 0, 2\pi, 4\pi \cdots 2n\pi,$

$$\therefore \mathbf{\phi = 2n\pi} \quad \ldots (9) \quad (n = 0,1,2,3 \ldots\ldots)$$

If **x** is the path difference between two wavelets and when they reach at the same point then

$$x = \frac{\lambda}{2\pi} \times \phi$$

$$\text{or } x = \frac{\lambda}{2\pi} \times 2n\pi$$

$$\text{or } \mathbf{x = n\lambda} \quad \dots (10)$$

$$\text{then } R = \sqrt{a^2 + b^2 + 2ab}$$

$$\text{or } R = \sqrt{(a+b)^2} = a + b$$

$$\mathbf{R = a + b} \quad \dots (11)$$

$$\text{and intensity } \mathbf{I_R = I_1 + I_2 + 2\sqrt{I_1 I_2}}$$

$$\text{or } \mathbf{I_R = \left(\sqrt{I_1} + \sqrt{I_2}\right)^2} \quad \dots (12) \text{ (max)}$$

$$\text{Also } I_R = kR^2$$

$$\therefore \mathbf{I_R = k(a+b)^2}$$

- **For destructive interference -**

I_R is minimum

i.e. $\cos\phi = -1$ (min.)

$$\text{or } \phi = \pi, 3\pi, 5\pi \dots (2n-1)\pi$$

$$\mathbf{\phi = (2n-1)\pi} \quad \dots (13)$$

$$(n = 1,2,3 \dots)$$

& path difference between the two waves

$$x = \frac{\lambda}{2\pi} \times \phi = \frac{\lambda}{2\pi}(2n-1)\pi$$

$$\text{or } \mathbf{x = (2n-1)\frac{\lambda}{2}} \quad \dots (14)$$

$$\text{then } R = \sqrt{a^2 + b^2 - 2ab}$$

$$\text{or } R = \sqrt{(a-b)^2}$$

$$\text{or } \mathbf{R = (a-b)} \quad \dots (15)$$

$$\text{and intensity } I_R = I_1 + I_2 - 2\sqrt{I_1 I_2}$$

$$\text{or } \mathbf{I_R = \left(\sqrt{I_1} - \sqrt{I_2}\right)^2} \quad \dots (16)$$

(minimum)

$$\text{Also } I_R = kR^2$$

$$\therefore \mathbf{I_R = k(a-b)^2} \quad \dots (17)$$

Please note -

$$\frac{R_{max}}{R_{min}} = \frac{(a+b)}{(a-b)} \quad \dots (18)$$

$$\text{And } \frac{I_{max}}{I_{min}} = \frac{(a+b)^2}{(a-b)^2} = \frac{\left(\sqrt{I_1}+\sqrt{I_2}\right)^2}{\left(\sqrt{I_1}-\sqrt{I_2}\right)^2}$$

If ω_1 and ω_2 are widths of two slits from which intensities of light I_1 and I_2 are coming then

$$\frac{\omega_1}{\omega_2} = \frac{I_1}{I_2} = \frac{a^2}{b^2}$$

101. Proof of conservation of energy during interference

In interference pattern obtained on the screen

$$I_{max} = k(a+b)^2 \quad \dots (1)$$

$$\& \ I_{min} = k(a-b)^2 \quad \dots (2)$$

Then average intensity of light in interference pattern will be

$$I_{av} = \frac{I_{max} + I_{min}}{2}$$

$$I_{av} = \frac{k(a+b)^2 + k(a-b)^2}{2}$$

$$\text{or } I_{av} = \frac{2k(a^2+b^2)}{2}$$

$$\therefore \mathbf{I_{av} = k(a^2 + b^2)} \quad \dots (3)$$

Now if there is no interference and light is incident on screen from two sources of intensities $\mathbf{I_1}$ & $\mathbf{I_2}$ respectively, then

$$I = I_1 + I_2 = ka^2 + kb^2$$

$$\text{or } \mathbf{I = k(a^2 + b^2)} \ ..(4)$$

Equation (3) represents average intensity when interference takes place and equation (4) represents average intensity when no interference is there. We see both are equal. This proves that during interference the energy gets redistributed.

102. Positions of maxima and minima on the screen and fringe width

Bright fringes are called maxima and dark fringes are called minima. The distance of

n^{th} bright fringe from the centre of the screen is

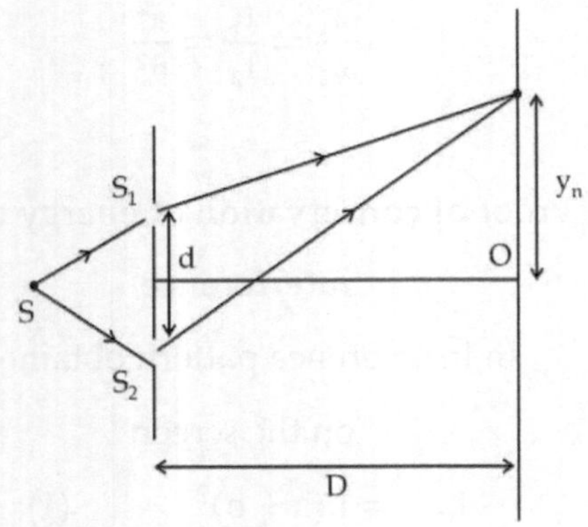

$$y_n = n\frac{\lambda D}{d} \quad \dots (1)$$

[For maxima or bright fringe]

[Here n = 1, 2, 3, 4]

Here λ = Wavelength of light used

D = Distance between slits and screen

d = Distance between two slits

The distance of nth dark fringe or minima from the centre is

$$y_n = (2n-1)\frac{\lambda D}{2d} \quad \dots (2)$$

[For minima or dark fringe]

[Here n = 1, 2, 3, 4]

Formula for fringe width

$$\beta = \frac{\lambda D}{d} \quad \dots (3)$$

All the fringes are of same width.

If we immerse **YDSE** apparatus in a liquid of refractive index μ then fringe width becomes $\frac{1}{\mu}$ times.

$$\therefore \beta' = \frac{\beta}{\mu} \quad \dots (4)$$

Angular width of fringes

$$\theta = \frac{\beta}{D}\left(\frac{\text{Arc}}{\text{Radius}}\right)$$

$$\therefore \theta = \frac{\lambda D}{d} \times \frac{1}{D} = \frac{\lambda}{d}$$

$$\theta = \frac{\beta}{D} = \frac{\lambda}{d} \quad \dots (5)$$

Intensity vs phase difference & intensity vs path difference graph -

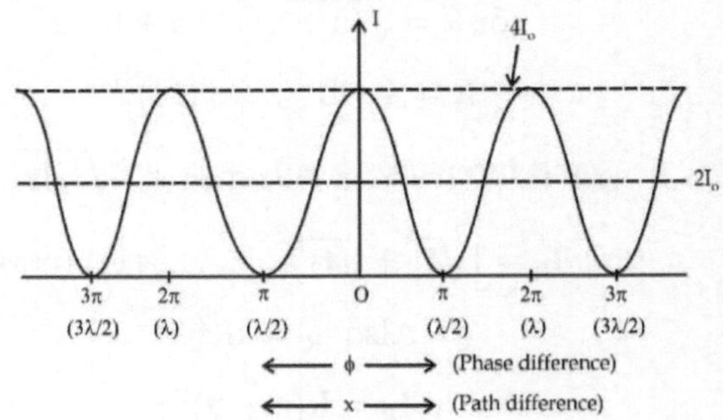

103. Diffraction of light and positions of secondary minima and secondary maxima

Definition: *The phenomenon of bending of light around the corners of an obstacle or an aperture which has size of the order of wavelength of light used is called diffraction.*

It's shown by single slit and based on Huygens's principle. Here, the plane wavefront incident on the slit emits secondary wavelets in all possible directions. The **bright bands** on the screen are called **secondary maxima** and **dark bands** on the screen are called **secondary minima**. At the centre of the screen, we get central maxima. Distance of nth **secondary minima** will be

$$y_n = \frac{n\lambda D}{a} \quad \text{[Here, n = 1,2,3,4 ...]}$$

λ = wavelength of light used

D = Distance between slit and screen

a = slit width

If 'x' is the path difference between the secondary wavelets reaching at a point on the screen from two ends of slit diameter for minima is $\mathbf{x = \pm n\lambda}$ (opposite to interference pattern)

sometimes

D = focal length of converging lens

$$\therefore\ y_n = \frac{n\lambda D}{a} = \frac{n\lambda f}{a}$$

- Distance of $\mathbf{n^{th}}$ **secondary maxima** from the central maxima will be

$$\mathbf{y_n = (2n+1)\frac{\lambda D}{2a}} \quad \text{[Here, n = 1,2,3,4 ...]}$$

And path difference $\mathbf{x = (2n+1)\frac{\lambda}{2}}$

- Angular position of $\mathbf{n^{th}}$ **secondary maxima** $\mathbf{\theta = \frac{y_n}{D}}$

$$\mathbf{\theta = \frac{(2n+1)\lambda}{2a}}$$

- Angular position of $\mathbf{n^{th}}$ **secondary minima** $\mathbf{\theta = \frac{n\lambda}{a}}$

- Angular width of central maxima

$$2\theta = \frac{2\lambda}{a}\ \text{(Double than other maxima)}$$

- Width of any maxima or minima -

$$y = \frac{\lambda D}{a}$$

- Width of central maxima

$$\mathbf{2y = \frac{2\lambda D}{a}}$$

- Intensity vs path difference graph -

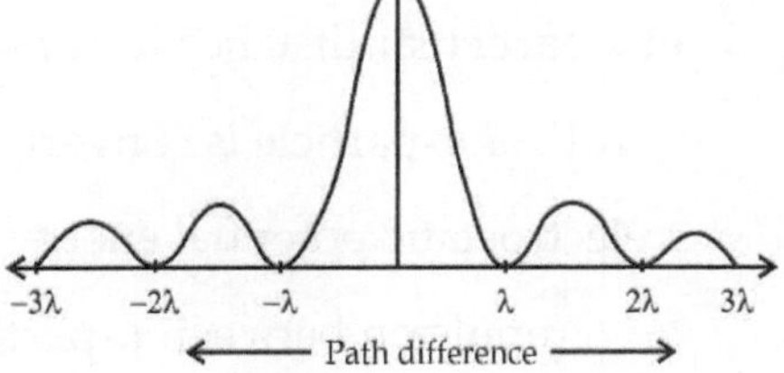

Intensity vs phase difference graph -

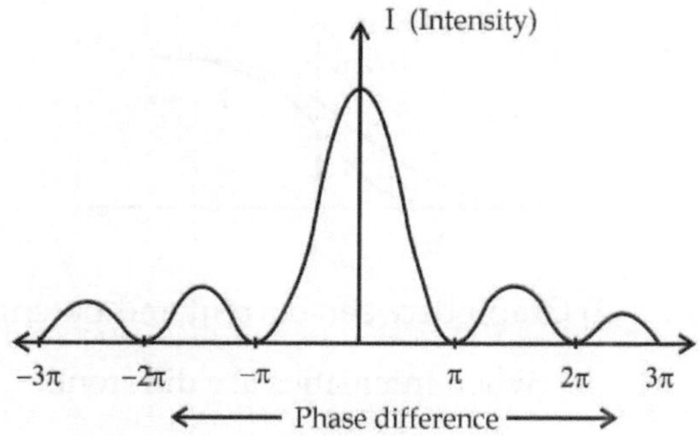

Chapter - 11

Modern Physics

104. Photo electric effect stopping potential equation

Photo electric effect Equation -

$$\mathbf{h\upsilon = h\upsilon_0 + \frac{1}{2}mv_{max}^2}$$

The minimum potential required to stop the electron just before reaching the collector plate is called stopping potential ($\mathbf{V_0}$)

$$\therefore \frac{1}{2}mv_{max}^2 = eV_0$$

$$h\upsilon = h\upsilon_0 + eV_0$$

$$\text{or}\ eV_0 = h\upsilon - h\upsilon_0$$

$$\text{or}\ \mathbf{V_0 = \frac{h\upsilon}{e} - \frac{h\upsilon_0}{e}}$$

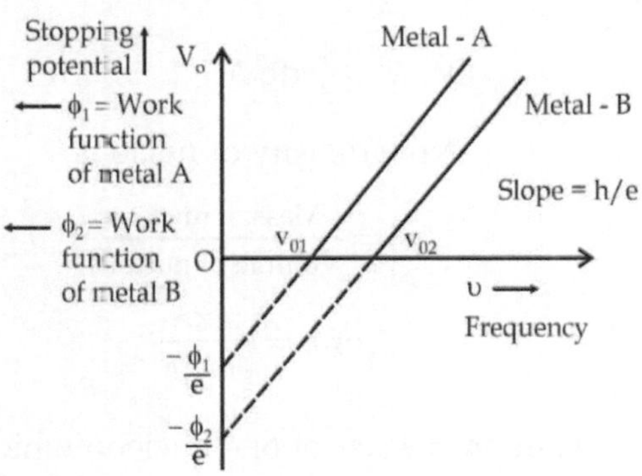

Note- (1) Graph between current and potential when different energy radiations are incident.

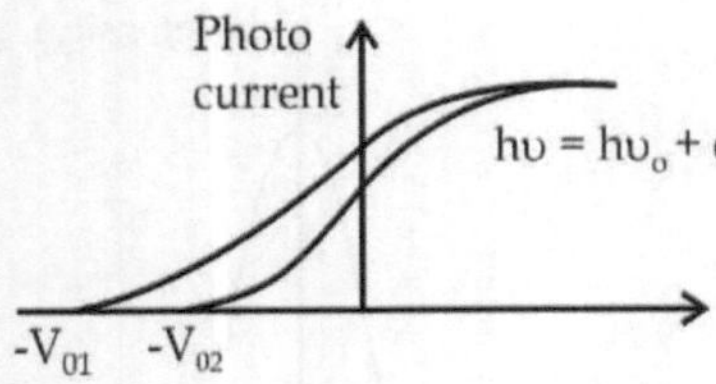

(2) Graph between current and potential when intensities are different.

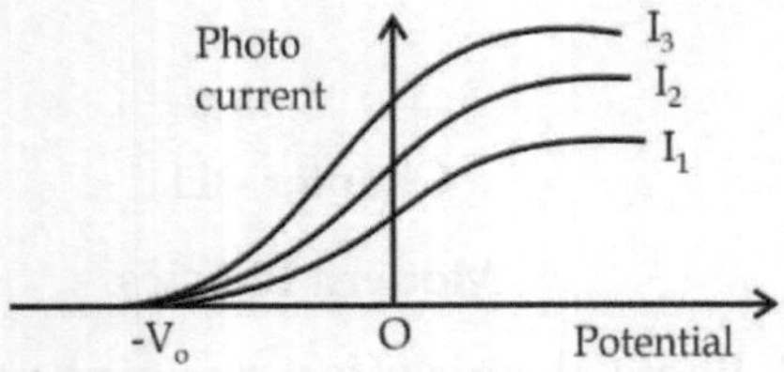

105. Proof for same nuclear density for all elements

Through experiments we have figured out that volume of a nucleus is directly proportional to its mass number **A**. i.e.

$$\frac{4}{3}\pi R^3 \propto A \quad \dots (1)$$

$$\text{or } R \propto A^{1/3}$$

$$\text{or } R = R_0 A^{1/3} \quad \dots (2)$$

Here $R_0 = 1.2 \times 10^{-15}$ m

Now volume of nucleus can be written as

$$V = \frac{4}{3}\pi R^3 = \frac{4}{3}\pi\left(R_0 A^{1/3}\right)^3$$

$$\text{Or } \; V = \frac{4}{3}\pi R_0^3 A \quad \dots (3)$$

Now density of nucleus

$$(\rho) = \frac{\text{Mass of nucleus}}{\text{Volume of nucleus}}$$

$$\text{or } \rho = \frac{m \times A}{\frac{4}{3}\pi R_0^3 A}$$

Here **m** = Mass of one nucleon which is 1.67×10^{-27} kg

$$\therefore \rho = \frac{m}{\frac{4}{3}\pi R_0^3}$$

$$\text{Or } \boldsymbol{\rho} = \frac{\mathbf{3m}}{\mathbf{4\pi R_0^3}} \quad \dots (4)$$

And everything in equation (4) is constant thus density of nucleus is constant for all elements.

106. Einstein's Mass-energy equivalence

When a certain mass **m** disappears in any process, an equivalent amount of energy is released which is equal to $\mathbf{\Delta E = \Delta mc^2}$

107. De Broglie wavelength of an electron

$$\lambda = \frac{h}{mv}$$

If an electron is accelerated by electric field having potential **V** then kinectic energy of electron

$$\frac{1}{2}mv^2 = eV \text{ or } v = \sqrt{\frac{2eV}{m}}$$

$$\because \lambda = \frac{h}{mv} = \frac{h\sqrt{m}}{m\sqrt{2eV}}$$

$$\text{or } \boldsymbol{\lambda} = \frac{\mathbf{h}}{\mathbf{\sqrt{2meV}}}$$

substituting $h = 6.64 \times 10^{-34}$ Js

$m = 9.1 \times 10^{-31}$ kg, $e = 1.6 \times 10^{-19}$c

$$\text{We get } \boldsymbol{\lambda} = \frac{\mathbf{12.27}}{\mathbf{\sqrt{V}}} \text{ Å}$$

108. Distance of closest approach

In Rutherford's experiment when **α**-particles bombarded on gold foil, for head on collision with gold nucleus, at a particular distance it stops because K.E. of **α**-particle is converted into electrostatic potential energy due to repulsion between **α**-particle and gold nucleus.

$$\therefore \frac{1}{2}mv^2 = \frac{k(Ze)e}{r}$$

$$\text{or } \mathbf{r} = \frac{\mathbf{2kZe^2}}{\frac{1}{2}\mathbf{mv^2}}$$

Here r = Distance of closest approach

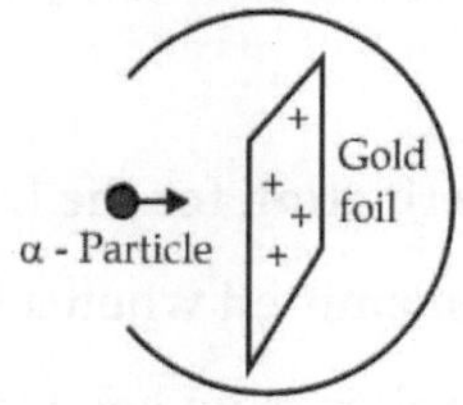

109. Energy of an electron in an atomic orbit.

When electron revolves around the nucleus, the electrostatic attraction force of nucleus provides necessary centripetal force which is given as

$$\frac{mv^2}{r} = \frac{k e \times e}{r^2} \text{ [In Hydrogen atom]}$$

Here m = mass of electron

v = speed of electron in the orbit

r = radius of orbit

e = charge an electron and proton

$$\text{Or } mv^2 = \frac{ke^2}{r} = \frac{e^2}{4\pi\varepsilon_0 r}$$

$$\text{Or kinetic energy } \frac{1}{2}mv^2 = \frac{1}{2}\frac{e^2}{4\pi\varepsilon_0 r} = \frac{e^2}{8\pi\varepsilon_0 r}$$

$$\therefore \mathbf{K.E.} = \frac{\mathbf{e^2}}{\mathbf{8\pi\varepsilon_0 r}}$$

$$\text{And potential energy} = \frac{k(e)(-e)}{r}$$

$$\text{or } \mathbf{P.E.} = -\frac{\mathbf{e^2}}{\mathbf{4\pi\varepsilon_0 r}}$$

Total mechanical engery (M.E) = Kinetic engery + potential energy

$$M.E. = \frac{e^2}{8\pi\varepsilon_0 r} - \frac{e^2}{4\pi\varepsilon_0 r}$$

$$\mathbf{M.E.} = -\frac{\mathbf{e^2}}{\mathbf{8\pi\varepsilon_0 r}}$$

Negative sign means that electron is bound to the nucleus. When we want to remove the electron we need to supply equal amount of external energy which is called **binding energy** [Ionization potential].

$$\therefore \mathbf{B.E.} = +\frac{\mathbf{e^2}}{\mathbf{8\pi\varepsilon_0 r}}$$

Note: (1) When electron jumps from one shell to other, energy is either emitted or absorbed. For any electron, for any element the energy of electron in a specific shell is given by-

$$\mathbf{E_n = -13.6\frac{Z^2}{n^2}\ eV}$$

Here, Z = atomic number

n = shell number

(2) Wavelength associated with energy absorbed/released

$$\mathbf{\frac{1}{\lambda} = RZ^2\left[\frac{1}{n_1^2} - \frac{1}{n_2^2}\right]}$$

Here, R is Rydberg's constant

$$R = 1.1 \times 10^7\ m^{-1}$$

110. Radius of nth orbit of an atom

Radius of any Bohr orbit -

$$\mathbf{r = 0.529\frac{n^2}{Z}\ \text{Å}}$$

n = no. of shell, Z = Atomic no.

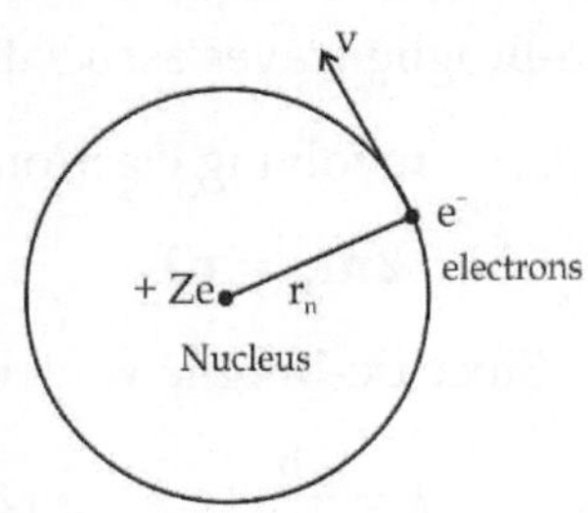

Suppose an electron is revolving in $\mathbf{n^{th}}$ orbit with speed $\mathbf{v_n}$. The radius of orbit is

r_n. Here centripetal force is provided by the electrostatic force of attraction, that means

$$\frac{mv_n^2}{r_n} = \frac{k(Ze)e}{r_n^2}$$

$$\text{or } mv_n^2 = \frac{kZe^2}{r_n} \quad ....(1)$$

Now from Bohr's **II** postulate

$$mv_nr_n = \frac{nh}{2\pi} \quad ....(2)$$

$$\text{or } v_n = \frac{nh}{2\pi mr_n} \quad ....(3)$$

substituting value of $\mathbf{v_n}$ in equation - (1), we get

$$m\,\frac{n^2h^2}{4\pi^2m^2r_n^2} = \frac{kze^2}{r_n}$$

$$\text{or } \mathbf{r_n} = \frac{\mathbf{n^2h^2}}{\mathbf{4\pi^2mkZe^2}}$$

111. Proof of Bohr's second postulate by de-Broglie explanation

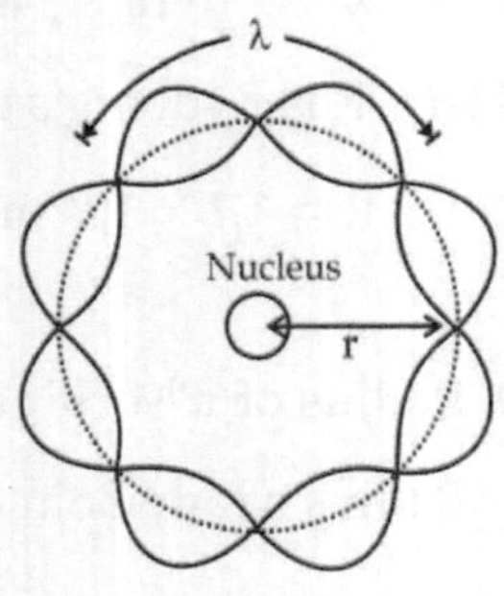

According to de-Broglie, a stationary orbit is that which contains an integral number of de-Broglie waves associated with the revolving electron.

$$\text{i.e. } \mathbf{2\pi r_n = n\lambda} \quad ...(1)$$

Since de-Brogile wavelength

$$\lambda = \frac{h}{mv_n} \quad ...(2)$$

n = shell number

substituting $\boldsymbol{\lambda}$ value in eq -(1)

$$2\pi r_n = \frac{nh}{mv_n}$$

$$\text{or } \mathbf{mv_nr_n} = \frac{\mathbf{nh}}{\mathbf{2\pi}}$$

This is Bohr's II postulate.

112. Derivation for the frequency of radiation emitted when a hydrogen atom de-excites from level 'n' to (n-1)

When electron revolves in an orbit, the necessary centripetal force is provided by electrostatic attraction force between electron and nuclei

$$\text{i.e. } \frac{mv^2}{r} = \frac{kZe^2}{r^2}$$

$$\text{or } \frac{1}{2}mv^2 = \frac{kZe^2}{2r} \quad ...(1)$$

$$\text{i.e. K.E. of electron} = \frac{kZe^2}{2r}$$

$$\text{\& potential energy of electron} = -\frac{kZe^2}{r}$$

$$\text{or P.E.} = -\frac{kZe^2}{r} \quad ...(2)$$

Total energy of electron in an orbit

$$E = K.E + P.E.$$

$$E = \frac{1}{2}\frac{kZe^2}{r} - \frac{kZe^2}{r}$$

$$\therefore E = (-)\frac{kZe^2}{2r} \quad ...(3)$$

$$\because r_n = \frac{n^2h^2}{4\pi^2mkZe^2}$$

substituting $\mathbf{r_n}$ value in equation - (3) we get

$$E_n = (-)\frac{kZe^2}{2} \times \frac{4\pi^2mkZe^2}{n^2h^2}$$

$$\text{Or } E_n = (-)\left(\frac{2\pi^2mk^2e^4}{h^2}\right)\frac{Z^2}{n^2} \quad ...(4)$$

For hydrogen atom Z = 1

$$\therefore E_n = -\frac{2\pi^2mk^2e^4}{h^2} \times \frac{1}{n^2} \quad ...(5)$$

When an electron de-excites from level $\mathbf{n_2}$ to level $\mathbf{n_1}$ then energy is released which is given by

$$h\upsilon = E_2 - E_1$$

$$\& \ \upsilon = \frac{E_2 - E_1}{h} \quad \dots (6)$$

From equation (5)

$$\upsilon = (-)\frac{2\pi^2 mk^2 e^4}{h^3}\left[\frac{1}{n_2^2} - \frac{1}{n_1^2}\right]$$

$$\text{or } \upsilon = \frac{2\pi^2 mk^2 e^4}{h^3}\left[\frac{1}{n_1^2} - \frac{1}{n_2^2}\right]$$

Given $n_1 = n - 1, n_2 = n$

$$\therefore \upsilon = \frac{2\pi^2 mk^2 e^4}{h^3}\left[\frac{1}{(n-1)^2} - \frac{1}{n^2}\right]$$

$$\text{or} \quad \boldsymbol{\upsilon = \frac{2\pi^2 mk^2 e^4}{h^3}\frac{(2n-1)}{(n-1)^2 n^2}}$$

Here k = Coulomb's constant

$$k = \frac{1}{4\pi\varepsilon_0} = 9 \times 10^9 \frac{Nm^2}{c^2}$$

113. Classical frequency of revolution of an electron in n^{th} orbit.

First we calculate speed of revolving electron in $\mathbf{n^{th}}$ orbit. Since angular momentum of a revolving electron **mvr** is integral multiple of $\frac{h}{2\pi}$

$$\therefore mvr = n\frac{h}{2\pi}$$

$$\therefore v = \frac{nh}{2\pi mr} \quad \dots (1)$$

$$\& \ r_n = \frac{n^2 h^2}{4\pi^2 mkZe^2} \quad \dots (2)$$

substituting value of $\mathbf{r_n}$ in equation-(1), we get

$$v = \frac{nh}{2\pi m} \times \frac{4\pi^2 mkZe^2}{n^2 h^2}$$

$$\text{or } v = \frac{2\pi kZe^2}{nh} \quad \dots (3)$$

since $v = r\omega$ (ω = Angular velocity)

$$\& \ \omega = 2\pi\upsilon$$

$$\therefore v = r \times 2\pi\upsilon$$

$$\text{or } \upsilon = \frac{v}{2\pi r}$$

substituting **v** value from equation - (3) we get

$$\upsilon = \frac{2\pi kZe^2}{nh \times 2\pi r} = \frac{kZe^2}{nhr}$$

substituting **r** value from equation (2) we get

$$\upsilon = \frac{kZe^2}{nh} \times \frac{4\pi^2 mkZe^2}{n^2 h^2}$$

$$\text{or } \boldsymbol{\upsilon = \frac{4\pi^2 k^2 mZ^2 e^4}{n^3 h^3}} \quad \dots (4)$$

$$\therefore \upsilon \propto \frac{1}{n^3}$$

Equation (4) represents classical frequency of revolution of an electron in n^{th} orbit.

114. Proof: When transition of electron takes place between higher orbitals then frequency of emitted radiation is equal to classical frequency of electron.

In previous derivation we obtained the frequency of emitted radiation when electron jumps from **n** to **(n − 1)**, which is

$$\upsilon = \frac{2\pi^2 mk^2 e^4 Z^2 (2n-1)}{h^3 \ (n-1)^2 n^2}$$

For large value of **n**

$$2n - 1 \approx 2n \ \& \ n - 1 \approx n$$

$$\text{Then } \upsilon = \frac{2\pi^2 mk^2 e^4 Z^2}{h^3} \times \frac{2n}{n^2 \times n^2}$$

$$\text{or } \upsilon = \frac{4\pi^2 mk^2 Z^2 e^4}{n^3 h^3} \quad \dots (1)$$

This is equal to classical frequency of a revolving electron in $\mathbf{n^{th}}$ orbit i.e.,

$$\boldsymbol{\upsilon = \frac{4\pi^2 k^2 mZ^2 e^4}{n^3 h^3}}$$

Hence Proved.

Question Bank for Entire Syllabus (PYQs)

I. Electric Charges and Fields

1. According to Coulomb's law, which is the correct relation for the following figure? (2022)

(a) $q_1q_2 > 0$ **(b) $q_1q_2 < 0$**
(c) $q_1q_2 = 0$ **(d) $1 > q_1/q_2 > 0$**

Ans. (b): As shown in the figure, there is a force of attraction that means the charges are unlike charges.

$\therefore F = \frac{kq_1q_2}{r^2}$

And for attractive force, $q_1q_2 < 0$, because

$q_1 = +$ ve and $q_2 = -$ ve

$\therefore q_1q_2 < 0$

2. A small object with charge q and weight mg is attached to one end of a string of length 'L' attached to a stationary support. The system is placed in a uniform horizontal electric field ' E ', as shown in the accompanying figure. In the presence of the field, the string makes a constant angle θ with the vertical. The sign and magnitude of q is (2021)

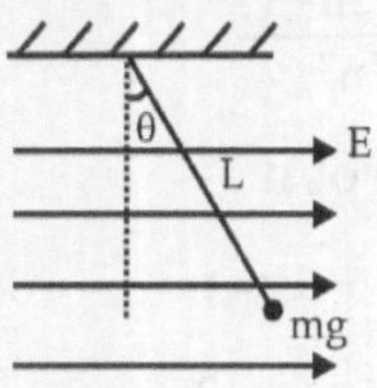

(a) positive with magnitude mg/E
(b) positive with magnitude (mg/E)tan θ
(c) negative with magnitude mg/Etan θ
(d) positive with magnitude E tan θ/mg

Ans. (b): In the figure at equilibrium, various forces acting on the object are shown.

From figure $T\cos\theta = mg$ ....(1)

and $T\sin\theta = qE$ .....(2)

Now $\frac{T\sin\theta}{T\cos\theta} = \frac{qE}{mg} = \tan\theta$

$\therefore qE = mg\tan\theta$

$q = \left(\frac{mg}{E}\right)\tan\theta$

$\tan\theta = \frac{qE}{mg}$ or $q = \frac{mg\tan\theta}{E}$ ..... (1)

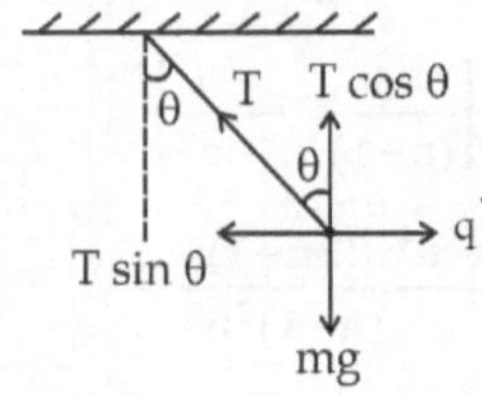

Since the force on the particle is in the direction of electric feld, the charge is positive.

3. Two point charges +8q and −2q are located at x = 0 and x = L respectively. The point on x axis at which net electric field is zero due to these charges is (2021)

(a) 8L **(b) 4 L**
(c) 2 L **(d) L**

Ans. (c): We consider point 'P' at a distance (L + r) from 8q and at r from −2q.

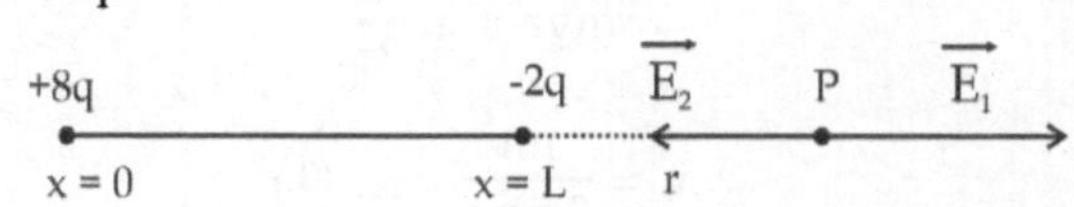

Here the magnitude of '8q' is greater than 2q. Point with zero electric field will not lie in the middle. Because electric field is in the same direction due to both the charges on the line joining the two charges.

Assume net electric field at P = 0

And $\vec{E}_1$ = Electric field at 'P' due to +8q

And $\vec{E}_2$ = Electric field at 'P' due to −2q.

Also $|\vec{E}_1| = |\vec{E}_2|$

$\therefore \frac{k(8q)}{(L+r)^2} = \frac{k(2q)}{r^2} \therefore \frac{4}{(L+r)^2} = \frac{1}{r^2}$

$\Rightarrow 4r^2 = (L+r)^2$

$\Rightarrow 2r = L + r$
$\therefore r = L$
$\therefore$ P is at $x = L + L = 2L$ from origin.

4. The magnitude of electric field due to a point charge 2q, at distance r is E. Then the magnitude of electric field due to a uniformly charged thin spherical shell of radius R with total charge q at a distance $\frac{r}{2}$ ($r \gg R$) will be (2021)

(a) $\frac{E}{4}$ **(b) 0**
(c) 2E **(d) 4E**

Ans. (c): Given Electric field due to a point charge (2q) at distance r is given by 'E'.

$E = \frac{k \cdot 2q}{r^2}$

Which is $E = \frac{2kq}{r^2}$ .....(1)

Now Electric field due to thin shell of radius 'R' and charge 'q' at a distance $\frac{r}{2}$ will be

$E' = \frac{k \cdot q}{(r/2)^2} = \frac{kq \times 4}{r^2}$ ....(2)

Dividing eq. (1) by (2),

$\frac{E}{E'} = \frac{k \cdot 2q}{r^2} \times \frac{r^2}{kq \times 4} = \frac{1}{2}$; $E' = 2E$

5. Three charges q, −q and q_0 are placed as shown in figure. The magnitude of the net force on the charge q_0 at point O is $\left[k = \frac{1}{(4\pi\varepsilon_0)}\right]$ (2022)

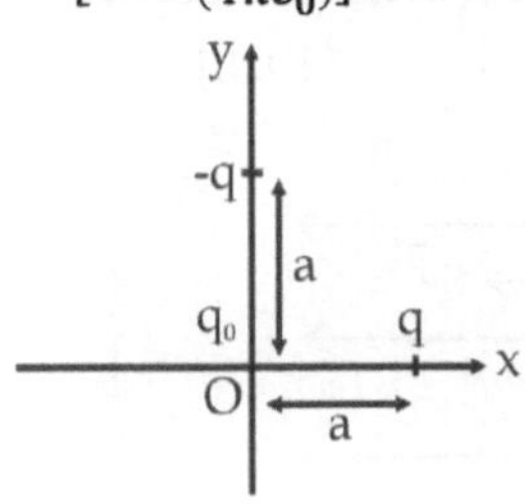

(a) 0 **(b) $\frac{2kqq_0}{a^2}$**
(c) $\frac{\sqrt{2}kqq_0}{a^2}$ **(d) $\frac{1}{\sqrt{2}}\frac{kqq_0}{a^2}$**

Ans. (c): From the diagram force on q_0 due to $-q$

$F_1 = \frac{kqq_0}{a^2}$ (along y-direction)

And force on q_0 due to $+q$

$F_2 = \frac{kqq_0}{a^2}$ (along negative x-direction)

Thus net force on q_0 will be -

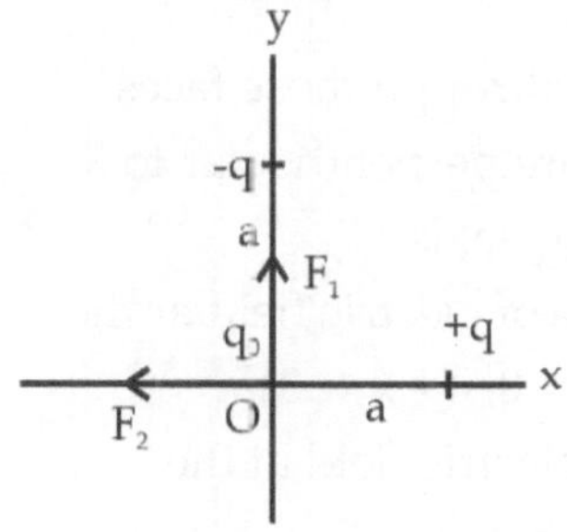

$F = \sqrt{F_1^2 + F_2^2} = \frac{kqq_0\sqrt{2}}{a^2}$

[$\because$ Angle between F_1 and F_2 is 90°].

6. In an experiment three microscopic latex spheres are sprayed into a chamber and became charged with charges +3e, +5e and −3e respectively. All the three spheres came in contact simultaneously for a moment and got separated. Which one of the following are possible values for the final charge on the spheres? (2022)

(a) + 5e, −4e, +5e
(b) + 6e, +6e, −7e
(c) − 4e, +3. 5e, +5. 5e
(d) + 5e, −8e, +7e

Ans. (b): Given $q_1 = 3e, q_2 = +5e$, $q_3 = -3e$

Thus total charge, $Q = q_1 + q_2 + q_3$
$= 3e + 5e + (-3e) = 5e$

Since charge is conserved, in option (b),
$Q = 6e + 6e - 7e = 5e$

We cannot take option (c) because charge cannot have a fractional value. It is integral multiple of 'e'.

7. Given the electric field in the region $\vec{E} = 2x\hat{i}$, find the electric flux through the cube and the charge enclosed by it. (2015)

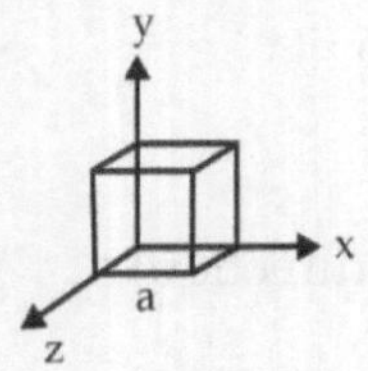

Ans. Given $\vec{E} = 2x\hat{i}$

Thus flux will pass through those faces of the cube which are perpendicular to X-axis, and parallel to y-axis.

And the magnitude of electric field at the left face will be $E_1 = 0$ $(\because x = 0)$

The magnitude of electric field at the right face will be -

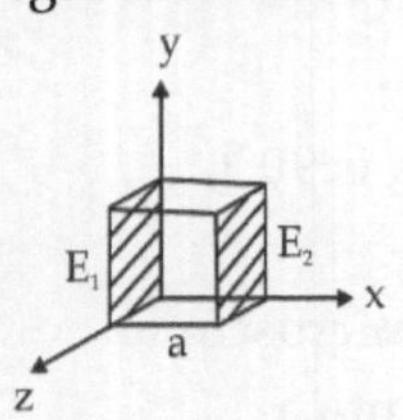

$E_2 = 2a$ $(\because x = a)$

So, net flux, $\phi = \vec{E} \cdot \Delta\vec{S}$

or $\phi = E_1 \Delta S \cos 180° + E_2 \Delta S \cos 0°$

$= 0 + 2a \times a^2 = 2a^3$

If the charge enclosed is 'q' then -

By Gauss's law, $\phi = \frac{q}{\varepsilon_0}$; $q = \varepsilon_0 \phi$

$\therefore q = 2a^3 \varepsilon_0$

8. Two small identical electric dipoles AB and CD, each of dipole moment $\vec{p}$ are kept at an angle of 120° to each other in an external electric field $\vec{E}$ pointing along the x-axis as shown in the figure. Find the

(a) dipole moment of the arrangement, and

(b) magnitude and direction of the net torque acting on it (2011)

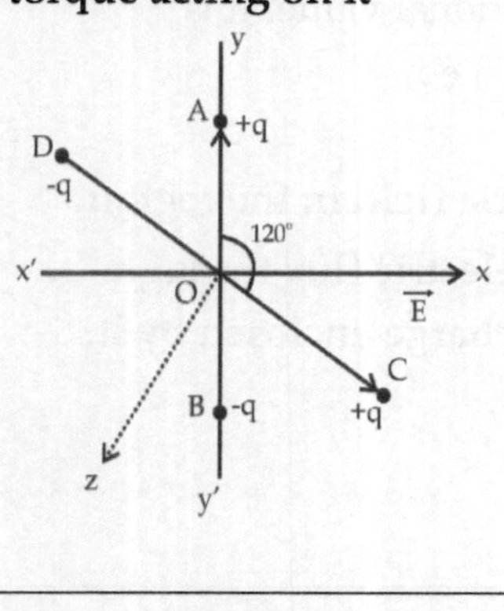

Ans. Here as shown in the figure, AB and CD are dipoles making an angle of 120° with respect to each other.

Thus, the resultant magnetic dipole moment can be calculated as -

$p_r = \sqrt{p^2 + p^2 + 2pp\cos 120°}$

$= \sqrt{2p^2 + 2p^2 \cos 120°}$

$= \sqrt{2p^2 + (2p^2) \times \left(-\frac{1}{2}\right)}$

$= \sqrt{2p^2 - p^2} = p$

The resultant dipole will make an angle 60° with Y-axis or 30° with x-axis.

Since torque is -

$\tau = \vec{p} \times \vec{E}$

or $\tau = pE\sin\theta$

or $\tau = pE\sin 30°$

$\therefore \tau = \frac{1}{2} pE$

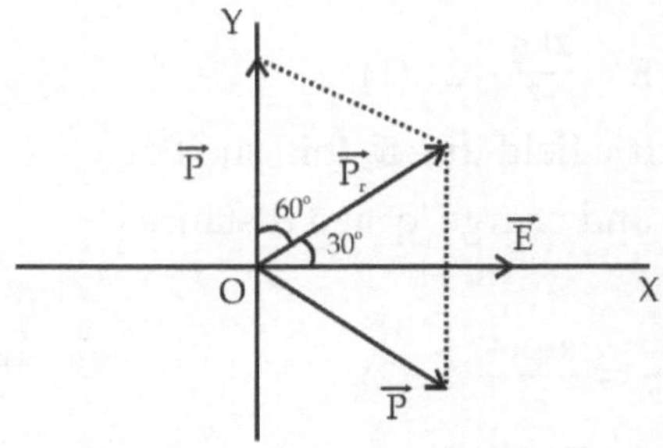

The direction of torque will be along negative Z-direction.

9. Derive the expression for the torque acting on an electric dipole, when it is held in a uniform electric field. Identify the orientation of the dipole in the electric field, in which it attains a stable equilibrium. (2020)

Ans.

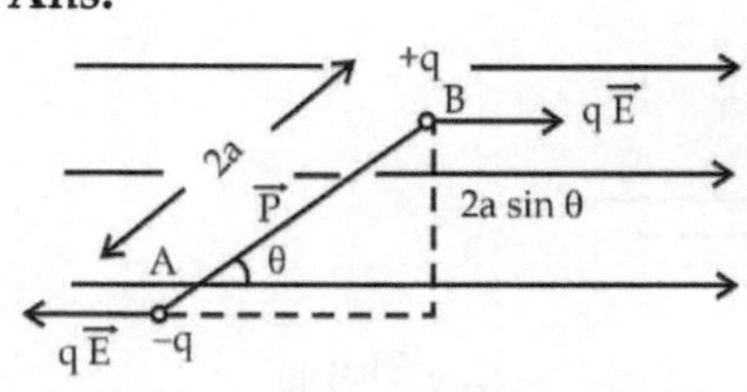

An electric dipole **AB** placed in uniform electric field $\vec{\mathbf{E}}$ and the dipole moment vector is making an angle $\boldsymbol{\theta}$ with the direction of uniform electric field $\vec{\mathbf{E}}$. Due to external electric field $\vec{\mathbf{E}}$, charge $+\mathbf{q}$

experiences force **qE** in the direction of electric field & charge $-\mathbf{q}$ experiences force in the opposite direction of electric field. Two equal and anti-parallel forces acting on a body form a couple and cause dipole moment (Torque), which is given by

$\tau =$ Force $\times \perp$ distance b/w two forces

$\tau = qE \times 2a \sin\theta$

$\because p = q \times 2a$

$\therefore \tau = pE \sin\theta$

In Vector form, $\vec{\tau} = \vec{p} \times \vec{E}$

Or $\tau = PE \sin\theta$

For stable equilibrium the torque acting on the dipole should be minimum i.e. $\theta = 0°$.

10. (a) Define electric flux. Is it a scalar or a vector quantity?

A point charge q is at a distance of d/2 directly above the centre of a square of side d, as shown in the figure. Use Gauss's law to obtain the expression for the electric flux through the square.

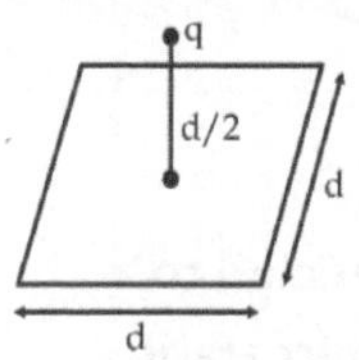

(b) If the point charge is now moved to a distance ' d ' from the centre of the square and the side of the square is doubled, explain how the electric flux will be affected. (2018)

Ans. (a) The electric flux linked with any surface is the number of electric lines of force passing through the surface normally and this can be measured as the surface integral of the electric field over that surface i.e., $\phi = \int \vec{E} \cdot d\vec{S}$

It is a scalar quantity.

Now we can assume that the given square is one face of a cube of edge 'd' cm. Since charge 'q' is at a distance of d/2 above the centre of the square, so it will be enclosed by the cube. Thus, electric flux linked with the cube is

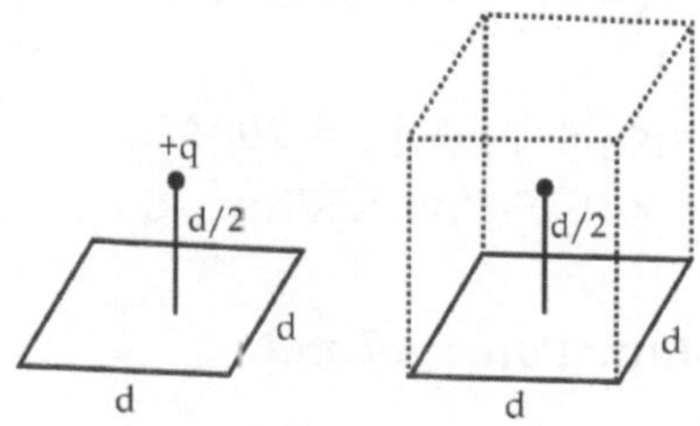

$\phi = \frac{q}{\varepsilon_0}$ (From Gauss' theorem)

Now, the magnitude of the electric flux through the square will be $\frac{1}{6}$ th of the total flux passing through the cube -

i.e., $\phi = \frac{\phi}{6} = \frac{q}{6\varepsilon_0}$

(b) Now, the distance of the point charge gets doubled and also the sides of the square also gets doubled. But electric flux through the square will not be changed because it does not depend on the distance. It only depends as the charge enclosed by the surface. i.e., $\phi = \frac{q}{6\varepsilon_0}$. Hence, there will be no change in electric flux.

11. An electric field along the x-axis is given by $\vec{E} = 100\hat{i}N/C$ for $x > 0$ and $\vec{E} = -100\hat{i}N/C$ for $x < 0$. A right circular cylinder of length 20 cm and radius 5 cm lies parallel to the x-axis, with its centre at the origin and one face at $x = +10$ cm, the other face at $x = -10$ cm. Calculate the net outward flux through the cylinder. (2019)

Ans. Flux through right circular cap,

$\phi_1 = \vec{E} \cdot d\vec{S} = 100\Delta S$

[Here $\Delta S = \pi r^2$]

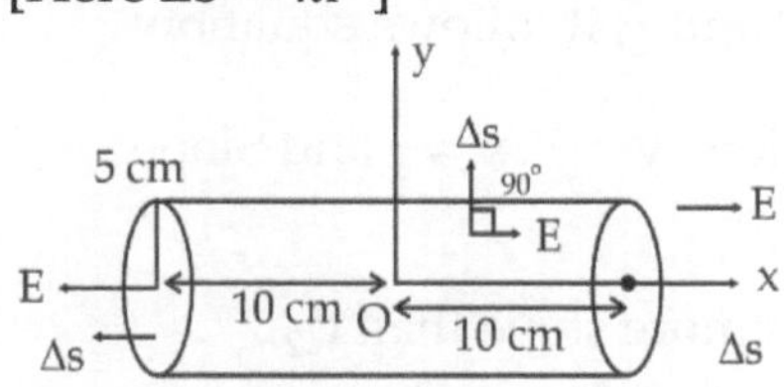

Now flux through left circular cap will be, $\phi_2 = \vec{E} \cdot \Delta\vec{S}$; $\phi_2 = 100\ \Delta S$

Flux through the curved surface will be,

$\phi_3 = \vec{E} \cdot \Delta\vec{S}$ (because angle between $\vec{E}$ and $\Delta\vec{S}$ is 90°)

Net flux, $\phi = \phi_1 + \phi_2 + \phi_3 = 200\Delta S$

$= 200 \times 3.14 \times (0.05)^2 = 1.57 Nm^2/C$

II - Electrostatic Potential and Capacitance

12. The electric potential V at any point (x, y, z) is given by $V = 3x^2$ where x is in metres and V in volts. The electric field at the point (1 m, 0, 2 m) is (2021)

(a) 6 $V m^{-1}$ along - x-axis

(b) 6 $V m^{-1}$ along +x-axis

(c) 1.5 $V m^{-1}$ along- x-axis

(d) 1.5 $V m^{-1}$ along +x-axis.

Ans. (a): Given Potential, $V = 3x^2$

$\because E = \frac{-dV}{dx} = -3 \times 2x = -6x$

$\because E(1,0,2) = -6 \times 1 = -6 V m^{-1}$

or $E = 6 V m^{-1}$ (along - x axis)

13. The figure shows the variation of potential V with $\frac{1}{r}$ for two point charges Q_1 and Q_2, where V is the potential at a distance r due to a point charge. Find $\frac{Q_1}{Q_2}$. (2023)

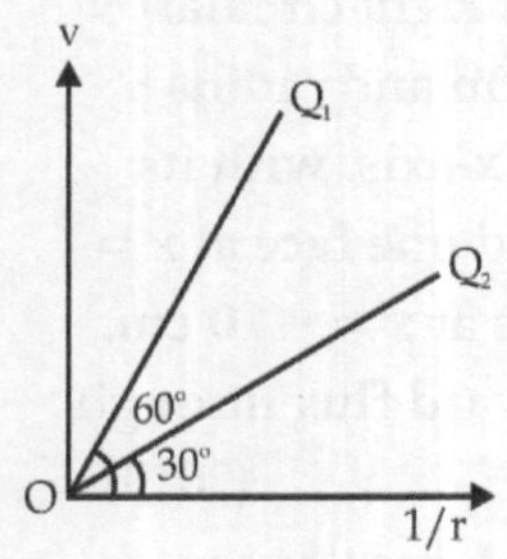

Ans. (b) Since $V = \frac{kQ}{r}$ and the graph is between 'V' and $\frac{1}{r}$, it follows equation

$y = mx$ [Here $y = V, x = \frac{1}{r}$ and Slope $m = kQ$]

Now Q_1 has more slope than Q_2.

So, $Q_1 > Q_2$ ......(i)

and $\frac{Q_1}{Q_2} = \frac{\tan 60°}{\tan 30°} = \frac{\sqrt{3}}{\frac{1}{\sqrt{3}}} \therefore \frac{Q_1}{Q_2} = \frac{3}{1}$

14. Three point charges Q, q and -q are kept at the vertices of an equilateral triangle of side L as shown in figure. What is (i) the electrostatic potential energy of the arrangement? And (ii) the potential at point D ? (2023)

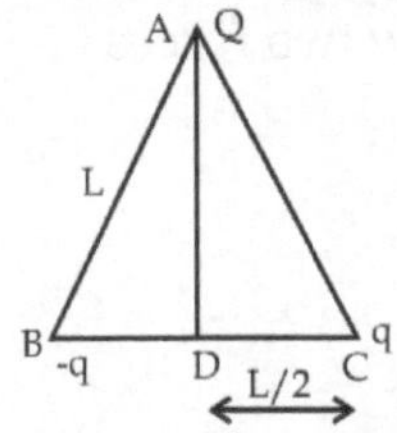

Ans. (i) We know that the potential energy of the arrangement is given by

$$U = \frac{kQ(-q)}{L} + \frac{kQq}{L} + \frac{k(-q)q}{L}$$

$$\text{Or } U = \frac{kq}{L}[-Q + Q - q]$$

$$\therefore U = -\frac{kq^2}{L}$$

(ii) $AD = \sqrt{L^2 - \frac{L^2}{4}} = \frac{\sqrt{3}}{2}L$

Now potential at D,

$$V = \frac{k(-q)\times 2}{L} + \frac{kq\times 2}{L} + \frac{kQ\times 2}{\sqrt{3}L}$$

$$\text{or } V = \frac{2kQ}{\sqrt{3}L}$$

15. A 12 pF capacitor is connected to a 50 V battery. How much electrostatic energy is stored in the capacitor? If another capacitor of 6 pF is connected in series with it with the same battery connected across the combination, find the charge stored and potential difference across each capacitor. (2017)

Ans. The Electrostatic energy stored in the capacitor, is -

$$U = \frac{1}{2}CV^2 = \frac{1}{2} \times 12 \times 10^{-12} \times (50)^2$$

Or $U = 1.5 \times 10^{-8} J$

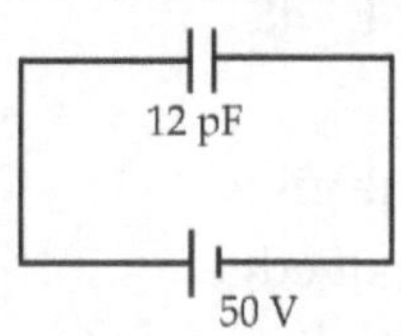

When 6 pF is connected in series the charge stored on each capacitor, is same and which is -

$Q = \frac{C_1C_2}{C_1+C_2}V = \frac{12\times6\times10^{-24}}{(12+6)\times10^{-12}} \times 50$

$= 200pC$

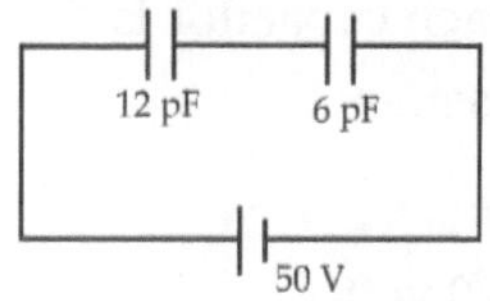

The, potential difference across 12 pF is,

$V_1 = \frac{Q}{C_1} = \frac{200\times10^{-12}}{12\times10^{-12}} = \mathbf{16.67\ V}$

and the potential difference across 6 pF is,

$V_2 = \frac{Q}{C_2} = \frac{200\times10^{-12}}{6\times10^{-12}} = \mathbf{33.33\ V}$

16. A cube of side 20 cm is kept in a region as shown in the figure. An electric field $\vec{E}$ exists in the region such that the potential at a point is given by V = 10x + 5, where V is in volt and x is in metre. (2020)

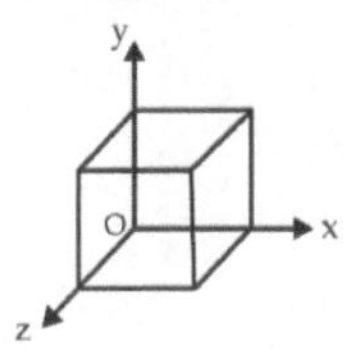

Find the (i) electric field $\vec{E}$ and
(ii) total electric flux through the cube.

Ans. (i) Now electric field,

$\bar{E} = \frac{-dV}{dx} = \frac{-d}{dx}(10x+5) = -10\hat{\imath}$

(only along −x axis)

(ii) The total electric flux through the cube, $\phi = \int E.dS$

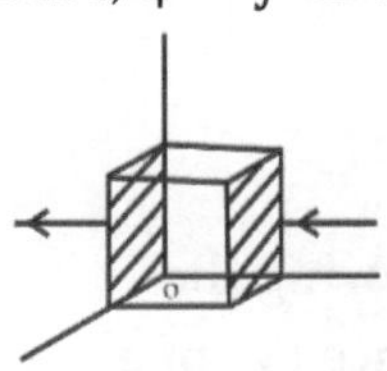

Now $E = -10\hat{\imath}$ only along −x axis. That means flux entering the right surface and comes out from left surface, hence total electric flux will be zero.

17. In the circuit shown in the figure, the charge on the capacitor of 4μ F is 16μC. Calculate the energy stored in the capacitor of 12μF capacitance. (2019)

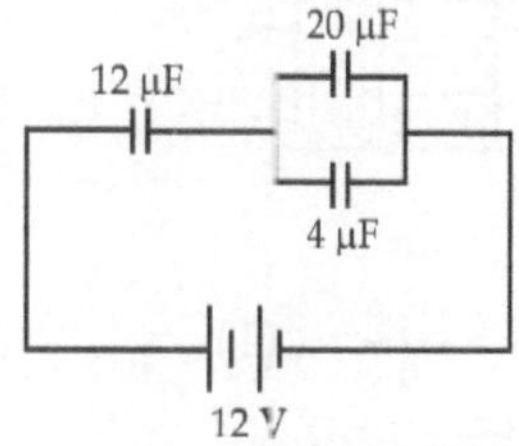

Ans. (b) Since 20 μF and 4 μF are in parallel.

Here, $C_p = 20 + 4 = 24\ \mu F$

Now $\frac{1}{C_s} = \frac{1}{C_p} + \frac{1}{12} = \frac{1}{24} + \frac{1}{12} = \frac{1}{8}$

$\therefore C_s = 8\ \mu F$

and $q = C_sV = 8 \times 12 = 96\mu C$

Now 'V' across 12 μF $= \frac{q}{C_1} = \frac{16}{12} = 8\ V$

and 'V' across 24 μF will be = 12 − 8 = 4 V

Charge on 4 μF will be (q) $= 4 \times 10^{-6} \times 4$ = 16 μC

And charge on 20 μF will be

$= 20 \times 10^{-6} \times 4$

= 80 μC

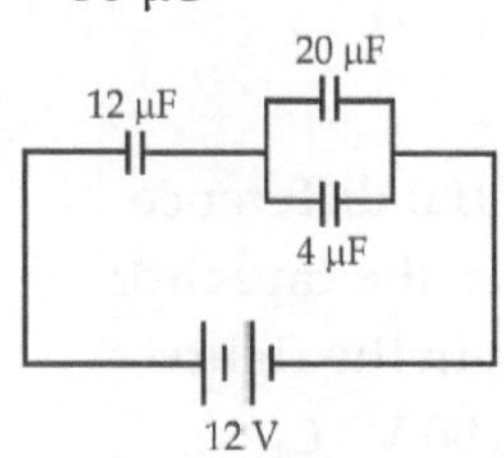

Energy stored in 12 μF

$U = \frac{1}{2} \times C \times V^2 = \frac{1}{2} \times 12 \times 8 \times 8 = 384\ \mu J$

18. The space between the plates of a parallel plate capacitor is completely filled in two ways. In the first case, it is filled with a slab of dielectric constant K. In the second case, it is filled with two slabs of equal thickness and dielectric constants K_1 and K_2 respectively as shown in the figure. The capacitance of the capacitor is same in the two cases. Obtain the relationship between K, K_1 and K_2. (2020)

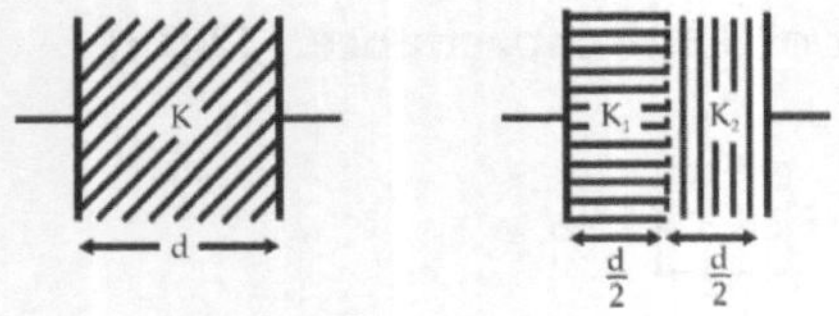

(Case 1) (Case 2)

Ans. (a): In case - 1 $C_1 = \frac{K\epsilon_0 A}{d}$ ...(1)

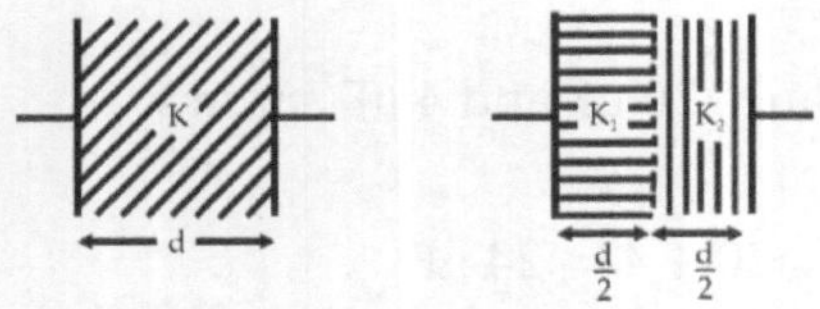

(Case 1) (Case 2)

And in case - 2

$$C_1' = \frac{K_1 \epsilon_0 A \times 2}{d}$$

$$C_2' = \frac{K_2 \epsilon_0 A \times 2}{d}$$

Please note that C_1' and C_2' are in series

$$\therefore \frac{1}{C_S} = \frac{d}{2\epsilon_0 K_2 A} + \frac{d}{2\epsilon_0 K_1 A}$$

$$\text{or } C_s = \left(\frac{2K_1K_2}{K_1+K_2}\right)\frac{\epsilon_0 A}{d} \quad ...(2)$$

So from equation (1) and (2)

$$C_1 = C_s$$

$$\text{Thus } K = \frac{2K_1K_2}{K_1+K_2}$$

19. Calculate the potential difference and the energy stored in the capacitor C_2 in the circuit shown in the figure. Given potential at A is 90 V , $C_1 = 20\mu$ F, $C_2 = 30\mu$ F and $C_3 = 15\mu$ F.

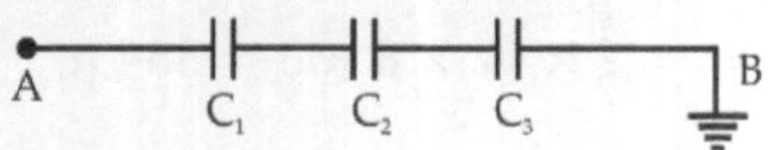

(2015)

Ans. The total capacitance (C_s) of the series circuit is given by

$$\frac{1}{C_S} = \frac{1}{20} + \frac{1}{30} + \frac{1}{15}$$

$$\frac{1}{C_S} = \frac{3+2+4}{60}$$

$$C_s = \frac{60}{9}\mu F$$

Charge on equivalent capacitor

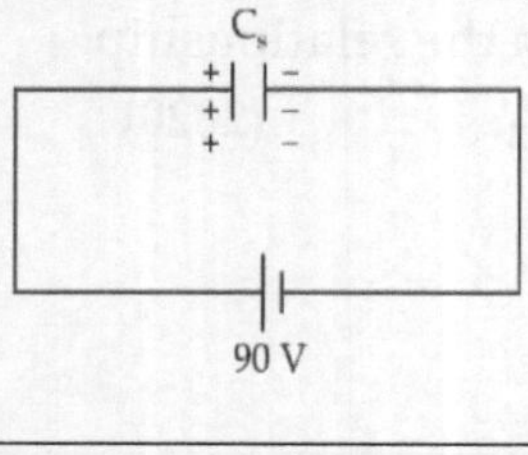

$$Q = C_sV = \frac{60}{9} \times 10^{-6} \times 90$$

$$Q = 600\mu C$$

Please note charge on each capacitor is same when they are in series.

Potential drop across C_2

$$V_2 = \frac{Q}{C_2} = \frac{600\times10^{-6}}{30\times10^{-6}} = 20 \text{ volt}$$

And Energy, $U = \frac{1}{2}C_2V_2^2$

$$U = \frac{1}{2} \times 30 \times 10^{-6} \times (20)^2$$

$$= 6 \times 10^{-3} \text{ joule}$$

20. Two slabs of dielectric constants 2K and K fill the space between the plates of a parallel plate capacitor of plate area A and plate separation d as shown in figure. Find an expression for capacitance of the system. (2023)

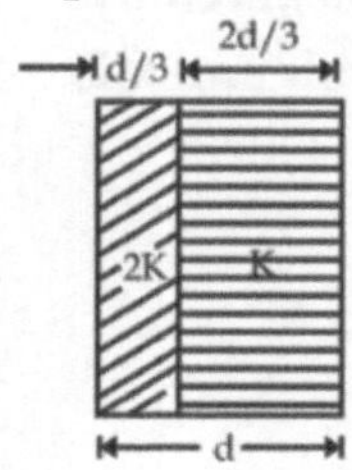

Ans. $C_1 = \frac{2K\epsilon_0 A \times 3}{d}$

And $C_2 = \frac{K\epsilon_0 A \times 3}{2d}$

Since both are in series

$$\frac{1}{C_S} = \frac{1}{C_1} + \frac{1}{C_2} = \frac{d}{6K\epsilon_0 A} + \frac{2d}{3K\epsilon_0 A}$$

$$\text{or } \frac{1}{C_S} = \frac{d+4d}{6K\epsilon_0 A}$$

$$\therefore C_S = \frac{6K\epsilon_0 A}{5d}$$

III. Current Electricity

21. A straight line plot showing the terminal potential difference (V) of a cell as a function of current (I) drawn from it, is shown in the figure. The internal resistance of the cell would be then (2021)

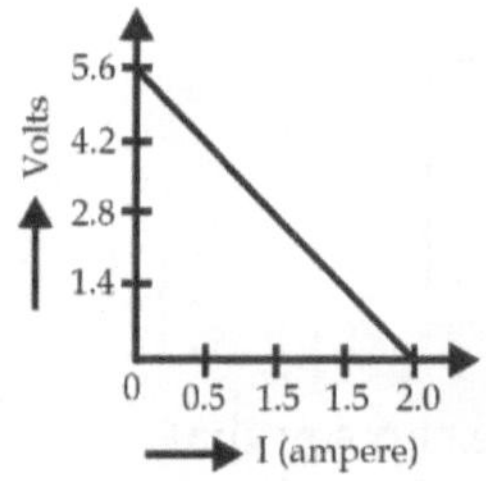

(a) 2.8Ω (b) 1.4Ω

(c) 1.2Ω (d) zero

Ans. We know that

$E = V + Ir$

Or $V = E - Ir$ [It follows $y = c - mx$]

Now Internal resistance,

$$r = \text{slope of V-I graph} = \frac{5.6-0}{2.0-0} = 2.8\Omega$$

22. The temperature (T) dependence of resistivity of material A and material B is represented by Fig. (i) and Fig. (ii), respectively. Identify material A and material B. (2021)

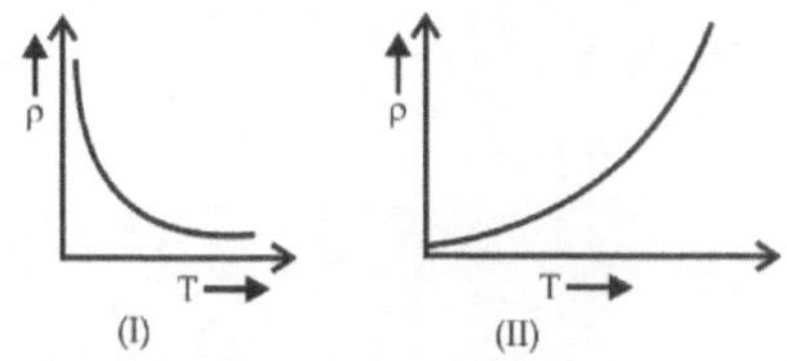

(a) Material A is copper and material B is germanium

(b) Material A is germanium and material B is copper

(c) Material A is nichrome and material B is germanium

(d) Material A is copper and material B is nichrome

Ans. (b) A is germanium and B is copper.

23. Two sources of equal emf are connected in series. This combination is in turn connected to an external resistance R. The internal resistance of two sources are r_1 and $r_2 (r_2 > r_1)$. If the potential difference across the source of internal resistance r_2 is zero, then R equals to (2022)

(a) $\frac{r_1+r_2}{r_2-r_1}$ (b) $r_2 - r_1$

(c) $\frac{r_1 r_2}{r_2-r_1}$ (d) $\frac{r_1+r_2}{r_1 r_2}$

Ans. (b) Here sources are in series

Thus net emf, $E_{net} = E + E = 2E$

And net resistance, $R_{net} = R + r_1 + r_2$

The current in the circuit, will be

$$I = \frac{E_{net}}{R_{net}} = \frac{2E}{R+r_1+r_2}$$

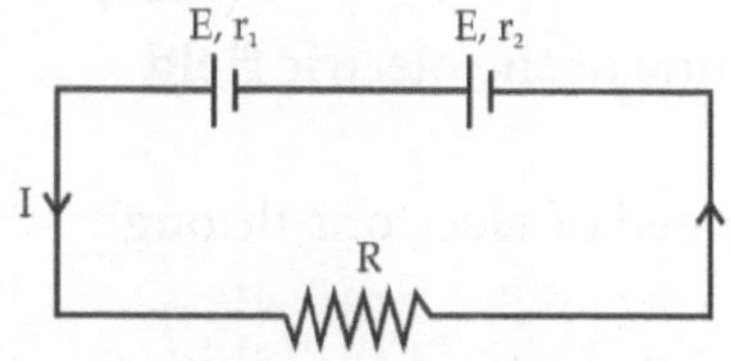

As given in the question, the potential difference across the source of internal resistance r_2 is zero.

i.e. $E - Ir_2 = 0$

or $E - \frac{2Er_2}{R+r_1+r_2} = 0$

or $R + r_1 - r_2 = 0$ or $R = r_2 - r_1$

24. If n, e, τ and m have their usual meanings, then the resistance of a wire of length l and cross-sectional area A is given by (2022)

(a) $\frac{ne^2A}{2m\tau l}$ (b) $\frac{ml}{ne^2\tau A}$

(c) $\frac{m\tau A}{ne^2 l}$ (d) $\frac{ne^2\tau A}{2ml}$

Ans. (b) $R = \frac{m\ell}{ne^2\tau A}$

25. Find the temperature at which the resistance of a conductor increases by 25% of its value at 27°C. The temperature coefficient of resistance of the conductor is $2.0 \times 10^{-4}\,°C^{-1}$. (2024)

Ans. Given, $\alpha = 2 \times 10^{-4}\,°C^{-1}$

We take $R_{27°} = R$,

$\therefore R_T = R + \frac{25}{100}R = 1.25R$ ..... (1)

Also $T_1 = 27 + 273 = 300$ K

Since $R_T = R_o(1 + \alpha\,\Delta T)$ .....(2)

We can say that

$1.25R = R[1 + 2.0 \times 10^{-4}(T_2 - 300)]$

$1 + 2.0 \times 10^{-4}(T_2 - 300) = 1.25$

or $2.0 \times 10^{-4}(T_2 - 300) = 0.25$

On solving the above equation, we get

$T_2 = 1550$ K

26. The following table gives the length of three copper wires, their diameters and the applied potential difference across their ends. Arrange the wires in increasing order according to the following (2017)
(i) The magnitude of the electric field within them
(ii) The drift speed of electrons through them, and
(iii) The current density within them.

Wire no.	Length	Diameter	Potential Difference
1	L	3d	V
2	2L	D	V
3	3L	2d	V

Ans. (i) If we take E_1, E_2 and E_3 are the magnitudes of electric field in wires 1,2 and 3, then

$E_1 = \frac{V}{L}, E_2 = \frac{V}{2L} = \frac{0.5V}{L}$

and $E_3 = \frac{2V}{3L} = \frac{0.66V}{L}$

Hence, $E_2 < E_3 < E_1$

(ii) If v_{d_1}, v_{d_2} and v_{d_3} are the drift velocities of electrons in wires 1, 2 and 3.

$\because v_d = \frac{e\tau}{m}E \quad \therefore v_d \propto E$

Thus $v_{d_2} < v_{d_3} < v_{d_1}$

(iii) J_1, J_2 and J_3 are the current densities in wires 1, 2 and 3.

Since $J = nev_d$

$\because J \propto v_d$

$\therefore J_2 < J_3 < J_1$

27. Three cells of emf E, 2E and 5E having internal resistances r, 2r and 3r respectively are connected across a variable resistance R as shown in the figure. Find the expression for the current. Plot a graph for variation of current with R. (2010)

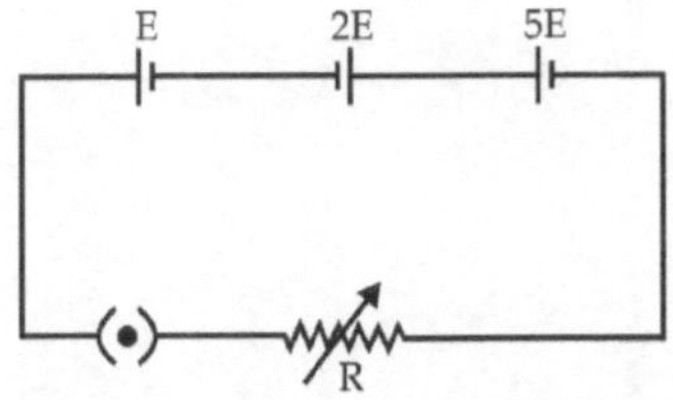

Ans. From the figure, we can say that

$E_1 = E, E_2 = -2E$

and $E_3 = 5E$

And $r_1 = r$

$r_2 = 2r$

and $r_3 = 3r$

Now net emf of the cell combination will be -

$E = E_1 + E_2 + E_3$

$= E - 2E + 5E = 4E$

And net resistance of the circuit will be -

$= r_1 + r_2 + r_3 + R$

$= r + 2r + 3r + R = 6r + R$

$\therefore$ Current, $I = \frac{4E}{6r+R}$

The graph between current I with resistance R will be -

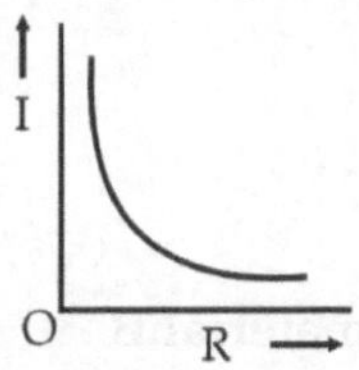

28. Two wires A and B of different metals have their lengths in ratio 1: 2 and their radii in ratio 2: 1, respectively. I − V graphs for them is shown in the figure. (2024)
Find the ratio of their
(i) Resistances (R_A/R_B) and
(ii) Resistivities (σ_A/σ_B)

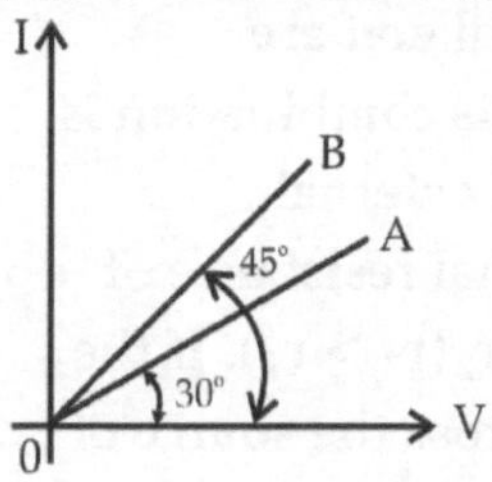

Ans. Given ratio of lengths = 1: 2

and ratio of radii = 2: 1

(i) Here Slope $\frac{\tan\theta_A}{\tan\theta_B} = \frac{R_B}{R_A} \quad \because \frac{I}{V} = \frac{1}{R}$

$$\therefore \frac{\tan 30°}{\tan 45°} = \frac{R_B}{R_A}$$

$$\left[\begin{array}{c}\theta_A = 30°, \theta_B = 45° \\ \text{(from figure given in question)}\end{array}\right]$$

$$\frac{R_B}{R_A} = \frac{1}{\sqrt{3}}$$

(ii) Since area $a = \pi r^2$

Thus $\frac{r_A^2}{r_B^2} = \frac{4}{1} \left[\because \frac{r_A}{r_B} = \frac{2}{1}\right]$

and $\frac{\ell_A}{\ell_B} >= \frac{1}{2}$

$\because \sigma_A = \frac{R_A a_A}{\ell_A}$

and $\sigma_B = \frac{R_B a_B}{\ell_B}$

$$\therefore \frac{\sigma_A}{\sigma_B} = \frac{R_A}{R_B} \cdot \frac{a_A}{a_B} \cdot \frac{\ell_B}{\ell_A} = \sqrt{3} \cdot \frac{4}{1} \cdot \frac{2}{1}$$

$$\frac{\sigma_A}{\sigma_B} = 8\sqrt{3}$$

29. Using Kirchhoff's rules calculate the potential difference between B and D in the circuit diagram as shown in the figure. (2018)

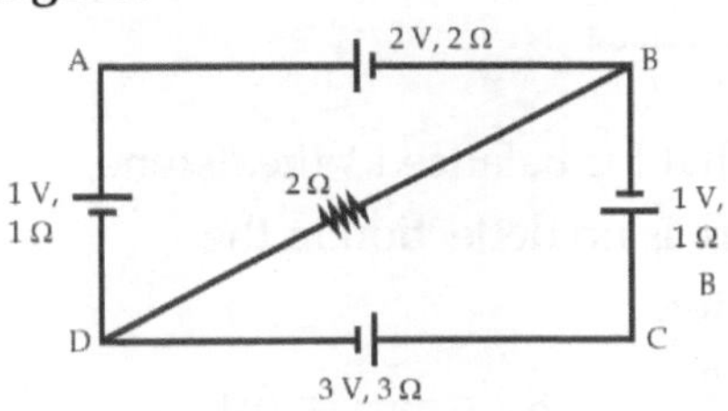

Ans. For loop BADB,

$$2 - 2i_1 - i_1 - 1 - 2(i_1 - i_2) = 0 \quad ....(1)$$

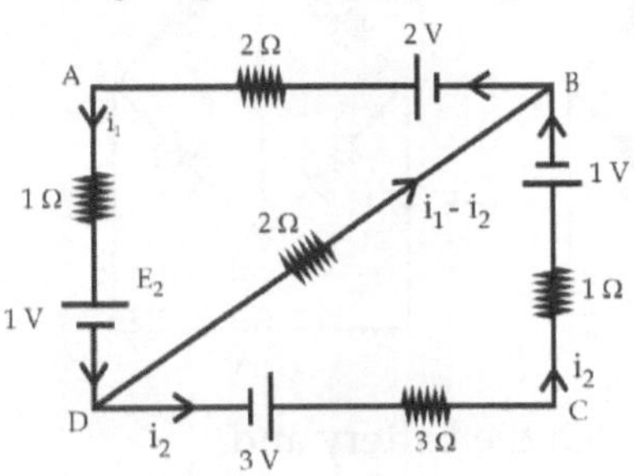

For loop BDCB,

$$2(i_1 - i_2) + 3 - 3i_2 - i_2 - 1 = 0 \quad .....(2)$$

Solving equations (1) and (2), we get

$$i_1 = \frac{5}{13} \text{A}, \quad i_2 = \frac{6}{13} \text{A}$$

and $i_1 - i_2 = -\frac{1}{13}$ A

Potential difference between B and D.

$$V_D - 2(i_1 - i_2) = V_B$$

$$\therefore \quad V_B - V_D = -2(i_1 - i_2) = \frac{2}{13} \text{V}$$

30. In the given network, calculate
(i) effective resistance between points A and M, and
(ii) power supplied by the battery.
(2024)

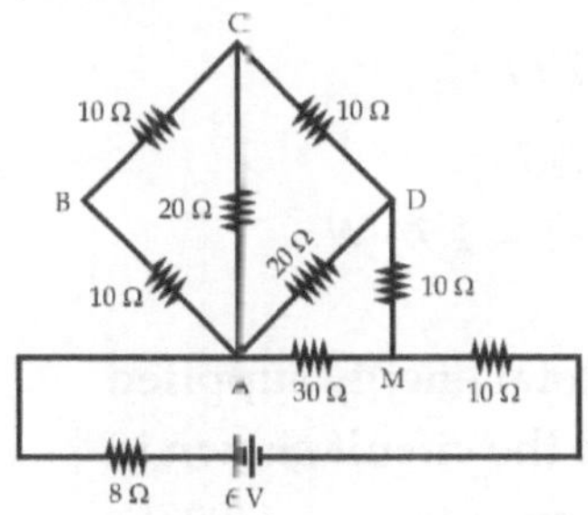

Ans. (i)

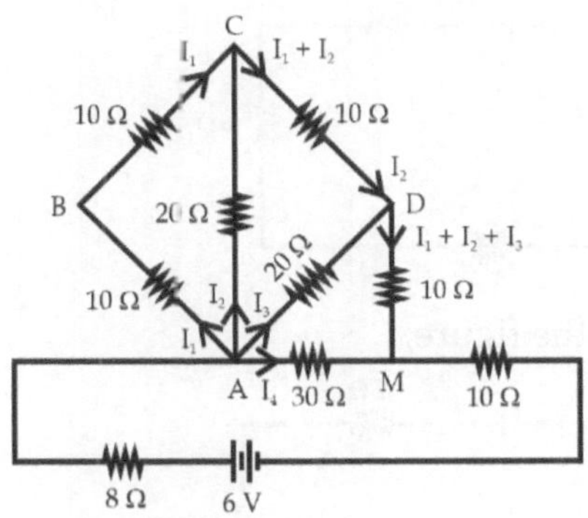

Here R_{AB} and R_{BC} are in series thus equivalent resistance, will be -

$R_1 = R_{AB} + R_{BC} = 10\Omega + 10\Omega = 20\Omega$

Now R_1 and R_{AC} are in parallel, thus net resistance, (R_2) will be -

$$R_2 = \frac{R_1 \times R_{AC}}{R_1 + R_{AC}} = \frac{20 \times 20}{40} = 10\Omega$$

Now R_2 and R_{CD} are in parallel, thus net resistance, (R_3) will be -

$$R_3 = \frac{R_2 \times R_{CD}}{R_2 - R_{CD}} = \frac{10 \times 10}{20} = 5\Omega$$

Now R_3, R_{AD} and R_{MD} are in parallel, thus net resistance (R_4) will be -

$$\frac{1}{R_4} = \frac{1}{R_3} + \frac{1}{R_{AD}} + \frac{1}{R_{MD}} = \frac{1}{5} + \frac{1}{20} + \frac{1}{10}$$

$$\text{or} \quad \frac{1}{R_4} = \frac{4+1+2}{20} = \frac{7}{20} \qquad R_4 = \frac{20}{7} \Omega$$

and this R_4 is in parallel with R_{AM}, then equivalent resistance (R_5) will be -

$$\Rightarrow \frac{1}{R_5} = \frac{1}{R_4} + \frac{1}{R_{AM}} = \frac{7}{20} + \frac{1}{30}$$

$$\Rightarrow \frac{1}{R_5} = \frac{210+20}{600} = \frac{230}{600} = \frac{23}{600}$$

$$R_5 = \frac{60}{23}\Omega = 2.60\Omega$$

Thus total resistance of across AM is 2.60Ω.

And total resistance of circuit,

$R = 2.60 + 10 + 8 = 20.60\Omega$

(ii) Current in the circuit will be -

$$I = \frac{V}{R} = \frac{6}{20.60} = 0.29 \text{ A}$$

And power supplied -

$P = VI = 6 \times (0.29) = 1.74$ W

31. Calculate the total energy supplied by the batteries to the circuit shown in the figure, in one minute. (2023)

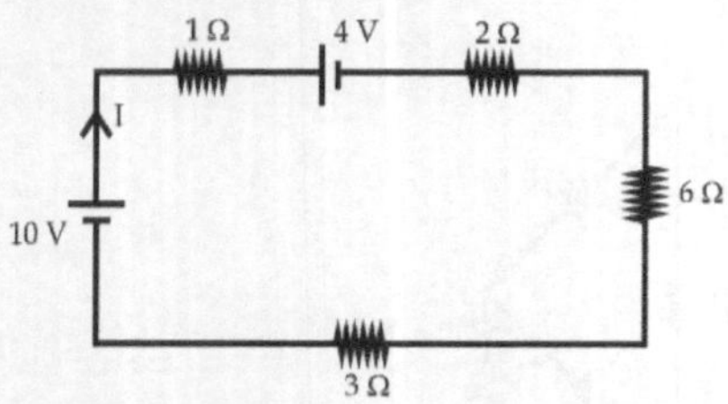

Ans. From the figure,

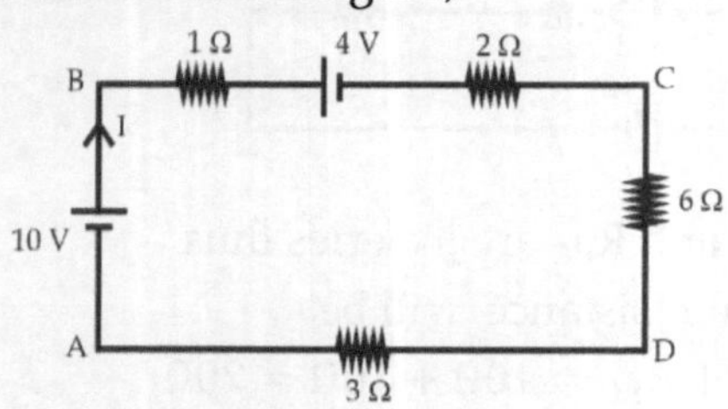

For loop ABCDA

$10 - I - 4 - 2I - 6I - 3I = 0$

$\therefore\ I = \frac{1}{2}$ A

And total resistance of the circuit

$(R) = 1 + 2 + 6 + 3$

$= 12\ \Omega$

And total energy supplied by the battery will be

$$H = I^2Rt = \frac{1}{2} \times \frac{1}{2} \times 12 \times 60 = 180 \text{ J}$$

32. Define the current sensitivity of a galvanometer. Write its SI unit. Figure shows two circuits each having a galvanometer and a battery of 3 V. When the galvanometer in each arrangement do not show any deflection, obtain the ratio R_1/R_2. (2013)

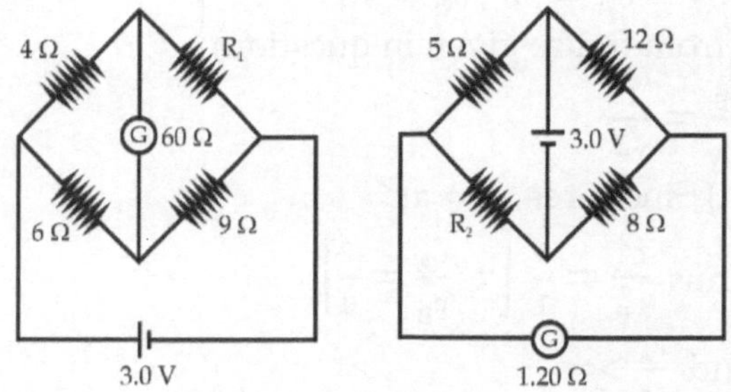

Ans. Current sensitivity (I_s) of a galvanometer is the deflection produed in the galvanometer per unit current flowing through it. Its SI unit is rad/ampere.

$$I_s = \frac{\theta}{I}$$

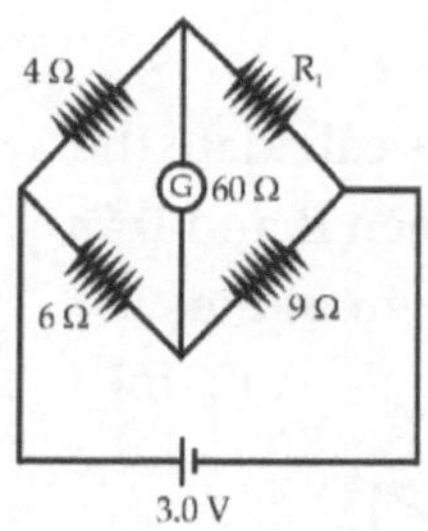

We know that for balanced Wheatstone bridge, there is no deflection in the galvanometer.

Thus $\frac{4}{R_1} = \frac{6}{9}$ $\quad\therefore R_1 = \frac{4 \times 9}{6} = 6\Omega$

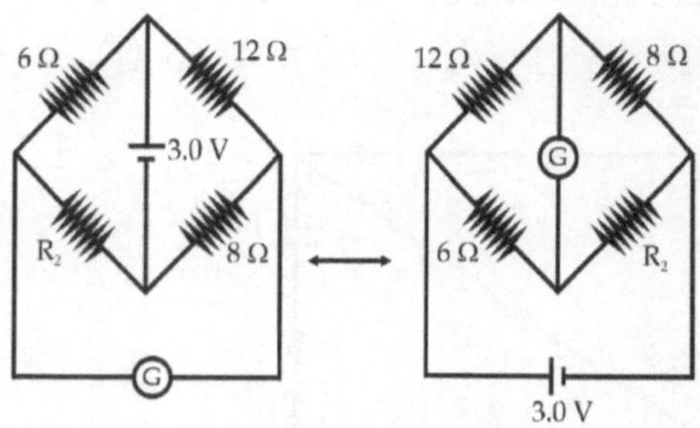

When we switch the battery and galvanometer position, the balanced condition is still applicable.

$$\therefore \frac{12}{8} = \frac{6}{R_2}$$

$$\text{or } R_2 = \frac{6 \times 8}{12} = 4\Omega$$

$$\therefore \frac{R_1}{R_2} = \frac{6}{4} = \frac{3}{2}$$

IV. Moving Charges and Magnetism

33. Two wires carrying currents I_1 and I_2 lie, one slightly above the other, in a horizontal plane as shown in figure. The region of vertically upward strongest magnetic field is (2022)

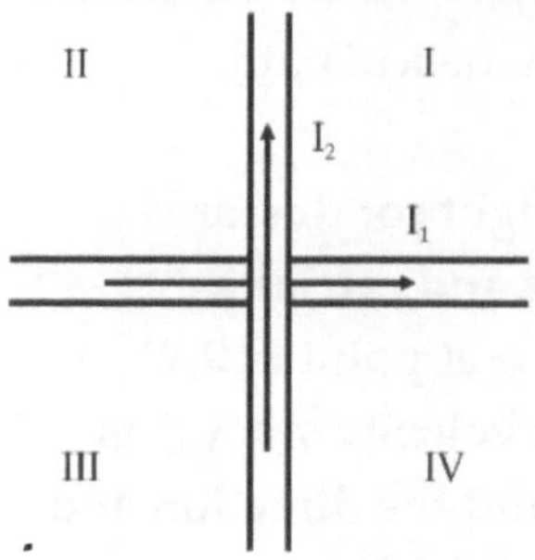

(a) I (b) II
(c) III (d) IV

Ans. (b) The region II due to both the wires the magnetic field is vertically upward, we can check this by using right hand thumb rule.

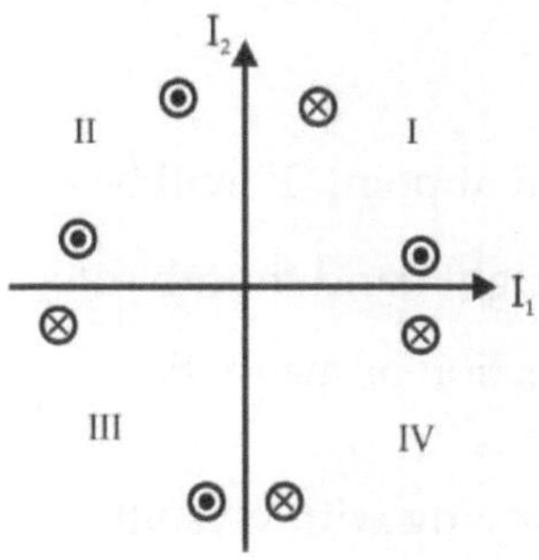

34. A current of 10 A is flowing from east to west in a long straight wire kept on a horizontal table. The magnetic field developed at a distance of 10 cm due north on the table is (2020)

(a) 2×10^{-5} T, acting downwards
(b) 2×10^{-5} T, acting upwards
(c) 4×10^{-5} T, acting downwards
(d) 4×10^{-5} T, acting upwards

Ans. (a) We know that the magnitude of magnetic field $|\mathbf{B}|$ for infinite length of wire is given by $= \frac{\mu_0 I}{2\pi r}$

$\therefore\ |\mathbf{B}| = \frac{4\pi \times 10^{-7} \times 10}{2 \times \pi \times 10 \times 10^{-2}}$

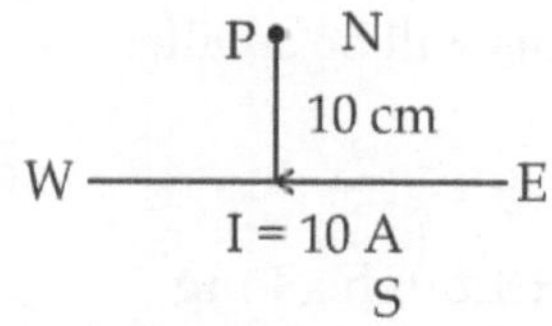

$= 2 \times 10^{-5}$ T

Magnetic field is acting downward.

35. Two long straight parallel conductors A and B, kept at a distance r, carry current I in opposite directions. A third identical conductor C, kept at a distance $\left(\frac{r}{3}\right)$ from A carry current I_1 in the same direction as in A. The net magnetic force on unit length of C is (2024)

(a) $\frac{3\mu_0 I I_1}{2\pi r}$**, towards A**
(b) $\frac{3\mu_0 I I_1}{2\pi r}$**, towards B**
(c) $\frac{3\mu_0 I I_1}{4\pi r}$**, towards A**
(d) $\frac{3\mu_0 I I_1}{4\pi r}$**, towards B**

Ans. (a) We know that the force per unit length between two parallel current carrying conductors is given by

$F = \frac{\mu_0}{4\pi} \cdot \frac{2 I_1 I_2}{r}$

As per the question we assume that the arrangement is as shown below -

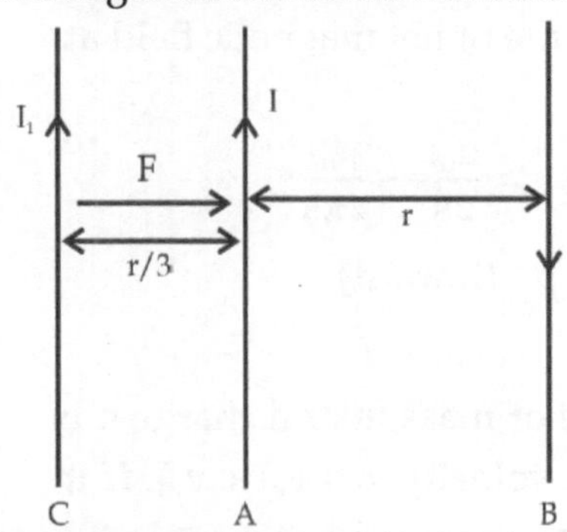

Net magnetic force per unit length of C,
Due to 'A'

$F = \frac{\mu_0}{4\pi} \cdot \frac{2 I I_1}{\frac{r}{3}} = \frac{3\mu_0 I I_1}{2\pi r}$, Attractive i.e., towards 'A' because current is flowing in same direction in both wires A and C.

Hence, option (a) is correct.

Wire 'B' is far thus the force per unit

length an 'C' due to this will be smaller than wire 'A'.

36. A current I flows through a long straight conductor which is bent into a circular loop of radius R in the middle as shown in the figure.

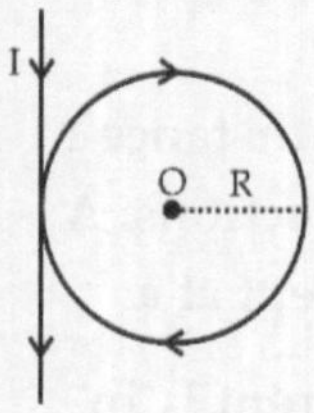

The magnitude of the net magnetic field at point O will be (2020)

(a) zero **(b) $\frac{\mu_0 I}{2R}(1+\pi)$**

(c) $\frac{\mu_0 I}{4\pi R}$ **(d) $\frac{\mu_0 I}{2R}\left(1-\frac{1}{\pi}\right)$**

Ans. (d) The magnetic field due to the straight conductor at O will be -

$B_1 = \frac{\mu_0 I}{2\pi R}$

(outward, using flemming's right hand thumb rule)

and due to the circular loop of radius R will be -

$B_2 = \frac{\mu_0 I}{2R}$

(Inward, using right hand screw rule)

$\therefore B_2 > B_1$

Thus magnitude of net magnetic field at point O is-

$B_{net} = B_2 - B_1 = \frac{\mu_0 I}{2R} - \frac{\mu_0 I}{2\pi R}$

$= \frac{\mu_0 I}{2R}\left(1-\frac{1}{\pi}\right)$ (inward)

37. A particle of mass m and charge q is moving with velocity $v = v_x\hat{i} + v_y\hat{j}$. If it is subjected to a magnetic field $B = B_0\hat{i}$, it will move in a (2004)

(a) straight line path **(b) circular path**

(c) helical path **(d) parabolic path**

Ans. (c) Whenever a charged particle moves in external magnetic field, it experiences a force called magnetic Lorentz force, which is given by

$\vec{F} = q(\vec{v} \times \vec{B})$

Here, the magnetic field is in X-direction and due to v_x component it covers linear distance and due to v_y component it moves in a circular path so the combined effect will produce a helical path.

38. An infinite straight conductor is kept along X'X-axis and carries a current I. A charge q at point P(0, r) starts moving with velocity $v = v_0\hat{j}$ as shown in figure. Find the direction and magnitude of force initially experienced by the charge (2024)

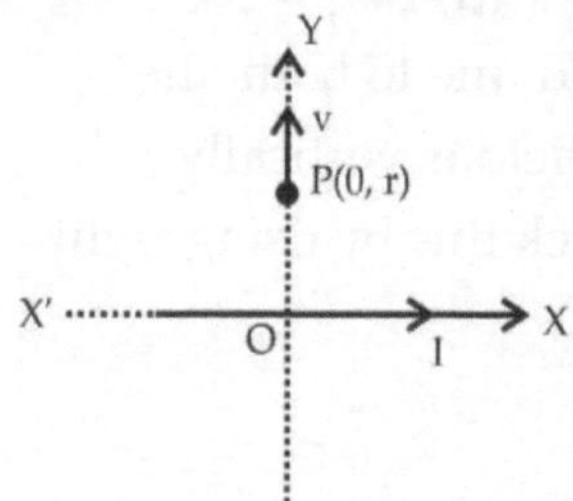

Ans. Magnetic field at point 'P' will be -

$\mathbf{B} = \frac{\mu_0 I}{2\pi r}\hat{\mathbf{k}}$ (From right hand thumb rule, we can get the direction of magnetic field)

Now charge q is moving with velocity $v = v_0\hat{\mathbf{j}}$.

Thus force of charge q due to straight wire, $\vec{F} = q(\vec{v} \times \vec{B})$

or $\vec{F} = q\left[(v_0\hat{j}) \times \left(\frac{\mu_0 I}{2\pi r}\hat{\mathbf{k}}\right)\right]$

or $\vec{F} = q\left[\frac{\mu_0 v_0 I}{2\pi r}(\hat{\mathbf{j}} \times \hat{\mathbf{k}})\right]$

$\therefore F = \frac{q\mu_0 v_0 I}{2p\pi r}\hat{\mathbf{i}}$ - Ans

39. A Circular loop of radius 10 cm carrying current of 1.0 A lies in X - Y plane. A long straight wire lies in the same plane parallel to X-axis at a distance of 20 cm as shown in figure.

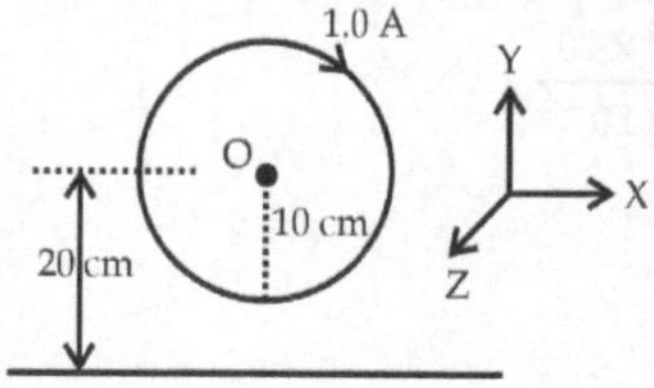

Find the direction and value of current that has to be maintained in the wire, so that the net magnetic field at O is zero. (2024)

Ans. From the figure -

Magnetic field at point O due to the circular loop, $B_1 = \frac{\mu_0 I_1}{2R}$

Now I = 1 A and radius = 10 cm = 0.1 m

$\therefore B_1 = \frac{\mu_0 \times 1}{2(0.1)}$

Now magnetic field at point O due to the straight wire,

$B_2 = \frac{\mu_0 I_2}{2\pi r}$

Here 'r' is the distance between wire and point 'O'

$r = 0.2$ m

$B_2 = \frac{\mu_0 I_2}{2\pi(0.2)}$

For net magnetic field at O to be zero.

$B_1 = -B_2$

or $B_1 + B_2 = 0$ and $|B_1| = |B_2|$

$\therefore \frac{\mu_0 \times 1}{2(0.1)} = \frac{\mu_0 \times I_2}{2\pi(0.2)}$

$\therefore I_2 = 6.28$ A

Due to the loop the magnetic field is into the plane of the paper (clockwise) thus due to straight wire at 'O' the magnetic field must be out of the plane of the paper which means from left to right.

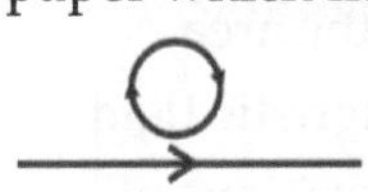

40. Two long insulated straight wires carrying currents of 3 A and 5 A are arranged in XY-plane as shown in figure. Find the magnitude and direction of the net magnetic field at points $P_1(2\text{ m}, 2\text{ m})$ and $P_2(-1\text{ m}, 1\text{ m})$. (2024)

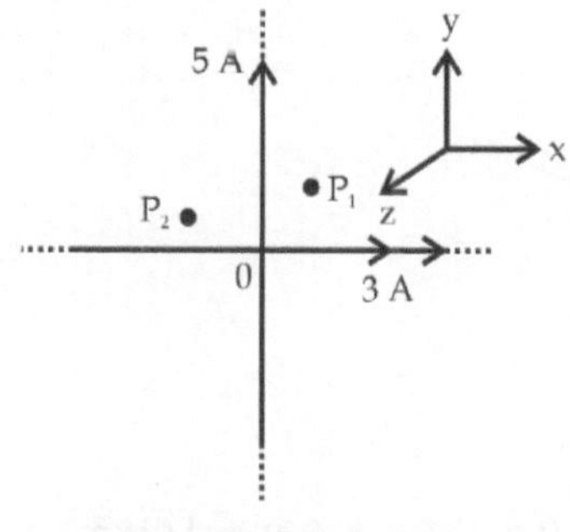

Ans. The directions of magnetic field due to both wires at P_1 and P_2 is shown in the figure which is determined by right hand thumb rule.

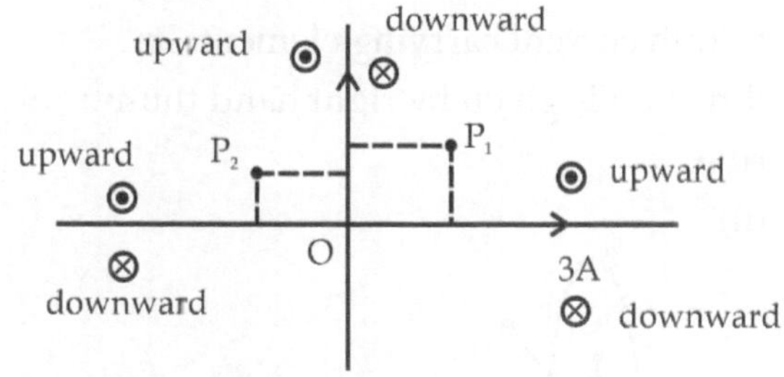

The net magnetic field, at point P_1 is

$\mathbf{B}_{net} = \mathbf{B}_x - \mathbf{B}_y = \frac{\mu_0 I_1}{2\pi r} - \frac{\mu_0 I_2}{2\pi r}$

$= \frac{\mu_0 5}{2\pi \times 2} - \frac{\mu_0 3}{2\pi \times 2} = \frac{+\mu_0}{2\pi}$ (Downward)

Because $I_1 > I_2$

And at point $P_2(-1\text{ m}, 1\text{ m})$,

The net magnetic field,

$\mathbf{B}_{net} = \frac{\mu_c \times 5}{2\pi \times 1} + \frac{\mu_0 \times 3}{2\pi \times 1} = \frac{4\mu_0}{\pi}$ (upward)

41. (i) State Biot-Savart's law and express this law in the vector form.

(ii) Two identical circular coils P and Q each of radius R, carrying currents 1 A and $\sqrt{3}$ A respectively, are placed concentrically and perpendicular to each other lying in the XY and YZ-planes. Find the magnitude and direction of the net magnetic field at the centre of the coils. (2017)

Ans. Biot-Savart's Law : According to this law, the magnetic field due to any current carrying element $I\,\vec{dl}$ at a nearby point is given by

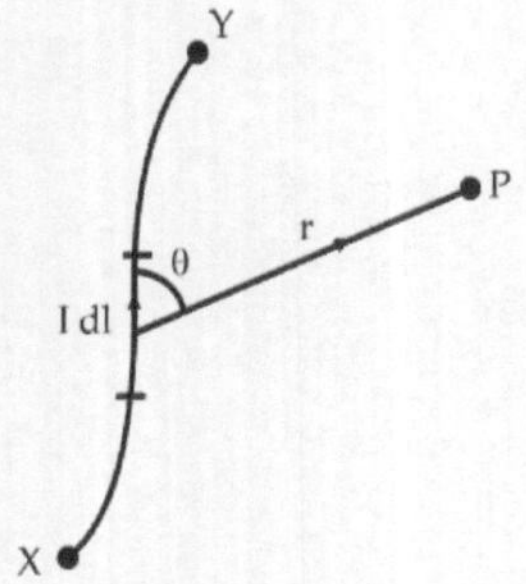

$dB = \frac{\mu_0}{4\pi} \cdot \frac{Idl\sin\theta}{r^2}$ or $\overrightarrow{dB} = \frac{\mu_0}{4\pi} \cdot \frac{I\,\overrightarrow{dl}\times\vec{r}}{r^3}$

here, $\mu_0/4\pi = 10^{-7}\ TA^{-1}m$

Here, μ_0 = permeability of free space or vacuum and $\vec{r}$ = position vector of point P from current carrying element. Its direction is given by right hand thumb rule.

(ii)

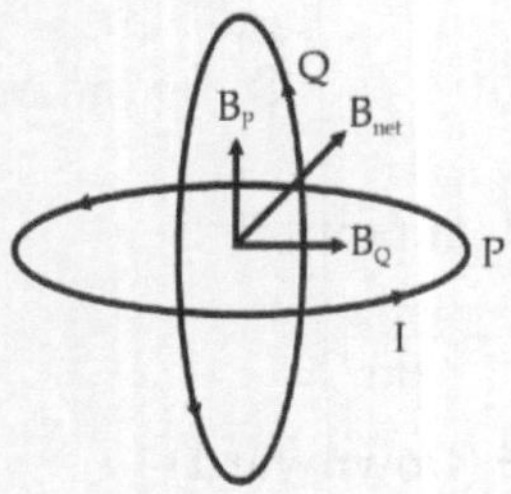

Magnetic field due to loop P,

$B_P = \mu_0 I_1/2R$ (vertically upwards considering direction of current anticlockwise from top)

and Magnetic field due to loop Q,

$B_Q = \mu_0 I_2/2R$ (Towards +X axis considering direction anticlockwise from right hand side)

Net Magnetic field at 'O' due to both the loops

$$B = \sqrt{B_P^2 + B_Q^2}$$

$$\therefore B = \sqrt{\left(\frac{\mu_0 I_1}{2R}\right)^2 + \left(\frac{\mu_0 I_2}{2R}\right)^2}$$

$$\text{Or } B = \sqrt{\left(\frac{\mu_0}{2R}\right)^2 (I_1^2 + I_2^2)}$$

$$\text{Or } B = \frac{\mu_0}{2R}\sqrt{I_1^2 + I_2^2}$$

$$\text{Or } B = \frac{4\pi\times10^{-7}}{2\times R}\sqrt{(1)^2 + (\sqrt{3})^2}$$

$\therefore B = 4\pi \times 10^{-7}/R$ Tesla

If resultant magnetic field makes an angle θ with direction of B_Q, then

$\tan\theta = B_P/B_Q = 1/\sqrt{3}$

$\Rightarrow \theta = 30°$

42. (a) Write the principle and explain the working of a moving coil galvanometer. A galvanometer as such cannot be used to measure the current in a circuit. Why?

(b) Why is the magnetic field made radial in a moving coil galvanometer? How is it achieved? (2023)

Ans. (a) **Principle of moving coil galvanometer** - When a current carrying coil is placed in a magnetic field, it experiences a torque which is given by

τ = NIAB

Here, N = number of turns,

I = current, A = area of a coil

and B = magnetic field.

The galvanometer cannot be used to measure the current because it is too sensitive, giving full scale deflection for very small current and its coil has significant resistance which affects original current being measured.

(b) Due the uniform radial magnetic field, the plane of the coil is always parallel to the direction of the magnetic field i.e., the angle between the area vector of the coil and the magnetic field is always 90° for all the orientations of the coil and deflection is directly proportional to the current magnitude and thus we can use a linear scale. We can achieve radial magnetic field by using cylindrical pole pieces of permanent magnet and a soft iron core placed inside the coil of the galvanometer.

43. (i) Briefly describe how the current sensitivity of a moving coil galvanometer can be increased.

(ii) A galvanometer shows full scale

deflection for current I. A resistance R_1 is required to convert it into a voltmeter of range (0 – V) and a resistance R_2, to convert it into a voltmeter of range (0 – 2 V). Find the resistance of the galvanometer. (2023)**

Ans. (i) Current Sensitivity of a moving coil galvanometer can be increased by increasing the number of turns in this coil, the area of coil and magnetic field, also by decreasing the torsion constant 'K' of the suspension wire.

$\left[I_s = \frac{\theta}{I} = \frac{NBA}{K}\right]$

(ii) A galvanometer shows full deflection for current I_g and for range (0-V), required resistance is R_1 and for range (0-2 V), required resistance is R_2

Resistance of galvanometer = ?

Now for voltmeter (range 0-V),

$1V = (G + R_1)I_g$ ....(1)

where, I_g is the current in galvanometer and for voltmeter range (0-2V)

$2V = (G + R_2)I_g$ ......(2)

Dividing Equation (1) by (2), we get

$\frac{1}{2} = \frac{(G+R_1)}{(G+R_2)}$

$G + R_2 = 2(G + R_1)$

$G + R_2 = 2G + 2R_1$

Or thus, $G = R_2 - 2R_1$ - Ans

V. Magnetism and Matter

44. The magnetic field lines near a substance are as shown in the figure. The substance is (2023)

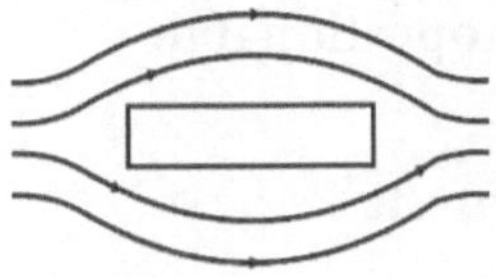

(a) copper **(b) iron**
(c) sodium **(d) aluminium**

Ans. (a) The behaviour of magnetic field lines shown in the figure is for a diamagnetic substance. So, the answer is copper.

45. A bar magnet has magnetic dipole moment M. Its initial position is parallel to the direction of uniform magnetic field B. In this position, the magnitudes of torque and force acting on it, respectively are - (2022)

(a) 0 and $\mathbf{M} \cdot \mathbf{B}$

(b) $\mathbf{M} \cdot \mathbf{B}$ and $\mathbf{M} \cdot \mathbf{B}$

(c) 0 and 0

(d) $|\mathbf{M} \times \mathbf{B}|$ and 0

Ans. (c) The torque on a bar magnet in external magnetic field is given by -

$\tau = |\mathbf{M} \times \mathbf{B}| = MB\sin\theta$

When the bar magnet is parallel to the direction of magnetic field, i.e., $\theta = 0°$

$\therefore \tau = MB\sin 0° = 0$

And the force acting at the poles of bar magnet are equal and opposite in direction thus net force acting is zero.

46. The magnetic susceptibility of χ of a given material is -0.5 . Identify the magnetic material. (2019)

Ans. For diamagnetic substances, the value of χ is small and negative. Thus, it is a diamagnetic substance.

47. Draw the magnetic field lines for a current carrying solenoid when a rod made of (i) copper, (ii) aluminium and (iii) iron are inserted within the solenoid as shown. (2017)

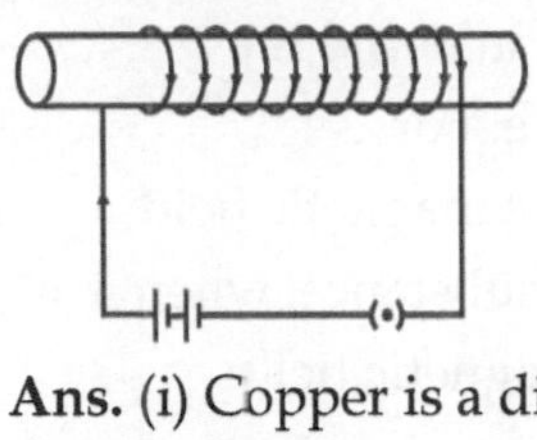

Ans. (i) Copper is a diamagnetic substance.

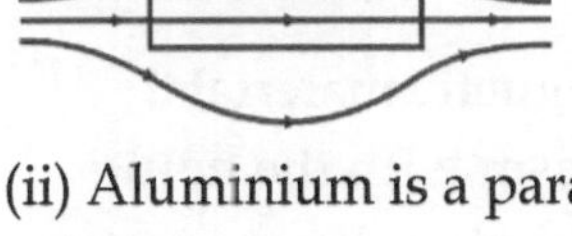

(ii) Aluminium is a paramagnetic substance.

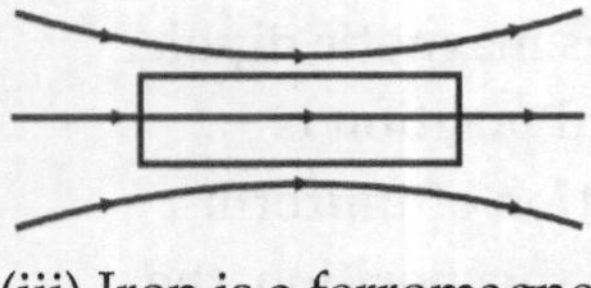

(iii) Iron is a ferromagnetic substance.

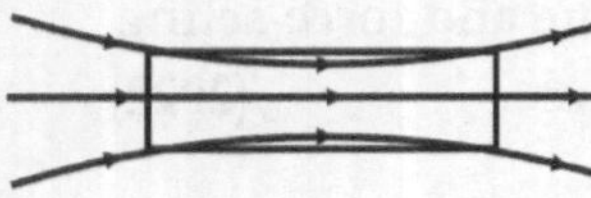

48. A steady current of 2A flows through a circular coil having 5 turns of radius 7 cm . The coil lies in XY-plane with its centre at the origin. Find the magnitude and direction of the magnetic dipole moment of the coil. (2020)

Ans. Given,

$I = 2$ A, and $r = 7$ cm or 0.07 m

Now the Magnetic dipole moment of a coil,

$M = NIA = 5 \times 2 \times \pi(0.07)^2$

$= 0.154\ \text{Am}^2$

Its direction is perpendicular to the plane of coil, i.e. it is along Z-axis.

49. The susceptibility of a magnetic material is 0.9853 . Identify the type of magnetic material. Draw the modification of the field pattern on keeping a piece of this material in a uniform magnetic field. (2018)

Ans. Here, susceptibility, $\chi_m = 0.9853$

Since, the susceptibility of material is small and positive, the material is a paramagnetic substance. For paramagnetic substance magnetic field lines pass through the substance, when it is placed in external magnetic field.

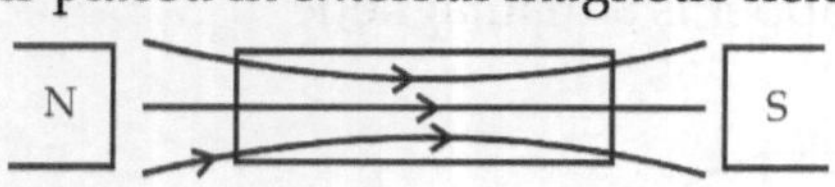

50. What are ferromagnetic materials? Explain ferromagnetism with the help of suitable diagrams, using the concept of magnetic domain. (2024)

Ans. Those substances which get strongly magnetised in an external magnetic field are called ferromagnetic substances. They move from weak magnetic field region to a strong magnetic field region, i.e. they get strongly attracted towards a magnet.

Reason for Ferromagnetism: The atoms in a ferromagnetic substance have a net a dipole moment aligned in one direction over a small volume which is called domain. Each domain has net magnetic moment.

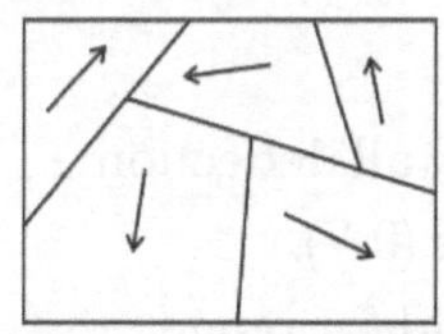

Randomly aligned domains

When we place them in external magnetic field B, the domains orient themselves in the direction of B.

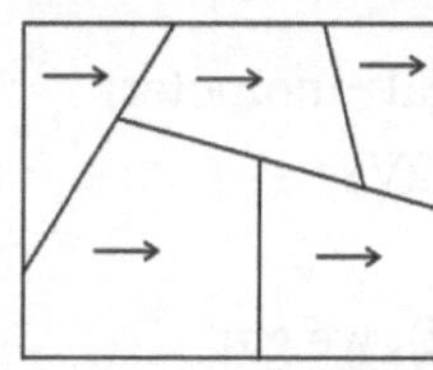

Aligned domains

E.g. alnico (an alloy of Fe, Al, Ni, Co and Cu).

VI. Electromagnetic Induction

51. A circular loop A of radius R carries a current I. Another circular loop B of radius $r\left(=\frac{R}{20}\right)$ is placed concentrically in the plane of A. The magnetic flux linked with loop B is proportional to (2024)

(a) R **(b) $\sqrt{R}$**

(c) $R^{\frac{3}{2}}$ **(d) R^2**

Ans. (a) The magnetic field produced by circular loop A will be –

$B_A = \frac{\mu_0 I}{2R}$

Now the magnetic flux linked with loop B, $\phi_B = B_A A_B = \frac{\mu_0 I}{2R} \cdot \pi\left(\frac{R}{20}\right)^2$

or $\phi_B = \frac{\mu_0 I\pi R}{800}$

$\therefore \ \phi_B \propto R$

52. An iron cored coil is connected in series with an electric bulb with an AC source as shown in figure. When iron piece is taken out of coil, the brightness of the bulb will (2022)

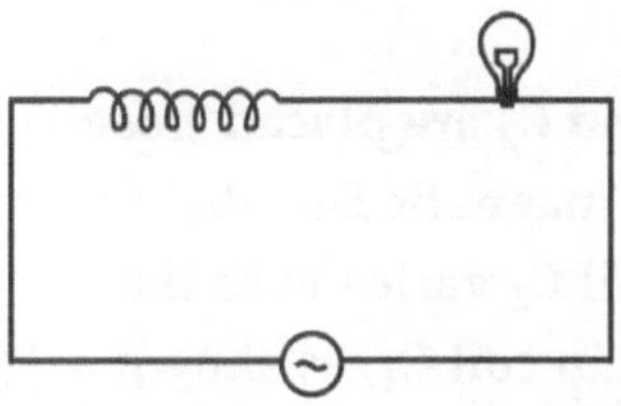

(a) decrease (b) increase
(c) remain unaffected (d) fluctuate

Ans. (b) When an iron cored coil is connected in series with an electric bulb and with an AC source, the resistance of the coil increases that means inductive reactance (X_L) increases and current decreases. When we take out the core, X_L decreases and current increases thus the brightness will increase.

53. The direction of induced current in the loop abc is (2023)

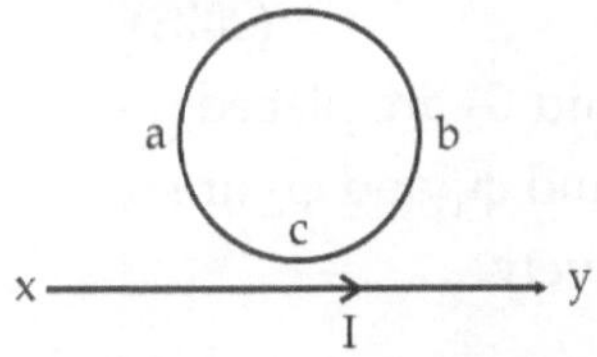

(a) along abc if I decreases
(b) along acb if I increases
(c) along abc if I is constant
(d) along abc if I increases

Ans. (d) From lenz's law, the direction of induced current in loop abc is along abc, if current increases.

54. Predict the polarity of the capacitor in the situation as shown in the figuregiven below. Explain the reason too. (2011)

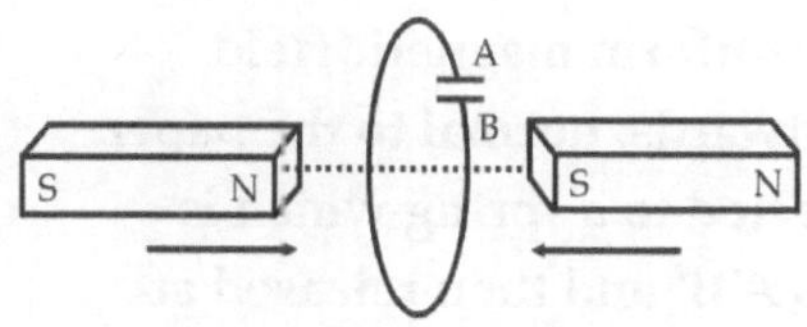

Ans. As shown in the figure, induced current in the loop will be clockwise when seen from the right hand side. That means the plate A will acquire higher potential and positively charged and plate B is negatively charged.

55. (i) The figure shows the variation of induced emf as a function of rate of change of current for two identical solenoids XY, one is air cored and the other is iron cored. Which one of them is iron cored? Why?

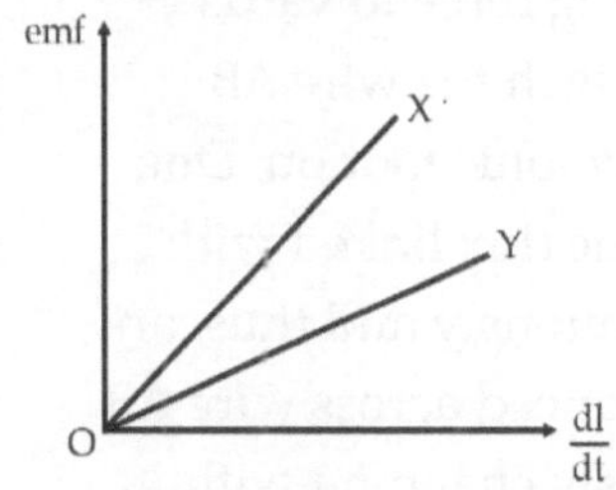

(ii) Obtain an expression for self-inductance of a long solenoid of length L and cross-sectional area A having N turns. (2023)

Ans. (i) Since induced emf,

$e = L\frac{dI}{dt}$ ...... (1)

it follows $y = mx$ .....(2)

From equation (1) and Eq. (2),

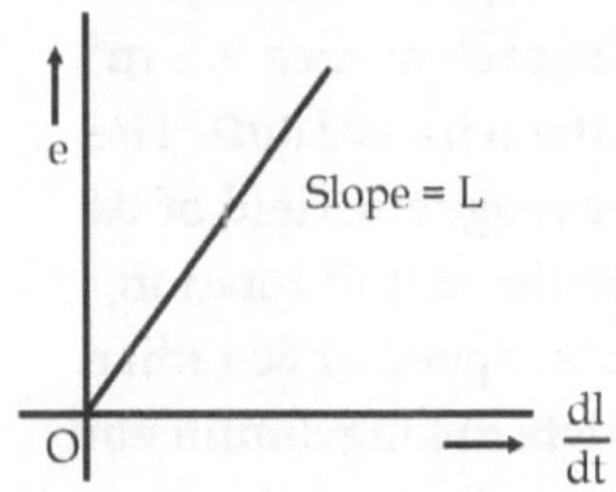

Since, Since slope of X > slope of y

$\therefore L_x > L_y$

And $L \propto \mu_r$ $\therefore (\mu_r)_x > (\mu_r)_y$

Thus, x has iron-core.

56. A rectangular frame of wire is

placed in a uniform magnetic field directed outwards, normal to the paper. AB is connected to a spring which is stretched to A′B′ and then released at time t = 0. Explain qualitatively how induced electromotive force in the coil would vary with time. (Neglect damping of oscillations of spring) (2018)

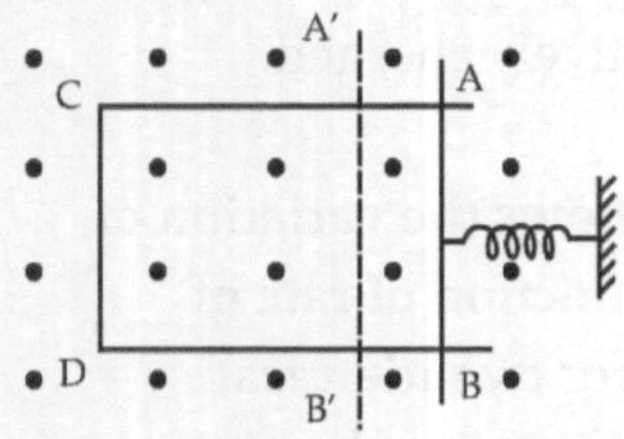

Ans. As shown in the figure, when wire AB is stretched to A′B′, then spring experience a restoring force towards right side, due to which the wire AB executes simple harmonic motion. Due to SHM the magnetic flux linked with wire changes continuously and thus, an induced emf is produced across wire AB which is continuously changing with time sinusoidally.

$x = -a\cos\omega t$

And $\phi = B.A = B\ell x$

Or $\phi = -B\ell a\cos\omega t$

And $e = -\frac{d\phi}{dt}$, $\therefore e = -\frac{d}{dt}(-B\ell a\cos\omega t)$

Or $e = -B\ell a\omega\sin\omega t$

57. The coil of an AC generator consists of 100 turns of wire each of area $0.5\ m^2$. The resistance of the wire is 100Ω. The coil is rotating in a magnetic field of 0.8 T perpendicular to the axis of rotation, at a constant angular speed of 60 radian per second. Calculate the maximum emf generated and power dissipated to the coil. (2023)

Ans. Given, N = 100

$A = 0.5\ m^2$

$R = 100\Omega$

$B = 0.8\ T$

$\omega = 60 rad/s$

$\because (e_0) = NBA\omega$

$\therefore e_0 = 100 \times 0.8 \times 0.5 \times 60$

$e_0 = 2400\ V$ – Ans

And Induced current,

$I_0 = \frac{e_0}{R} = \frac{2400}{100} = 24\ A$

Now Power dissipated,

$P = e_0 \times I_0 = 2400 \times 24 = 57600$

$P = 576 \times 10^2\ W$

58. Two coils C_1 and C_2 are placed close to each other. The magnetic flux ϕ_2 linked with the coil C_2 varies with the current I_1 flowing in coil C_1, as shown in figure.

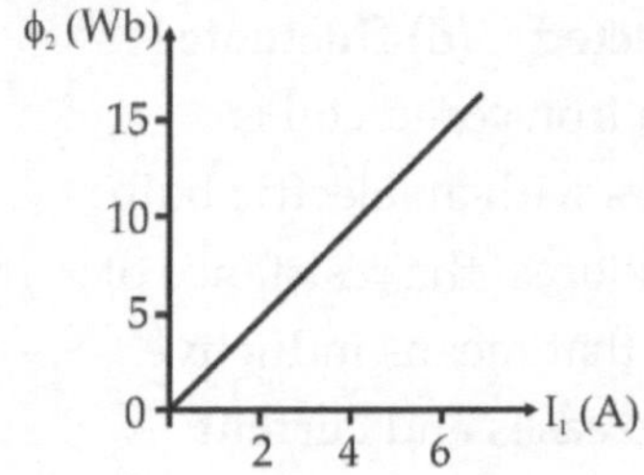

Find

(i) the mutual inductance of the arrangement and

(ii) the rate of change of current $\left(\frac{dI_1}{dt}\right)$ that will induce an emf of 100 V in coil C_2. (2023)

Ans. Here coils C_1 and C_2 are placed close to each other and ϕ_1 and ϕ_2 are flux of coils respectively.

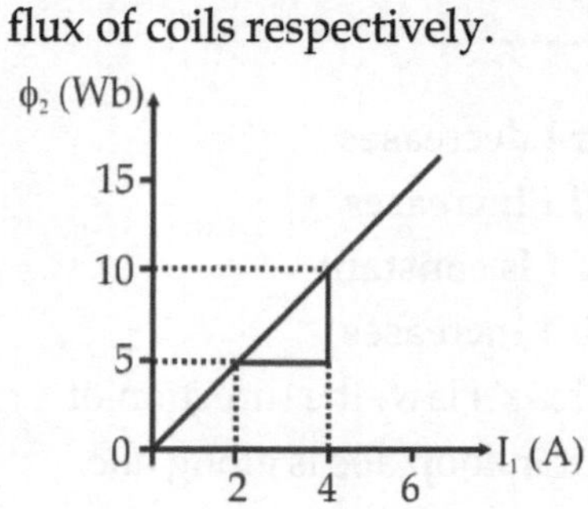

(i) Since $\phi = M \times I$

$\therefore \phi_2 = M_{21} I_1$

$\therefore M_{21} = \frac{\phi_2}{I_1} = 2.5\ H$

(ii) $\frac{dI_1}{dt} = ?$

$\because e = M \cdot \frac{dI_1}{dt}$

$\therefore\ e = 2.5 \times \frac{dI_1}{dt}$

$100 = 2.5 \times \frac{dI_1}{dt}$

$\therefore \frac{dI_1}{dt} = 40\ A$

VII. Alternating Current

59. A 300Ω resistor and a capacitor of $\left(\frac{25}{\pi}\right)$ μF are connected in series to a 200 V – 50 Hz AC source. The current in the circuit is (2022)

(a) 0.1 A **(b) 0.4 A**

(c) 0.6 A **(d) 0.8 A**

Ans. (b) Here, $R = 300\Omega$,

$C = \frac{25}{\pi}\ \mu F = \frac{25}{\pi} \times 10^{-6}\ F$,

$E = 200\ V$ and $\upsilon = 50\ Hz$

The capacitive reactance, (X_C) will be

$X_C = \frac{1}{2\pi\upsilon C} = \frac{1}{2\pi \times 50 \times \frac{25}{\pi} \times 10^{-6}} = 400\Omega$

Also impedance of circuit, (Z) -

$Z = \sqrt{R^2 + X_C^2}$

$= \sqrt{(300)^2 + (400)^2}$

$= 500\Omega$

And current in circuit, $I = \frac{E}{Z} = \frac{200}{500}$

$= 0.4\ A$

60. Figure shows the variation of inductive reactance X_L of two ideal inductors of inductances L_1 and L_2, with angular frequency ω. The value of $\frac{L_1}{L_2}$ is (2024)

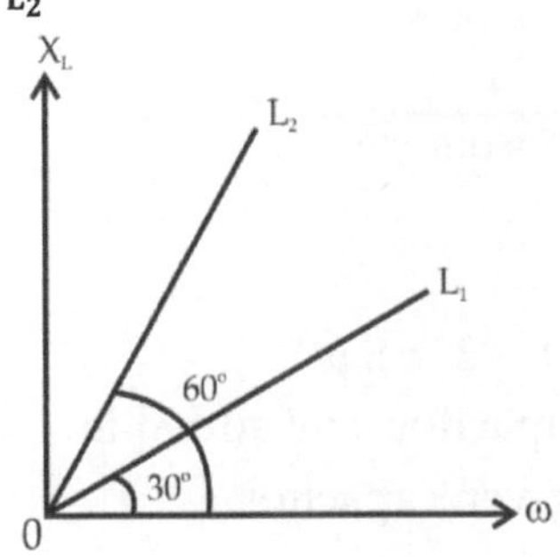

(a) $\sqrt{3}$ **(b) $\frac{1}{\sqrt{3}}$**

(c) 3 **(d) $\frac{1}{3}$**

Ans. (d) $\because\ X_L = \omega L$ or $\frac{X_L}{\omega} = L$

i.e., slope of $X_L - \omega$ graph gives self inductance -

$\therefore L_1 = \tan 30° = \frac{1}{\sqrt{3}}$

and $L_2 = \tan 60° = \sqrt{3}$

Thus, $\frac{L_1}{L_2} = \frac{\frac{1}{\sqrt{3}}}{\sqrt{3}} = \frac{1}{3}$

61. The rms current in a circuit connected to a 50 Hz AC source is 15 A. The value of the current $\left(\frac{1}{600}\right)$ s after its value becomes zero is (2022)

(a) $\frac{15}{\sqrt{2}}$ A **(b) $15\sqrt{2}$ A**

(c) $\frac{\sqrt{2}}{15}$ A **(d) 8 A**

Ans. (a) The instantaneous current is given by -

$I = I_0 \sin \omega t = I_0 \sin 2\pi\upsilon t$

$\therefore I = (I_{rms} \times \sqrt{2}) \sin\left(2\pi \times 50 \times \frac{1}{600}\right)$

$= 15 \times \sqrt{2} \times \sin\frac{\pi}{6} = 15 \times \sqrt{2} \times \frac{1}{2} = \frac{15}{\sqrt{2}}$ A

62. In a circuit, the phase difference between the alternating current and the source voltage is $\frac{\pi}{2}$. Which of the following cannot be the element(s) of the circuit? (2022)

(a) Only C **(b) Only L**

(c) L and R **(d) L or C**

Ans. (c)

63. An AC voltage $V = V_0 \sin\ \omega t$ is applied to a series combination of a resistor R and an element X. The instantaneous current in the circuit is $I = I_0 \sin\left(\omega t + \frac{\pi}{4}\right)$. Then, which of the following is correct? (2023)

(a) X is a capacitor and $X_C = \sqrt{2}R$

(b) X is an inductor and $X_L = R$

(c) X is an inductor and $X_L = \sqrt{2}R$

(d) X is a capacitor and $X_C = R$

Ans. (d) Given $V = V_0 \sin\ \omega t$

And instantaneous current -

$I = I_0 \sin\left(\omega t + \frac{\pi}{4}\right)$

Since current is leading thus element 'X' is a capacitor

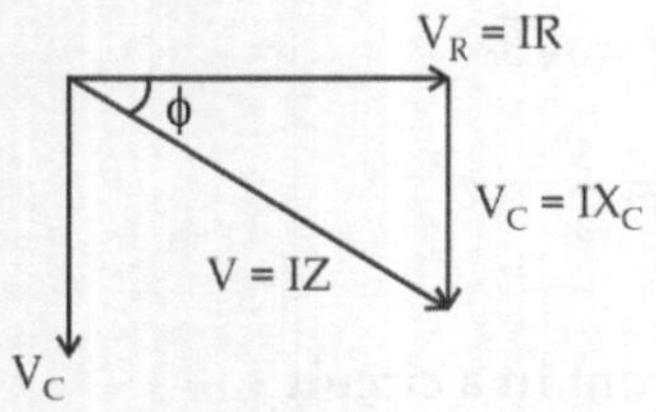

Now $\phi = \frac{\pi}{4}$

$\therefore \tan\frac{\pi}{4} = \frac{IX_C}{IR} = 1$

$\therefore X_C = R$

64. Mention the factors on which the resonant frequency of a series LCR circuit depends. Plot a graph showing variation of impedance of a series LCR circuit with the frequency of the applied AC source. (2024)

Ans. In LCR series circuit, the resonant frequency is given by -

$\upsilon_0 = \frac{1}{2\pi\sqrt{LC}}$

The resonant frequency depends on self inductance L and capacitance C.

The graph between impedance and frequency is -

Impedance is minimum at υ_r.

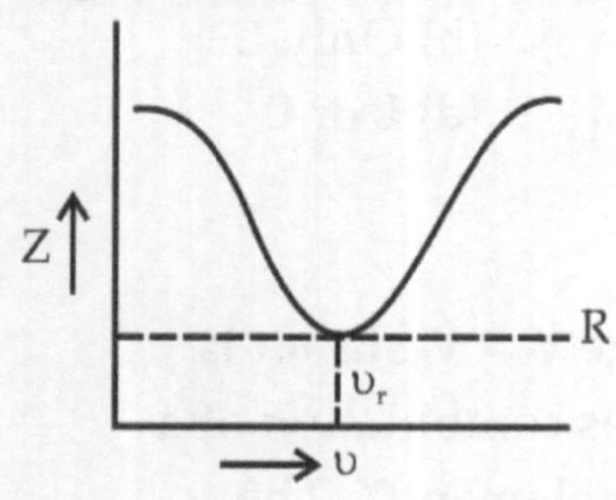

65. An electric lamp having coil of negligible inductance connected in series with a capacitor and an AC source is glowing with certain brightness.

How does the brightness of the lamp change on reducing the (i) capacitance, and (ii) the frequency ? (2010)

Justify your answer.

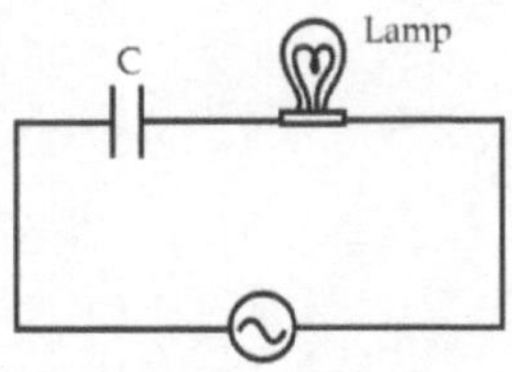

Ans. In AC circuit capacitative reactance is -

$X_C = \frac{1}{\omega C} = \frac{1}{2\pi\nu C}$.

It allows AC to pass through it.

(i) On reducing capacitance C, reactance increases due to which current decreases and brightness of the bulb reduces.

(ii) When frequency υ is reduced, X_C increases thus current decreases and brightness of the bulb reduces.

66. Calculate the value of the additional capacitor which may be joined suitably to the capacitor C that would make the power factor of the circuit unity. (2015)

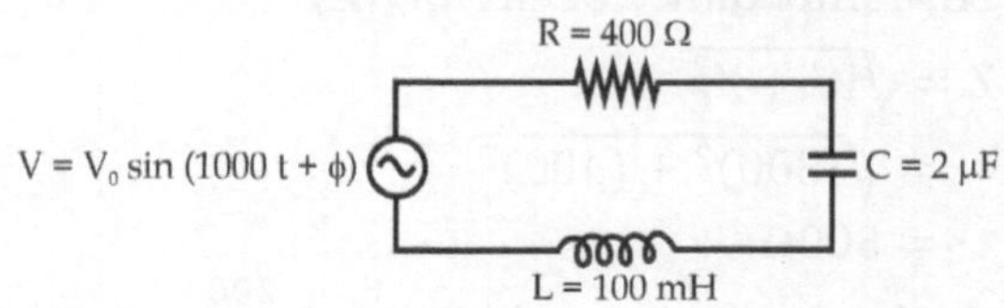

Ans. For, $\cos\phi = 1$

$\therefore \frac{R}{Z} = 1$ or $Z = R$

i.e. $X_L = X_C$

$\omega L = 1/\omega C'$ ...(1)

Here, $C' = C + C_1$

(For parallel combination and C_1 is the capacitance required)

Now from eq (1)

$C' = \frac{1}{\omega^2 L} = \frac{1}{(1000)^2 \times 100 \times 10^{-3}}$

$= 10^{-5}\ F = 10\ \mu F$

$\because C' = C + C_1$

$\therefore C_1 = C' - C = 10 - 2 = 8\ \mu F$

Thus we need a capacitor 8 μF added in parallel with the given capacitor.

67. A series RL circuit with R = 10Ω and $L = \left(\frac{100}{\pi}\right)$ mH is connected to an AC source of voltage, V = 141sin (100πt), where V is in volts and t is in seconds.

Calculate **(2023)**
(i) impedence of the circuit
(ii) phase angle and
(iii) voltage drop across the inductor

Ans. Given $R = 10\Omega$

And $L = \left(\frac{100}{\pi}\right)$ mH

Also voltage, $V = 141\sin(100\pi t)$

$\because V = V_0 \sin \omega t$

$\therefore \omega = 100\pi$

(i) Impedence,

$Z = \sqrt{R^2 + X_L^2}$

$\because X_L = \omega L$

$\therefore Z = \sqrt{(10)^2 + (100\pi)^2 \times \left(\frac{100}{\pi}\right)^2}$

or $Z = \sqrt{(10)^2 + (100)^2 \times (100)^2}$

$Z = 10000\Omega$ - Ans

(ii) Now phase angle ϕ,

$\because \cos\phi = \frac{R}{Z}$

$\therefore \cos\phi = \frac{10}{10000}$

$\phi = \cos^{-1}(0.001)$

or $\phi = 90°$

(iii) And voltage drop across inductor,

$V_{rms} = \frac{V_0}{\sqrt{2}}$

$V_{rms} = \frac{141}{\sqrt{2}} = \frac{141}{1.41} = 100$ V

VIII. Electromagnetic Waves

68. The electromagnetic radiations used to kill germs in water purifiers are called **(2023)**

(a) Infrared waves **(b) X-rays**
(c) Gamma rays **(d) Ultraviolet rays**

Ans. (d) Ultraviolet rays.

69. The phase difference between electric field E and magnetic field B in an electromagnetic wave propagating along Z-axis is - **(2024)**

(a) zero **(b) π**
(c) $\frac{\pi}{2}$ **(d) $\frac{\pi}{4}$**

Ans. (a) The phase difference between electric field **E** and magnetic field **B** is zero.

70. The electromagnetic waves used in RADAR systems are - **(2023)**

(a) Infrared waves
(b) Ultraviolet rays
(c) Microwaves
(d) X-rays

Ans. (c) Microwaves.

71. The ratio of the magnitude of the electric field and magnetic field of a plane electromagnetic wave is - **(2023)**

(a) 1 **(b) $\frac{1}{c}$**
(c) c **(d) $\frac{1}{c^2}$**

Ans. (c) It gives speed of EM waves -

i.e., $c = \frac{E_0}{B_C}$

72. The magnetic field in a plane electromagnetic wave is given by
$B_y = 12 \times 10^{-8} \sin(1.20 \times 10^7 z + 3.60 \times 10^{15} t)$
T.
Calculate the
(i) energy density associated with the electromagnetic wave and
(ii) speed of the wave **(2017)**

Ans. Given,

$B_y = 12 \times 10^{-8} \sin(1.20 \times 10^7 z + 3.60 \times 10^{15} t)$

$\therefore B_0 = 12 \times 10^{-8}$ $\quad \because C = \frac{E_0}{B_0}$

Thus $E_0 = c \times B_0 = 3 \times 10^8 \times 12 \times 10^{-8}$

$\therefore E_0 = 36 \frac{N}{C}$

(i) Energy density, $u = \frac{1}{2}\varepsilon_0 E_0^2$ ......(2)

And $E_0 = 36$ N/C

$\therefore u = \frac{1}{2} \times 8.85 \times 10^{-12} \times (36)^2$

or $u = 5.74 \times 10^{-15}$ J/m^3

(ii) Speed of the wave $= \frac{\omega}{k}$

$\because B_y = 12 \times 10^{-8} \sin(1.20 \times 10^7 z + 3.60 \times 10^{15} t)$

$\therefore k = 1.20 \times 10^7$

and $\omega = 3.6 \times 10^{15}$

$\therefore$ Speed of the wave $= \frac{3.6\times10^{15}}{1.20\times10^{7}}$

$= 3 \times 10^8$ m/s

73. Name the electromagnetic waves that are widely used as a diagnostic tool in medicine. (2018)

Ans. X-ray.

74. Electromagnetic waves of wavelengths λ_1, λ_2, and λ_3 are used in RADAR systems, in water purifiers and in remote switches of TV, respectively.

(i) Identify the electromagnetic waves and (2022)

(ii) Write one source of each of them.

Ans. In RADAR system Microwaves are used

In water purifiers UV rays and

in remote switches of TV Infrared waves are used.

(ii) Microwaves - Magnetron valve

UV rays - Inner shell electrons of atoms moving from one energy level to another level.

Infrared waves - Vibration of atoms and molecules in hot bodies.

75. How is the equation for Ampere's circuital law modified in the presence of displacement current? Explain. (2019)

Ans. Ampere's circuital law was modified to Ampere-Maxwell's circuital law,

$\oint \mathbf{B} \cdot \mathbf{dl} = \mu_0(I_c + I_d)$

Here, I_c = conduction current

and I_d = displacement current.

$\therefore \oint \mathbf{B} \cdot \mathbf{dl} = \mu_0\left(I_c + \varepsilon_0 \frac{d\phi_E}{dt}\right)$

In order to overcome the inconsistency of Ampere's circuital law, Maxwell introduced a term which is called displacement current (I_d).

Displacement current comes into play in a region where electric field and electric flux is changing with time.

76. Arrange the following in descending order of wavelength.

X-rays, radio waves, blue light, infrared light. (2010)

Ans. Radio wave > Infrared > Blue light > X-ray.

77. Name the current which can flow even in the absence of electric charge. (2018)

Ans. Displacement current between the plates of capacitor.

78. A capacitor of capacitance C is being charged by connecting it across a DC source along with an ammeter. Will the ammeter show a momentary deflection during the process of charging? If so, how would you explain this momentary deflection and the resulting continuity of current in the circuit? Write the expression for the current inside the capacitor. (2012)

Ans. The ammeter will show the momentary deflection during charging of capacitor only and when fully charged, the current in the circuit becomes zero. Because during the charging of capacitor, due to change in electric field between the plates of capacitor the displacement current is produced (I_D).

(Here I_C = conduction current through the wires)

The current inside the capacitor, $I_C = I_D$

where, I_D is displacement current,

$I_D = \varepsilon_0 \frac{d\phi_E}{dt}$

79. Name the electromagnetic waves which (2012)

(i) maintain the earth's warmth and

(ii) are used in aircraft navigation.

Ans. (i) Infrared rays.
(ii) Microwaves.

80. Identify the electromagnetic wave whose wavelengths range is from about
(i) 10^{-12} m to about 10^{-8} m
(ii) 10^{-3} m to about 10^{-1} m
Write one use of each. (2023)

Ans. (i) The wave is X-ray. It is used in taking images of bones.
(ii) Radio wave. It is used in TV and radio communication.

81. How is the speed of electromagnetic waves in vacuum determined by the electric and magnetic fields? (2017)

Ans. We use the formula, $c = E_0/B_0$
Here, E_0 and B_0 are maximum electric field and magnetic field component respectively.

IX. Ray Optics and Optical Instruments

82. A biconcave lens of power P verically splits into two identical plano-concave parts. The power of each part will be (2020)

(a) 2P **(b) $\frac{P}{2}$**

(c) P **(d) $\frac{P}{\sqrt{2}}$**

Ans. (b) When a symmetrical biconcave lens is splitted into two plano concave parts as given here, each part's focal length becomes 2f.

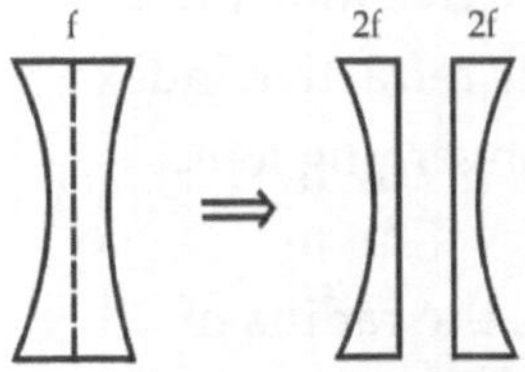

(You can check using the lens maker's formula)

$\therefore \frac{1}{f} = (\mu - 1)\left(\frac{1}{R_1} - \frac{1}{R_2}\right)$

Please note, that the plane part's focal length is ∞)

$\because P = \frac{1}{f}$

Now $P' = \frac{1}{2f}$ or $P' = \frac{P}{2}$

83. A ray of light of wavelength 600 nm propagates from air into a medium. If its wavelength in the medium becomes 400 nm . The reflactive index on the medium is (2023)

(a) 1.4 **(b) 1.5**

(c) 1.6 **(d) 1.8**

Ans. (b) Given Wavelength of light,
$\lambda_a = 600$ nm (in air)
and $\lambda_m = 400$ nm (in medium)
∴ Refractive index of the medium, μ_m

$\mu_m = \frac{\lambda_a}{\lambda_m} = \frac{600}{400}$

$\mu_m = 1.5$

84. (a) Write two necessary conditions for total internal reflection.
(b) Two prisms ABC and DBC are arranged as shown in figure. (2022)

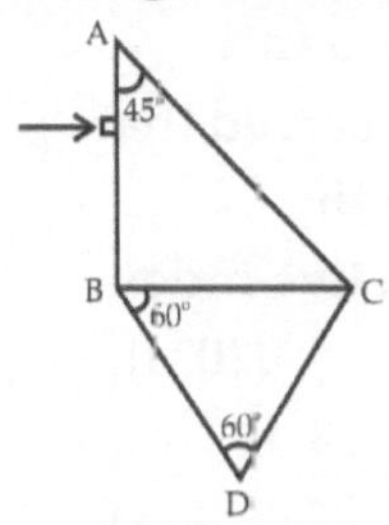

The critical angle for the two prisms with respect to air are 41.1° and 45°, respectively. Trace the path of the ray through the combination.

Ans. (a) The two necessary conditions for total internal reflection are –
(i) Light must travel from optically denser to optically rarer medium.
(ii) The angle of incidence in denser medium must be greater than critical angle for the given media.
(b) The ray incident normally on the face AB, it passes undeviated and falls at an angle 45° on face AC as shown in the figure. Since, the critical angle for prism ABC is 41.1°, the ray undergoes total internal reflection at face AC and falling

normally at BC and goes undeviated.

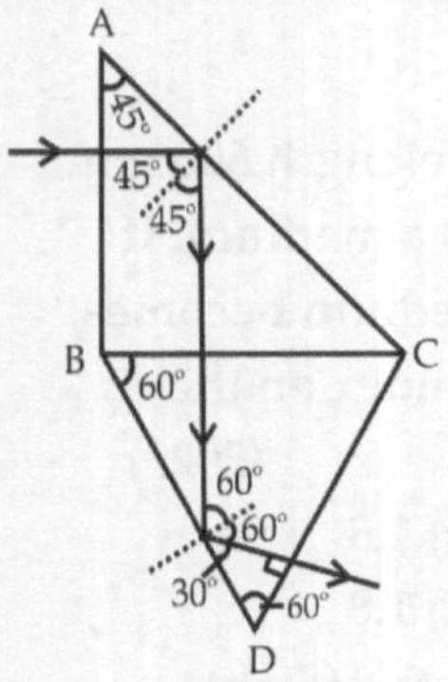

Now, it falls at face BD and we can see that, the angle of incidence is 60° which is greater than the critical angle of the prism (i.e., 45°), thus it again shows total internal reflection. Now at CD it again falls normally and emerges out undeviated.

85. A ray of light is incident normally on one face of an equilateral glass prism of refractive index μ. When the prism is completely immersed in a transparent medium, it is observed that the emergent ray just grazes the adjacent face. Find the refractive index of the medium. (2024)

Ans. From Snell's law,

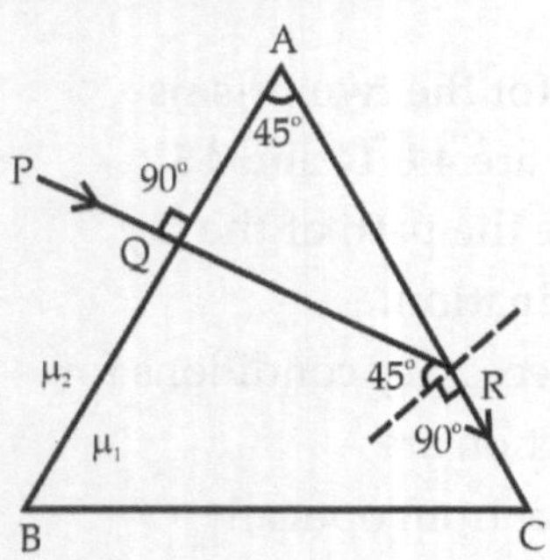

At face AC, light goes from glass to air, thus using snell's law -

$\frac{\sin i}{\sin r} = \frac{\mu_1}{\mu_2}$

Or $\frac{\sin 45°}{\sin 90°} = \frac{\mu_1}{\mu_2}$

Or $\mu_1 = \mu_2 \sin 45°$

Or $\mu_1 = \frac{\mu_2}{\sqrt{2}}$

Thus, refractive index of the outer medium is

$\mu_1 = \frac{\mu}{\sqrt{2}}$ - Ans

86. A ray of light is incident on a medium with angle of incidence i and is refracted into a second medium with angle of refraction r. The graph of sin i versus sin r is as shown. Find the ratio of the velocity of light in the first medium to the velocity of light in the second medium. (2018)

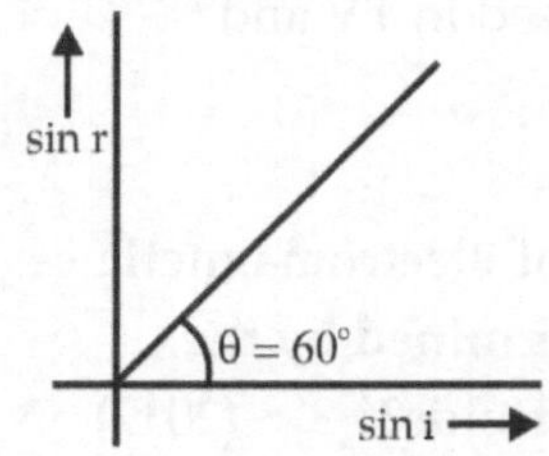

Ans. Here slope $\frac{\sin i}{\sin r} = \tan 60° = \sqrt{3}$

$\because \mu = \frac{\sin i}{\sin r} \quad \therefore \mu = \sqrt{3}$

Also $\mu = \frac{c}{v} = \sqrt{3}$ - Ans

87. A concave lens of refractive index 1.5 is immersed in a medium of refractive index 1.65. What is the nature of the lens? (2015)

Ans. A concave lens acts as a diverging lens when it is placed in a medium of lesser refractive index and acts as converging lens when placed in a medium of greater refractive index that means the nature changes if outer medium has more refractive index. Here outer medium has refractive index 1.65 which is greater than its refractive index 1.5 thus it acts like a converging lens.

88. In the given figure, the radius of curvature of curved face in the plano-convex and the plano-concave lens is 15 cm each. The refractive index of the material of the lenses is 1.5. Find the final position of the image formed. (2023)

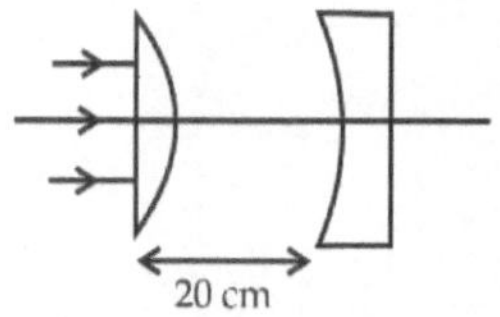

Ans. The radius of curvature of both the lenses is same, thus the focal length of plano-convex will be -

$\frac{1}{f} = (1.5-1)\left(\frac{1}{15}-\frac{1}{\infty}\right) = \frac{1}{30}$

$\therefore f = +30$ cm

Now focal length of plano-concave lens will be -

$\frac{1}{f} = (1.5-1)\left(\frac{1}{\infty}-\frac{1}{15}\right) = -\frac{1}{30}$

$\therefore f = -30$ cm

Parallel beam is incident on plano-convex lens thus the image will be formed at its focal point i.e., +30 cm from the lens. It acts like an object for plano-concave lens. Since both the lenses are at distance 20 cm, thus the distance of image from plano concave lens will be +10 cm thus 'u' for plano concave lens is 10 cm.

$\therefore \frac{1}{f} = \frac{1}{v} - \frac{1}{u}$

$\frac{1}{-30} = \frac{1}{v} - \frac{1}{10}$

$\frac{1}{v} = \frac{1}{10} - \frac{1}{30} = \frac{3-1}{30} = \frac{2}{30}$

or $v = +15$ cm

The final image is at a distance of 15 cm from plano concave lens.

89. (i) An object is placed in front of a converging lens. Obtain the conditions under which the magnification produced by the lens is (a) negative and (b) positive

(ii) A point object is placed at O in front of a glass sphere as shown in figure.

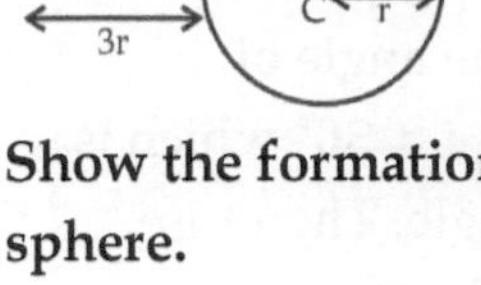

Show the formation of image by the sphere. **(2022)**

Ans. (i) Convex lens or converging lens produces both kind of images.

(a) When image is real, the magnification is negative because real image is always inverted.

$m = \frac{\text{Height if image}}{\text{Height of object}}\left(\frac{I}{O}\right)$

(b) When image is virtual, the magnification is positive because virtual image is erect.

(ii) Here the object is placed at a distance of 3r from left face of the sphere.

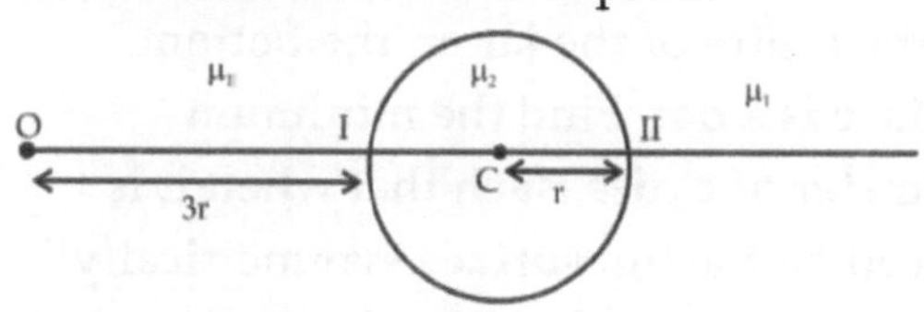

When refraction takes place from I surface, light goes from rarer (μ_1) medium to denser medium.

$\therefore$ We use formula

$\frac{\mu_2}{v} - \frac{\mu_1}{u} = \frac{\mu_2-\mu_1}{R}$

Given

$u = -3r$, $R = r$, $v = ?$, $\mu_1 = 1$, $\mu_2 = 1.5$

$\therefore \frac{1.5}{v} - \frac{1}{-3r} = \frac{1.5-1}{r}$

$\frac{1.5}{v} + \frac{1}{3r} = \frac{0.5}{r}$

Solving this we get v = 9r

That means from surface I the image I' is formed at a distance 9r.

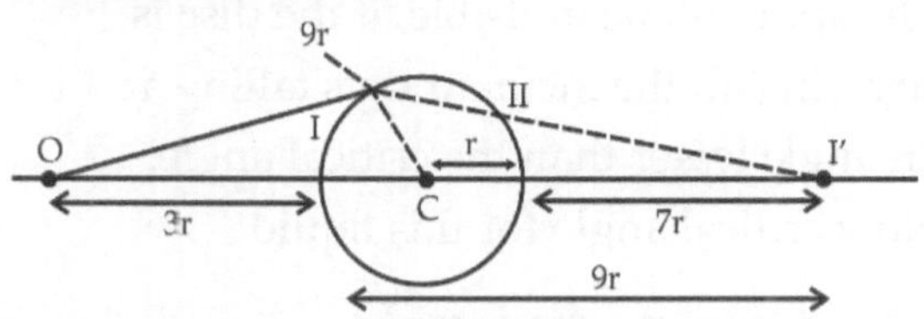

Here I' acts like a virtual object for II surface. In refraction from surface II. Light goes from denser to rarer medium. Here 'u' is the distance of virtual object I' from the surface II.

$u = +7r$, $v = ?$, $R = -r$, $\mu_1 = 1$, $\mu_2 = 1.5$

The formula for refraction from denser to rarer medium is -

$\frac{\mu_1}{v} - \frac{\mu_2}{u} = \frac{\mu_1-\mu_2}{R}$

$\frac{1}{v} - \frac{1.5}{7r} = \frac{1-1.5}{-r}$

Or $\frac{1}{v} = \frac{0.5}{r} + \frac{1.5}{7r}$

Solving this we get $v = \frac{7r}{5}$

Or $v = 1.4\,r$ - Ans

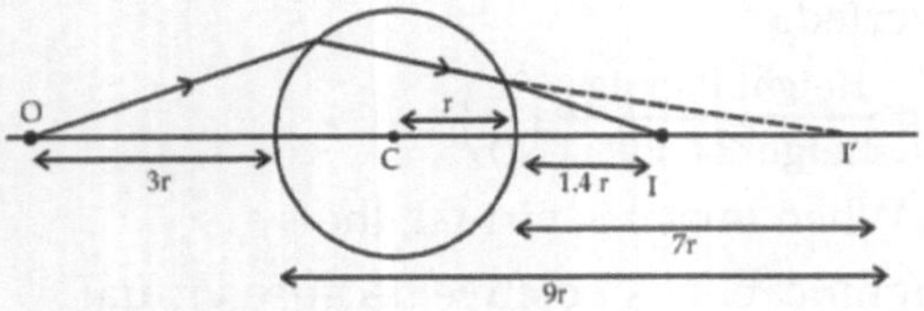

90. A jar of height h is filled with a transparent liquid of refractive index μ. At the centre of the jar on the bottom surface is a dot. Find the minimum diameter of a disc, such that when it is placed on the top surface symmetrically about the centre, then the dot is invisible. (2018)

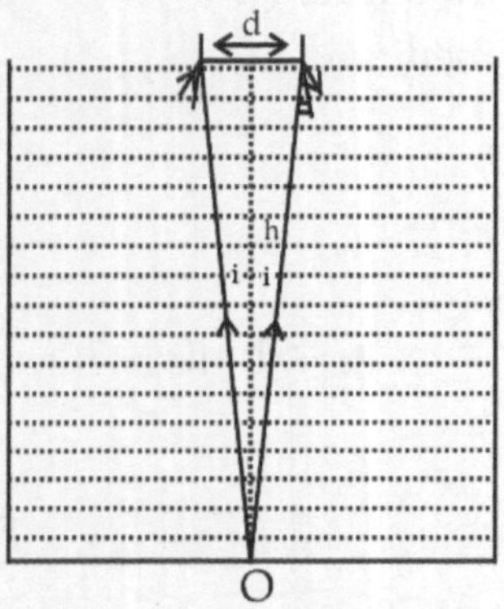

Ans. Assume that diameter of the disc is d.

The spot will be invisible, if the disc is covering all the incident rays falling at an angle lesser than the critical angle.

Now critical angle for this liquid

$\sin i_c = \frac{1}{\mu}$ or $\sin i_c = \frac{d/2}{x}$

From figure

or $x = \frac{d/2}{\sin i_c} = \frac{\mu d}{2}$ ....(1)

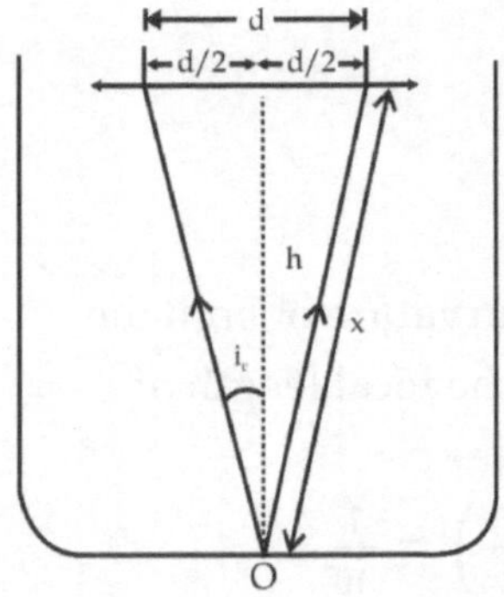

Also $x = \sqrt{h^2 + \left(\frac{d}{2}\right)^2}$ ...... (2)

∴ From equation (1)

$\frac{\mu d}{2} = \sqrt{h^2 + \left(\frac{d}{2}\right)^2}$

Or $\frac{\mu^2 d^2}{4} = h^2 + \frac{d^2}{4}$

∴ $\frac{\mu^2 d^2}{4} - \frac{d^2}{4} = h^2$

Or $\frac{d^2}{4}(\mu^2 - 1) = h^2$

$\frac{d}{2} = \frac{h}{\sqrt{\mu^2 - 1}}$

Or $d = \frac{2h}{\sqrt{\mu^2 - 1}}$ - Ans

91. A ray PQ incident normally on the refracting face AB is refracted in the prism BAC made of material of refractive index 1.5. Complete the path of ray through the prism. From which face will the ray emerge? Justify your answer. (2016)

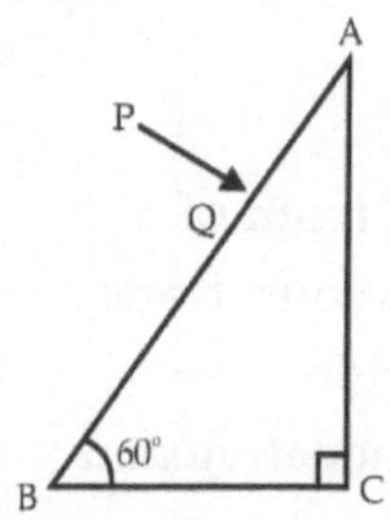

Ans. Given, $\mu = 1.5$

and Critical angle for the prism will be –

$\sin C = \frac{1}{\mu} = \frac{1}{1.5} = \frac{2}{3}$

or $C = \sin^{-1}(2/3) \simeq 42°$

From the ray diagram, the angle of incidence, At face AC is $i = 30°$ which is lesser than the critical angle. Therefore, the ray at the face AC will not experience total internal reflection and emerges out from the surface.

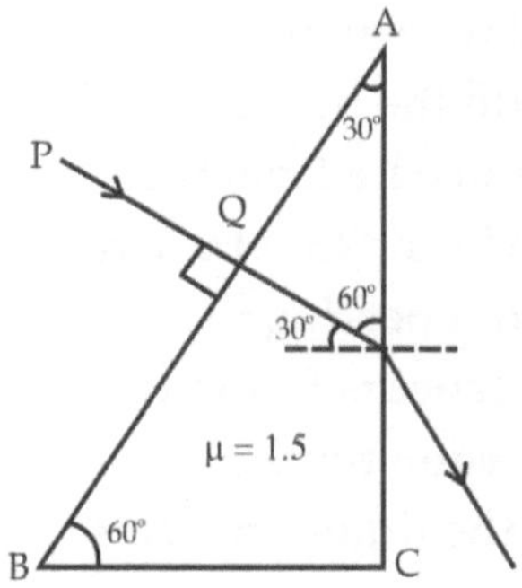

92. An air bubble is trapped at point B (CB = 20 cm) in a glass sphere of radius 40 cm and refractive index 1.5 as shown in figure. Find the nature and position of the image of the bubble as seen by an observer at point P. (2024)

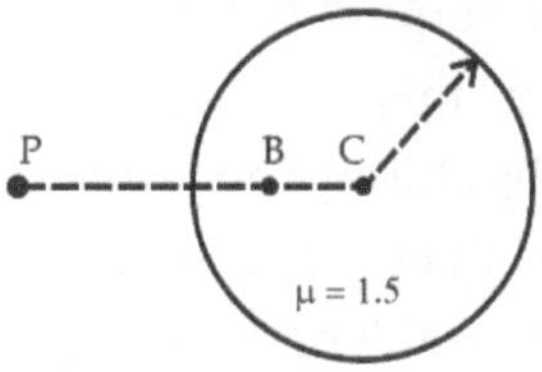

Ans.

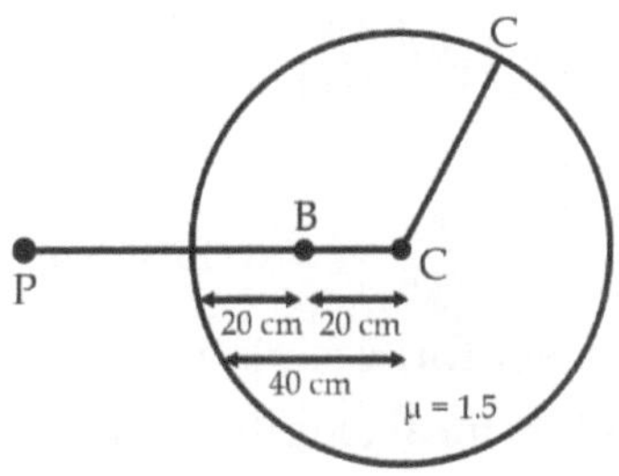

Given

R = 40 cm and $\mu_1 = 1.5$

Distance of the bubble from the centre of the sphere u = −20 cm

Refractive index of air $\mu_2 = 1$

Please note the object is in denser medium thus we need to use the formula of refraction when light travels from denser to rarer.

$$\therefore \frac{\mu_2}{v} - \frac{\mu_1}{u} = \frac{(\mu_2-\mu_1)}{R}$$

$$\frac{1}{v} - \frac{1.5}{-20} = \frac{1-1.5}{-40}$$

$$\frac{1}{v} = \frac{05}{40} - \frac{1.5}{20} = \frac{1}{80} - \frac{1.5}{200}$$

$$\frac{1}{v} = \frac{5-30}{400} = -\frac{25}{400}$$

$$v = -\frac{400}{25} = -16 \text{ cm}$$

Thus, image formed is at v = −16 cm to the right of the left surface.

93. A ray of light incident on the face AB of an equilateral triangular prism makes an angle of incidence i and is deviated by angle β as shown in the figure. Show that in the position of minimum deviation ∠β = ∠α. Also find out the condition, when the refracted ray QR suffer total internal reflection. (2019)

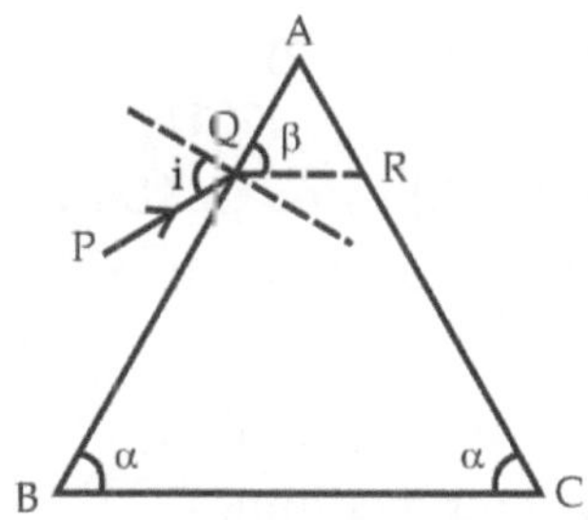

Ans. Here, From the figure

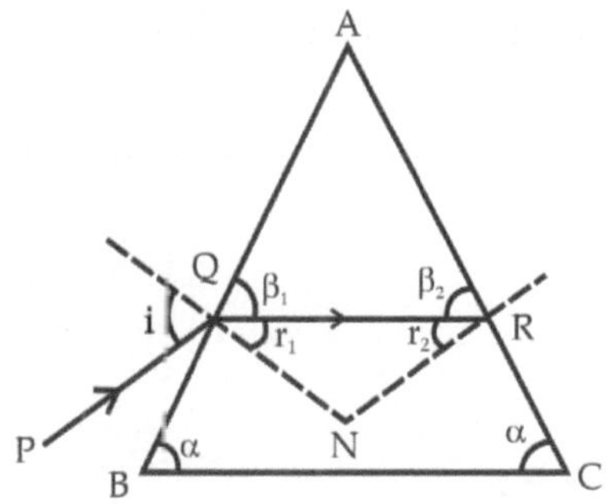

$180° - A = 2\alpha$ ....(1)

Now for the minimum deviation -

$r_1 + r_2 = A$ ......(2)

And $r_1 = r_2$

Also $r_1 = 90° - \beta$

Thus from equation (2)

$(90° - \beta) + (90° - \beta) = A$

Or $180° - 2\beta = A$

Or $180° - A = 2\beta$ ......(3)

From equation (1) and (3)

$2\alpha = 2\beta$

Or $\angle\alpha = \angle\beta$ Hence proved.

For total internal reflectin at face AC we need tc find the value of i_C

$\because r_1 + r_2 = A$

Or $r_1 + i_C = A$ $\therefore i_C = A - r_1$

And $r_1 = 90° - \beta$ $\therefore i_C = A - (90° - \beta)$

Thus r_2 or i_C should be greater than the value $= A - (90° - \beta)$ $\because \angle\beta = \angle\alpha$

$\therefore r_2$ or $i_C = 60° - 90° + 60° = 30°$

$\therefore i_C = 30°$

Thus $\mu \geq \frac{1}{\sin 30°}$ $\therefore \mu \geq 2$ -- Ans

94. State the necessary conditions for producing total internal reflection of light. Draw ray diagrams to show how specially designed prisms make use of total internal reflection to obtain inverted image of the object by deviation of rays (2011)

(a) through 90° and (b) through 180°.

Ans. Following are the criteria for total internal reflection -

(i) Light must travel from optically denser to optically rarer medium.

(ii) Angle of incidence in denser medium must be greater than critical angle for that pair of media.

(a) Deviation of light rays through 90°

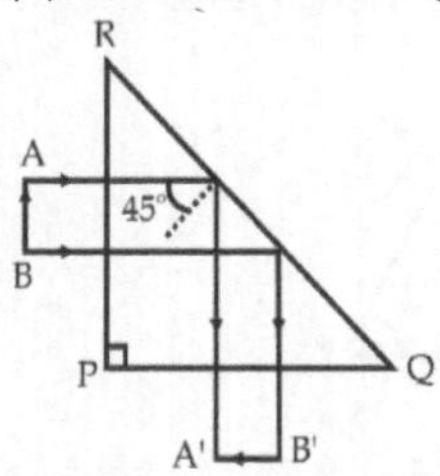

(b) Deviation of light rays through 180°

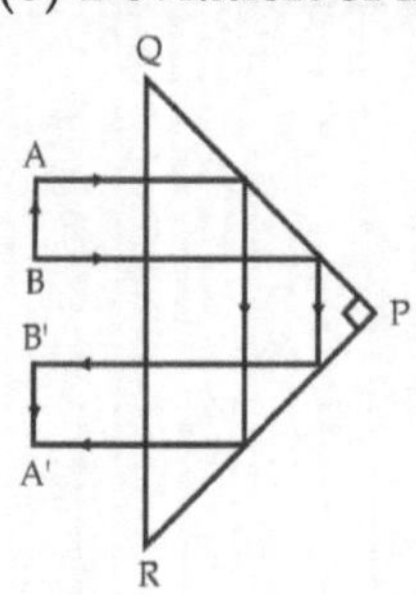

95. A symmetric biconvex lens of radius of curvature R and made of glass of refractive index 1.5 , is placed on a layer of liquid placed on the top of a plane mirror as shown in the figure. An optical needle with its tip on the principal axis of the lens is moved along the axis until its real, inverted image coincides with the needle itself. The distance of the needle from the lens is measured to be x. On removing the liquid layer and repeating the experiment, the distance is found to be y. Obtain the expression for the refractive index of the liquid in terms of x and y. (2018)

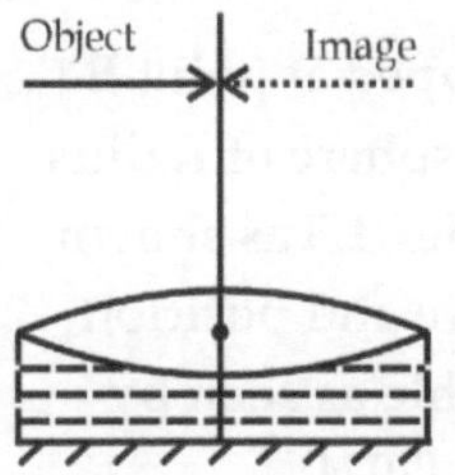

Ans. In the absence of liquid, the final image is formed at ' y ' this is only possible when the object is at f and after refraction by the lens the rays become parallel. If ' f_1 ' is the focal length of convex lens then $f_1 = y$.

and $\frac{1}{f_1} = (\mu - 1)\left[\frac{1}{R} - \left(-\frac{1}{R}\right)\right]$

or $\frac{1}{f_1} = (1.5 - 1)\left(\frac{2}{R}\right)$

or $f_1 = R$ ....(1)

Now the liquid layer forms a plano-concave lens and its focal length will be -

$\frac{1}{f_2} = (\mu_\ell - 1)\left[\frac{1}{-R} - \left(\frac{1}{\infty}\right)\right]$

Or $\frac{1}{f_2} = (\mu_\ell - 1)\left[-\frac{1}{R}\right]$

or $f_2 = \frac{R}{1-\mu_\ell}$ .....(2)

Now $\frac{1}{f_{eq}} = \frac{1}{f_1} + \frac{1}{f_2} = \frac{1}{R} + \frac{(1-\mu_\ell)}{R}$

Or $\frac{1}{f_{eq}} = \frac{2-\mu_\ell}{R}$ or $f_{eq} = \frac{R}{2-\mu_\ell} = x$ ....(3)

Now from equation -(1) $f_1 = R = y$

$\therefore f_{eq} = x = \frac{y}{2-\mu_\ell}$ (from equation (3))

$\therefore 2 - \mu_\ell = \frac{y}{x}$

or $\mu_\ell = 2 - \frac{y}{x}$ - Ans

96. A biconvex lens made of a transparent material of refractive index 1.5 is immersed in water of refractive index 1.33. Will the lens behave as a converging or a diverging lens? Give reason. (2014)

Ans. A biconvex lens acts as a converging lens when outer medium has lesser refractive index and acts like a diverging lens when outer medium has greater refractive index. Here outer medium is water which has lesser refractive index thus it acts like a converging lens only.

97. You are given three lenses L_1, L_2 and L_3 each of focal length 10 cm . An object is kept at 15 cm in front of L_1 as shown. The final real image is formed at the focus I of L_3. Find the separation between L_1, L_2 and L_3. (2012)

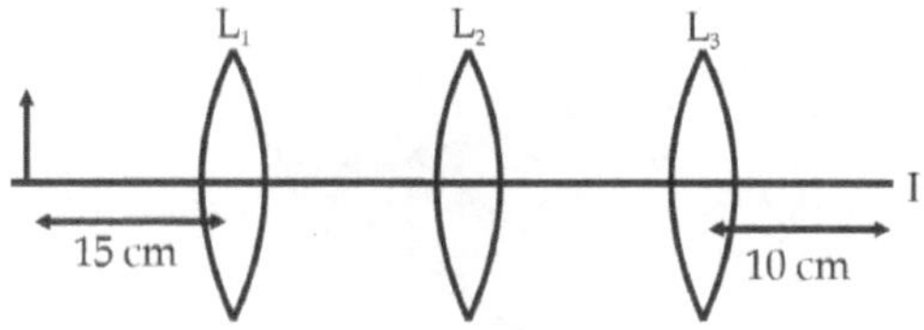

Ans. For lens $L_1, \frac{1}{f_1} = \frac{1}{v_1} - \frac{1}{u_1}$

Here $u_1 = -15$ cm, $v_1 = ?, f_1 = +10$ cm

$\frac{1}{10} = \frac{1}{v_1} + \frac{1}{15}$

Thus distance of image from lens L_1,

$\Rightarrow v_1 = 30$ cm

And for lens L_3, $\frac{1}{f_3} = \frac{1}{v_3} - \frac{1}{u_3}$

Now distance of image from lens L_3,

$v_3 = 10$ cm (given)

$\therefore \frac{1}{10} = \frac{1}{10} + \frac{1}{u_3}$ Or $\frac{1}{u_3} = \frac{1}{10} - \frac{1}{10}$

Or $u_3 = \infty$

This means the refracted rays from lens L_2 become parallel to principal axis, which is possible only, when image formed by L_1 lies at focus of L_2 i.e. at a distance of 10 cm from L_2.

$\therefore$ Separation between L_1 and L_2 is

$= 30 + 10 = 40$ cm

The distance between L_2 and L_3 can have any value.

98. You are given two converging lenses of focal lengths 1.25 cm and 5 cm to design a compound microscope. If it is desired to have a magnification of 30 , find out the separation between the objective and the eyepiece. (2015)

Ans. For compound microscope objective lens is of shorter focal length and eye piece is of higher focal length.

$\therefore f_e = +5$ cm

And $f_0 = 1.25$ cm

and Maximum magnification produced by compound microscope is

$m = \frac{v_0}{u_0}\left[1 + \frac{D}{f_e}\right]$

$\therefore 30 = \frac{v_0}{u_0}\left[1 + \frac{25}{5}\right]$ or $30 = \frac{v_0}{u_0}[6]$

$v_0 = 5u_0$ ....(1)

$\because f_0 = 1.25$ cm

$\therefore \frac{1}{v_0} - \frac{1}{u_0} = \frac{1}{f_0} \Rightarrow \frac{1}{5u_0} - \frac{1}{-u_0} = \frac{1}{1.25}$

[From Eq. (i)]

$\frac{6}{5u_0} = \frac{1}{1.25}$ or $5u_0 = +7.5$ cm or $u_0 =$ 1.5 cm

Thus $v_0 = +7.5$ cm

(Here $v_e = -25$ cm least distance of distinct vision because final image is formed 'D'.)

$\frac{1}{v_e} - \frac{1}{u_e} = \frac{1}{f_e}$ or $\frac{1}{-25} - \frac{1}{-u_e} = \frac{1}{5}$

$\frac{1}{u_e} = \frac{1}{5} + \frac{1}{25} = \frac{5+1}{25}$ or $u_e = \frac{25}{6}$ cm

Hence, separation between two lenses should be

$v_0 + u_e = 7.5 \text{ cm} + \frac{25}{6 \text{ cm}}$

$= 11.67$ cm - Ans

99. Three lenses of focal lengths +10 cm, −10 cm and +30 cm are arranged coaxially as in the figure given below. Find the position of the final image formed by the combination. (2019)

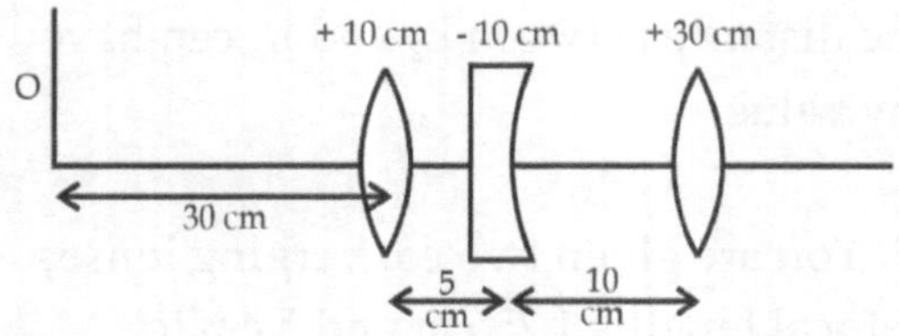

Ans. Since, $\frac{1}{v} - \frac{1}{u} = \frac{1}{f}$

Here, for first lens, $u = -30$ cm,

$f_1 = +10$ cm

$$\frac{1}{v_1} = \frac{1}{f_1} + \frac{1}{u_1} = \frac{1}{10} - \frac{1}{30}$$

$$\frac{1}{v_1} = \frac{2}{30}$$

$v_1 = 15$ cm

(To the right side of I lens)

∴ The image formed by first lens acts like a virtual object of II lens for second lens,

$u_2 = +10$ cm, $f_2 = -10$ cm

$$\frac{1}{v_2} = \frac{1}{f_2} + \frac{1}{u_2} = \frac{1}{10} - \frac{1}{10} \qquad \therefore v_2 = \infty$$

Thus, for III lens the object is at infinity, hence the final image is formed at the focus of the lens i.e. 30 cm.

$$\frac{1}{f_3} = \frac{1}{v_3} - \frac{1}{u_3} \qquad \frac{1}{v_3} = \frac{1}{f_3} + \frac{1}{u_3}$$

$v_3 = f_3 = +30$ cm

100. You are given three lenses of power 0.5 D, 4D and 10 D to design a telescope.
(a) Which lenses should be used as objective and eyepiece? Justify your answer.
(b) Why is the aperture of the objective preferred to be large? (2013)

Ans. (a) We know that, the objective lens of a telescope should be of larger focal length and eyepiece should be of smaller focal length, so lens of power 10 D can be used as eyepiece because it has smallest focal length and lens of power 0.5 D can be used as objective lens because it has largest focal length.

(b) The objective lens of a telescope must have larger aperture, so that it can gather maximum light rays from the distant object forms a bright image.

101. (i) Draw a ray diagram depicting the formation of the image by an astronomical telescope in normal adjustment.
(ii) You are given the following three lenses. Which two lenses will you use as an eyepiece and as an objective to construct an astronomical telescope? Give reason. (2017)

Lenses	Power (D)	Aperture (cm)
L_1	3	8
L_2	6	1
L_3	10	1

Ans. (i) In normal adjustment the final image is formed at ∞.

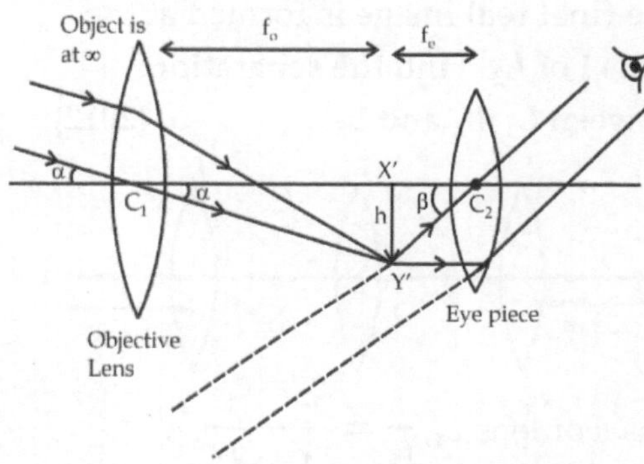

(ii) In Astronomical telescope the aperture of objective lens is larger than eyepiece. Therefore, we can use pair either (L_1 and L_3) or (L_1 and L_2). The focal length of the objective (f_o) must be greater than the focal length of eyepiece (f_e).

$$\therefore f_o > f_e \text{ or } P_o < P_e \text{ or } 1 < \frac{P_e}{P_o}$$

∴ Power of objective (P_o) must be lesser than power of eyepiece (P_e).

If we take L_1 and L_3 combination, then

$$\left(\frac{f_o}{f_e}\right) = \frac{P_e}{P_o} = \frac{10}{3}$$

and if we take L_1 and L_2 combination, then

$$\left(\frac{f_o}{f_e}\right)' = \frac{P_e}{P_o} = \frac{6}{3} = 2 < \left(\frac{f_o}{f_e}\right)$$

Hence the better pair combination will be L_1 and L_3.

102. A small telescope has an objective lens of focal length 140 cm and an eyepiece of focal length 5.0 cm . Find the magnifying power of the telescope for viewing distant objects when
(a) the telescope is in normal adjustment.
(b) the final image is formed at the least distance of distinct vision. (2021)

Ans. Given, $f_o = 140$ cm
And, $f_e = 5$ cm
Now least distance of distinct vision is
$D = 25$ cm
(a) ∴ In normal adjustment,
$$m = \frac{f_0}{f_e} = \frac{140}{5} = 28$$
(b) When final image is at the least distance of distinct vision (D) –
$$m = \frac{f_0}{f_e}\left[1 + \frac{f_e}{D}\right]$$
$$= \frac{140}{5}\left[1 + \frac{5}{25}\right] = 33.6 \text{ cm}$$

103. The magnifying power of an astronomical telescope in normal adjustment is 2.9 and the objective and the eyepiece are separated by a distance of 150 cm . Find the focal lengths of the two lenses. (2023)

Ans. In normal adjustment final image is formed at ∞
and tube length $L = f_0 + f_e$
$$\therefore f_0 + f_e = 150 \text{ cm} \quad ......(1)$$
And Magnification, $m = \frac{f_0}{f_e} = 2.9$
$$\therefore f_0 = 2.9 f_e \quad .....(2)$$
From Eq. (2), we get
$$2.9 f_e + f_e = 150$$
$$f_e = 38.46 \text{ cm}$$
$$\therefore f_0 = 150 - f_e = 111.53 \text{ cm}$$

104. A compound microscope uses an objective lens of focal length 4 cm and eyepiece lens of focal length 10 cm . An object is placed at 6 cm from the objective lens. Calculate the magnifying power of the compound microscope. Also, calculate the length of the microscope. (2011)

Ans. Given,
$f_0 = 4$ cm and $f_e = 10$ cm
$u_0 = -6$ cm, and $v_e = -D = -25$ cm
(We will calculate maximum magnifying power)
∴ for objective lens, $\frac{1}{f_0} = \frac{1}{v_0} - \frac{1}{u_0}$
$$\frac{1}{4} = \frac{1}{v_0} + \left(\frac{1}{6}\right)$$
$$\text{or} \quad \frac{1}{v_0} = \frac{1}{4} - \frac{1}{6} = \frac{1}{12}$$
$$\therefore v_o = 12 \text{ cm}$$
∴ Magnifying power, $m = \left(\frac{v_0}{u_0}\right)\left(1 + \frac{D}{f_e}\right)$
$$m = \left(-\frac{12}{6}\right)\left(1 + \frac{25}{10}\right)$$
Solving we get $m = 7$
Now for eyepiece, $v_e = -25$ cm,
$f_e = 10$ cm, $u_e = ?$
$$\frac{1}{f_e} = \frac{1}{v_e} - \frac{1}{u_e}$$
$$\text{or} \quad \frac{1}{u_e} = \frac{1}{v_e} - \frac{1}{f_e} = \frac{1}{-25} - \frac{1}{10}$$
$$\therefore \quad \frac{1}{u_e} = \frac{-2-5}{50} = -\frac{7}{50}$$
or $u_e = -50/7$ cm $= -7.14$ cm
∴ Length of microscope
$= |v_o| + |u_e| = 12 + 7.14$
$= 19.14$ cm – Ans

X. Wave Optics

105. A plane wavefront is incident on a concave mirror of radius of curvature R. The radius of the reflected wavefront will be (2023)

(a) 2R **(b) R**
(c) $\frac{R}{2}$ **(d) $\frac{R}{4}$**

Ans. (c) When light rays parallel to principle axis strikes the concave mirror, they meet at focal point after reflection.

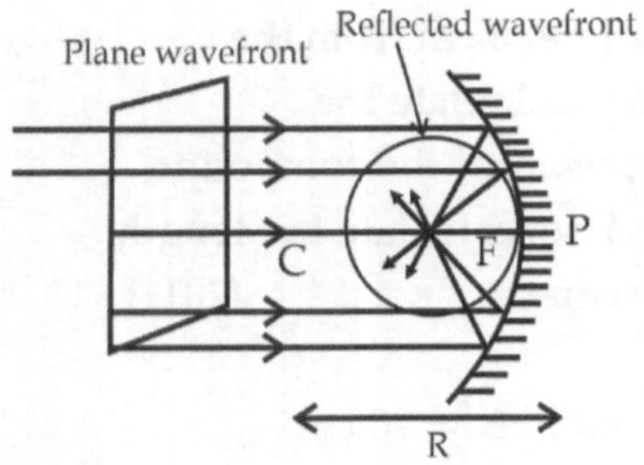

Here the reflected wavefront is spherical with focal length (f) and $f = \frac{R}{2}$

106. Write two points of difference between an interference pattern and a diffraction pattern. (2023)

Ans. Main differences between Interference and Diffraction

1. Fringe width - In interference pattern are the fringes are of same width but in diffraction the central bright fringe has double width than other secondary maxima or minima.

2. Intensity of fringes - In interference pattern all the bright fringes are of equal intensity but in diffraction, the intensity of secondary maxima (bright fringes) keeps on decreasing with increasing order of maxima.

3. Contrast between bright and dark fringes - There is a good contrast in interference pattern while it is not seen in diffraction.

4. Interference is produced by two different wavefronts coming from two coherent sources while diffraction is produced due to interaction of light coming from different parts of same wavefront.

107. If the slits in Young's double slit experiment are identical, then intensity at any point on the screen may vary between zero and four times to the intensity due to single slit. Justify the above statement through a relevant mathematical expression. (2022)

Ans. The intensity at any point is given by -

$I = I_1 + I_2 + 2\sqrt{I_1 I_2}\cos\phi$

When slits are identical $I_1 = I_2 = I_0$

$\therefore I = 2I_0 + 2I_0\cos\phi = 2I_0(1 + \cos\phi)$

→ The max. Intensity will be obtained when $\cos\phi = +1$ and $I_{max} = 4I_0$

The minimum Iintensity will be obtained when $\cos\phi = -1$ and $I_{min} = 0$

Thus the intensity at any point on the screen varies between zero and four times to the intensity due to single slit.

108. In a Young's double slit experiment using light of wavelength 600 nm , the slit separation is 0.8 mm and the screen is kept 1.6 m from the plane of the slits. Calculate the distance of

(i) third minimum and

(ii) fifth maximum, from the central maximum. (2022)

Ans. Given, $\lambda = 600\text{ nm} = 600 \times 10^{-9}$ m

And $d = 0.8\text{ mm} = 0.8 \times 10^{-3}$ m,

$D = 1.6$ m

(i) For nth minima, $y_n = (2n-1)\frac{\lambda D}{2d}$

and for third minima, $n = 3$

$\therefore y_3 = [2(3) - 1] \times \frac{1.6 \times 600 \times 10^{-9}}{2 \times 0.8 \times 10^{-3}}$

$= \frac{5 \times 1.6 \times 600 \times 10^{-6}}{2 \times 0.8} = 3$ mm

(ii) For fifth maximum,

$\because y_n = \frac{n\lambda D}{d}$

$\therefore y_5 = \frac{5 \times 1.6 \times 600 \times 10^{-9}}{0.8 \times 10^{-3}}$

$= 6 \times 10^{-3}$ m or 6 mm

109. When light travels from a rarer to a denser medium, the speed decreases. Does this decrease in speed imply a decrease in the energy carried by the light wave?

Justify your answer. (2010)

Ans. When light goes from a rarer medium to denser medium the speed decreases because of decreases in wavelegth. Energy of light depends on th frequency and number of protons n the beams of light. These two remain unchanged thus the energy remains unchanged.

110. What is the effect on the interference fringes in a Young's double slit experiment, if the monochromatic source S is replaced by a source of white light? (2023)

Ans. When we use white light only the central fringe will be white and other fringes will be of different colours because, the central fringe for all the colours are formed at the centre point of the screen thus it becomes white after mixing up. Since fringe width (β) is different for different wavelengths, we get fringes of all the colours on the screen.

111. Two coherent light waves of intensity $5 \times 10^{-2} Wm^{-2}$ each superimpose and produce the interference pattern on a screen. At a point where the path difference between the waves is $\frac{\lambda}{6}$, λ being wavelength of the wave. Find the
(i) phase difference between the waves
(ii) resultant intensity at the point
(iii) and resultant intensity in terms of the intensity at the maximum. (2020)

Ans. Given, intensity, $I = 5 \times 10^{-2} Wm^{-2}$

and path difference, $\Delta x = \frac{\lambda}{6}$

(i) We know that the path difference between the interfering waves is given as

$\Delta x = \frac{\lambda}{2\pi} \Delta\phi$

Here, $\Delta\phi$ = phase difference.

Given $\frac{\lambda}{2\pi}\Delta\phi = \frac{\lambda}{6} \quad \therefore \Delta\phi = \frac{\pi}{3}$

(ii) Resultant intensity,

$$I_R = I_1 + I_2 + 2\sqrt{I_1 I_2}\cos\phi$$

Here, $I_1 = I_2 = I = 5 \times 10^{-2} Wm^{-2}$

$$\therefore I_R = 2I + 2I\cos\frac{\pi}{3}$$

$$\text{or } 3I = 3 \times 5 \times 10^{-2}$$
$$= 15 \times 10^{-2} Wm^{-2}$$

It is 3 times the intensity due to single source.

$\because I_R = 3I \quad .....(1)$

Now for maximum intensity

$\Delta\phi = 2n\pi$

$$I_{max} = I + I + 2\sqrt{I \times I}\cos 2n\pi$$
$$= 2I + 2I \quad [\because \cos 2n\pi = 1]$$
$$= 4I \quad ...(2)$$

From Eq. (1) by Eq. (2), we get

$$\frac{I_R}{I_{max}} = \frac{3I}{4I} \text{ or } I_R = \frac{3}{4} I_{max}$$

XI. Dual Nature of Radiation and Matter

112. Which one of the following metals does not exhibit emission of electrons from its surface when irradiated by visible light? (2023)

(a) Rubidium **(b) Sodium**
(c) Cadmium **(d) Caesium**

Ans. (c) All alkali metals except lithium (Li) show photoelectric effect even with visible light. Here, cadmium is not an alkali metal.

113. The kinetic energy of a proton and that of an α-particle are 4 eV and 1 eV, respectively. The ratio of the de-Broglie wavelengths associated with them, will be (2020)

(a) 2:1 **(b) 1:1**
(c) 1:2 **(d) 4:1**

Ans. (b) The de-Broglie wavelength of a particle is given by $\lambda = \frac{h}{\sqrt{2mK}}$

Given here kinetic energy of proton $K_p = 4eV$ and

kinetic energy of α-particle $K_\alpha = 1eV$

Now $m_p = m$ and $m_\alpha = 4m$

$$\therefore \frac{\lambda_p}{\lambda_\alpha} = \sqrt{\frac{m_\alpha K_\alpha}{m_p K_p}} = \sqrt{\frac{4}{1} \times \frac{1}{4}}$$

$\left[\because \frac{m_\alpha}{m_p} = 4\right]$

Thus $\lambda_p : \lambda_\alpha = 1:1$

114. The quantum nature of light explains the observations on photoelectric effect as
(a) there is a minimum frequency of incident radiation below which no electrons are emitted
(b) the maximum kinetic energy of photoelectrons depends only on the frequency of incident radiation
(c) when the metal surface is illuminated, electrons are ejected from the surface after sometime
(d) the photoelectric current is independent of the intensity of incident radiation (2024)

Ans. (a) There is a minimum frequency of incident radiation below which no electrons are emitted.

115. Figure shows a plot of stopping potential (V_0) versus $\frac{1}{\lambda}$, where λ is the wavelength of the radiation causing photoelectric emission from a surface. The slope of the line is equal to (2023)

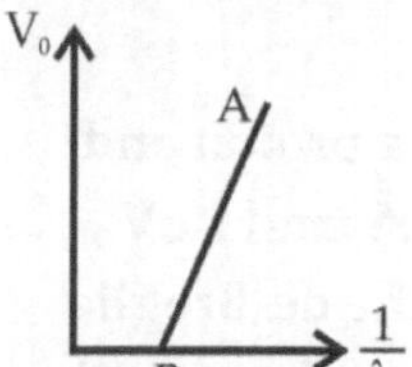

(a) ϕ_0 **(b) $\frac{h}{e}$**
(c) $\frac{hc}{e}$ **(d) $\frac{h^2c}{e^2}$**

Ans. (c) According to photoelectric equation,

$eV_0 = \frac{hc}{\lambda} - \phi_0$

or $V_0 = \left(\frac{hc}{e}\right) \cdot \frac{1}{\lambda} - \frac{\phi_0}{e}$ $(y = mx - c)$

Hence, the graph of V_0 versus $\frac{1}{\lambda}$ has slope $\left(\frac{hc}{e}\right)$ and intercept $\left(\frac{-\phi_0}{e}\right)$.

116. Light of frequency 6.4×10^{14} Hz is incident on a metal of work function 2.14 eV. The maximum kinetic energy of the emitted electrons is about (2023)

(a) 0.25 eV **(b) 0.51 eV**
(c) 1.02 eV **(d) 0.10 eV**

Ans. (b) Given,

$\upsilon = 6.4 \times 10^{14}$ Hz, and $\phi_0 = 2.14$eV

$\therefore KE_{max} = h\upsilon - \phi_0 = \frac{6.6\times10^{-34}\times6.4\times10^{14}}{1.6\times10^{-19}}$
-2.14
$= 0.51$eV

117. Photons of energy 3.2 eV are incident on a photosensitive surface. If the stopping potential for the emitted electrons is 1.5 V, the work function for the surface is (2023)

(a) 1.5 eV **(b) 1.7 eV**
(c) 3.2 eV **(d) 4.7 eV**

Ans. (b) Given, energy of photon,

$E = h\upsilon = 3.2$eV

Also stopping potential $V_0 = 1.5$ V and

$eV_0 = 1.5$ eV

$\because$ Work function, $\phi_0 = E - eV_0$

$\therefore \phi_0 = 3.2 - 15 = 1.7$eV

118. The figure shows v_m^2 versus $\frac{1}{\lambda}$ graph for photoelectrons emitted from a surface, where v_m is the maximum speed of electrons and λ is the wavelength of incident radiation. Using this graph and Einstein's photoelectric equation, obtain the expression for Planck's constant and work function of the surface. (2023)

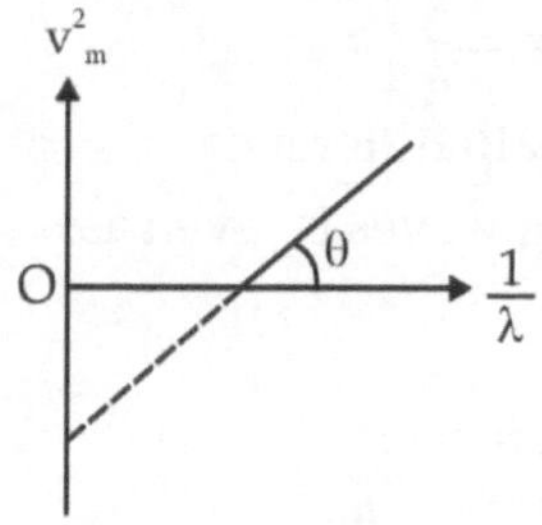

Ans. The photoelectric equation is

$$h\upsilon = \phi_o + \frac{1}{2}mv_m^2$$

or $\frac{1}{2}mv_m^2 = \frac{hc}{\lambda} - \phi_o$

or $v_m^2 = \frac{2hc}{\lambda m} - \frac{2\phi_o}{m}$

It follows $y = mx - c$

$\therefore$ Slope $= \frac{2hc}{m}$ .....(1)

Assume $\left(\frac{1}{\lambda_1}, v_{m_1}^2\right)$ and $\left(\frac{1}{\lambda_2}, v_{m_2}^2\right)$ are two points on the graph

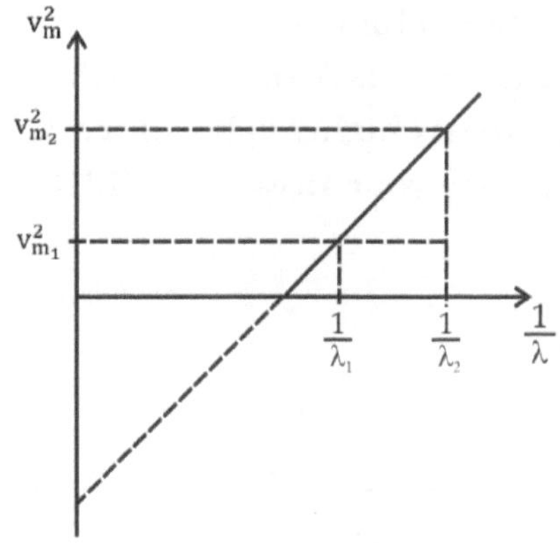

$\therefore$ Slope $m = \frac{v_{m_2}^2 - v_{m_1}^2}{\frac{1}{\lambda_2} - \frac{1}{\lambda_1}}$ ....(2)

$\therefore$ From Eq (1) and (2), we get

$$\frac{2hc}{m} = \frac{v_{m_2}^2 - v_{m_1}^2}{\frac{1}{\lambda_2} - \frac{1}{\lambda_1}}$$

$\therefore$ Planck's constant,

$$h = \frac{m(v_{m_2}^2 - v_{m_1}^2)}{2c\left(\frac{1}{\lambda_1} - \frac{1}{\lambda_2}\right)} \quad ....(3)$$

Also, the work function of the surface is

$$\phi_0 = \frac{hc}{\lambda_1} - \frac{1}{2}mv_{m_1}^2$$

$$= \frac{c}{\lambda_1} \times \frac{m(v_{m_2}^2 - v_{m_1}^2)}{2c\left(\frac{1}{\lambda_1} - \frac{1}{\lambda_2}\right)} - \frac{1}{2}mv_{m_1}^2$$

$$\therefore \phi_0 = \frac{m}{2}\left[\frac{v_{m_2}^2 - v_{m_1}^2}{\left(1 - \frac{\lambda_1}{\lambda_2}\right)} - v_{m_1}^2\right]$$

119. A photon and a electron have the same de-Broglie wavelength λ. Prove that the energy of the photon is (2mλc/h) times the kinetic energy of the electron. (2019)

Ans. Energy of photon, $E = \frac{hc}{\lambda}$

Energy of electron (moving particle),

$$K = \frac{1}{2}\frac{p^2}{m}$$

And de-Broglie wavelength for a moving particle is given by-

$\lambda = h/p$ or $p = h/\lambda$

$$\therefore K = \frac{1}{2}\frac{(h/\lambda)^2}{m} = \frac{1}{2}\frac{h^2}{\lambda^2 m}$$

$$\therefore \frac{E}{K} = \frac{hc/\lambda}{\frac{1}{2}\frac{h^2}{\lambda^2 m}} = \frac{2m\lambda c}{h}$$

$$\therefore E = \frac{2m\lambda c}{h}K$$

120. Explain how does (i) photoelectric current and (ii) kinetic energy of the photoelectrons emitted in a photocell vary, if the frequency of incident radiation is doubled but keeping the intensity same? Show the graphical variation in the above two cases. (2021)

Ans. (i) Photoelectric current depends on the number of electrons emitted per second by the metal surface. It does not depend on frequency of incident photons.

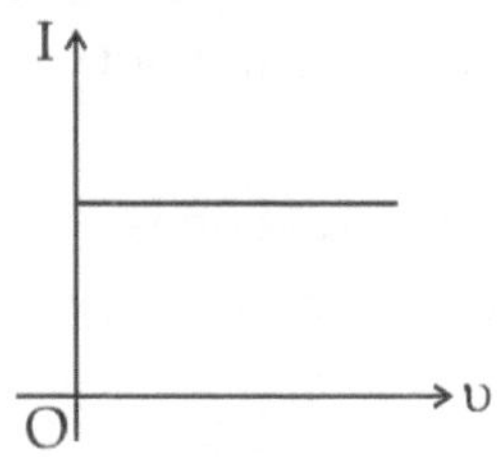

(ii) From photoelectric equation -

$$\frac{1}{2}mv^2 = h\upsilon - \phi_o$$

$\therefore KE \propto \upsilon$

Thus kinetic energy increases with increase in the frequency of incident radiation.

The between KE and υ is-

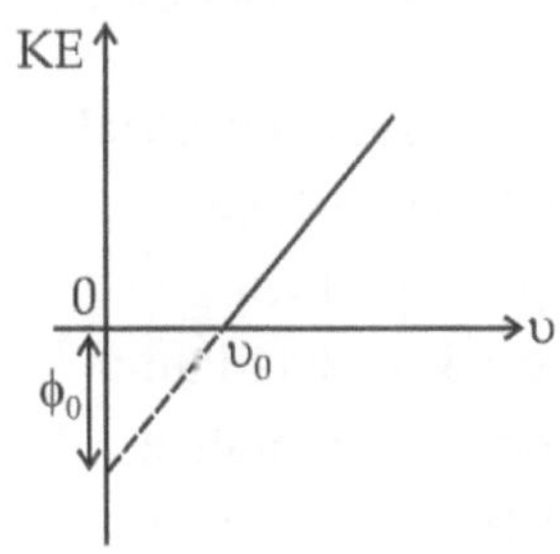

121. (i) Why photoelectric effect cannot

be explained on the basis of wave nature of light? Give reasons.

(ii) Write the basic features of photon picture of electromagnetic radiation on which Einstein's photoelectric equation is based. (2013)

Ans. (i) The photoelectric effect cannot be explained by the wave nature of light because wave nature of radiation cannot explain the following -

(a) Wave theory predicts some time delay for emission of electron which is not true.

(b) According to wave theory, any frequency of light, if intense enough should be able to eject electron but we have seen that there is threshold frequency associated with a given metal surface.

(c) According to wave theory, the kinetic energy of emitted electrons must depend on the intensity of light but we have observed that K.E. depends on the frequency of incident radiation and not on the intensity.

(ii) Einstein's photoelectric equation is based on particle nature of light. Its basic features are as follows -

(a) The radiation is made of particles called photons.

(b) Each photon has energy and momentum.

(c) All photons of a particular frequency υ and wavelength λ have same energy and momentum irrespective of the intensity of radiation.

122. How would the stopping potential for a given photosensitive surface change, if (i) the frequency of the incident radiation were increased and (ii) the intensity of incident radiation were decreased? Justify your answer. (2023)

Ans. From photoelectric equation,

$eV_0 = h\upsilon - \phi_0$

From the above equation we can say that stopping potential increases with the frequency of incident radiationis.

(ii) The stopping potential does not depend upon the intensity of incident light.

123. The graph shows the variation of stopping potential with frequency of incident radiation for two photosensitive metals A and B. Which one of the two has higher value of work function? Justify your answer. (2014)

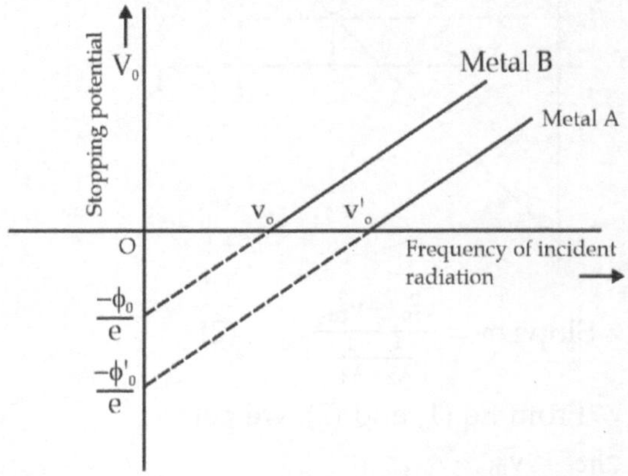

Ans. Metal A has a higher value of work function because the intercept of the line of $V_o - \upsilon$ graph depends on work function.

$eV_o = \frac{hc}{\lambda} - \phi_o$ or $V_o = \frac{hc}{e\lambda} - \frac{\phi_o}{e}$

Thus, higher is the intercept of the graph, higher is the work function.

XII. Atoms

124. A photon beam of energy 12.1 eV is incident on a hydrogen atom. The orbit to which electron of H -atom be excited is (2019)

(a) 2nd **(b) 3rd**

(c) 4th **(d) 5th**

Ans. (b) The energy of electron in n[th] orbit of hydrogen atom is given by -

$E_n = -\frac{13.6}{n^2}$ eV

Given -

$E_n - E_1 = 12.1$eV

thus $-\frac{13.6}{n^2} - (-13.6) = 12.1$

$\therefore\ n = 3$

Thus the electron excited to 3rd energy level.

125. In Balmer series of hydrogen atom, as the wavelength of spectral lines decreases, they appear

(a) equally spaced and equally intense.

(b) further apart and stronger in intensity.

(c) closer together and stronger in intensity.

(d) closer together and weaker in intensity. (2024)

Ans. (d) When the wavelength of spectral lines decreases, they appear closer and have weaker intensity because with increase in energy level, the gap between successive energy levels decreases.

126. The diagram shows four energy level of an electron in Bohr model of hydrogen atom. Identify the transition in which the emitted photon will have the highest energy. (2023)

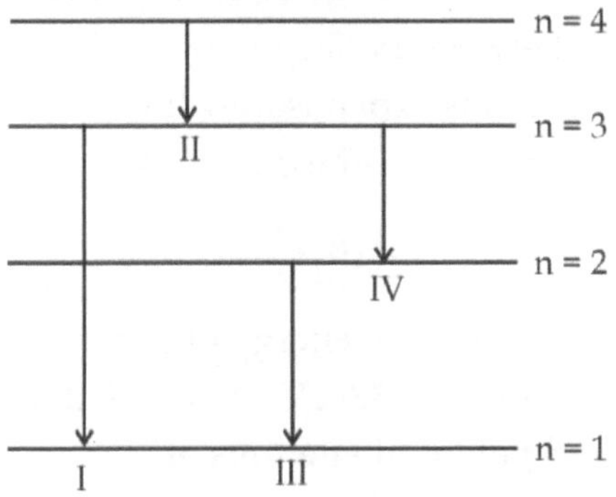

(a) I **(b) II**

(c) III **(d) IV**

Ans. (a) The energy of electron in n^{th} orbit of H-atom is, given by -

$$E_n = -\frac{13.6}{n^2}$$

From the figure -

$$E_3 - E_1 = -\frac{13.6}{9} - (-13.6) = 12.08\ eV$$

$$E_2 - E_1 = -\frac{13.6}{4} - (-13.6) = 10.2\ eV$$

$$E_3 - E_2 = -\frac{13.6}{9} - \left(-\frac{13.6}{4}\right) = 1.89\ eV$$

$$E_4 - E_3 = -\frac{13.6}{14} - \left(-\frac{13.6}{9}\right) = 0.66\ eV$$

Thus the transition $(E_3 - E_1)$ has highest energy.

127. The potential energy of an electron in the second excited state in hydrogen atom is (2023)

(a) -3.4 eV **(b) -3.02 eV**

(c) -1.51 eV **(d) -6.8 eV**

Ans. (b) Second excited state means electron is in energy level three, and Total energy in second excited state in hydrogen atom will be -

$$E_3 = -\frac{13.6}{(3)^2} = -1.51 eV$$

And potential energy in the second excited state will be,

$$PE_3 = 2E_3 = 2 \times (-1.51) = -3.02 eV$$

128. (i) State Bohr's quantisation condition for defining stationary orbits. How does de-Broglie's hypothesis explain the stationary orbits?

(ii) Find the relation between the three wavelengths λ_1, λ_2 and λ_3 from the energy level diagram shown below. (2014)

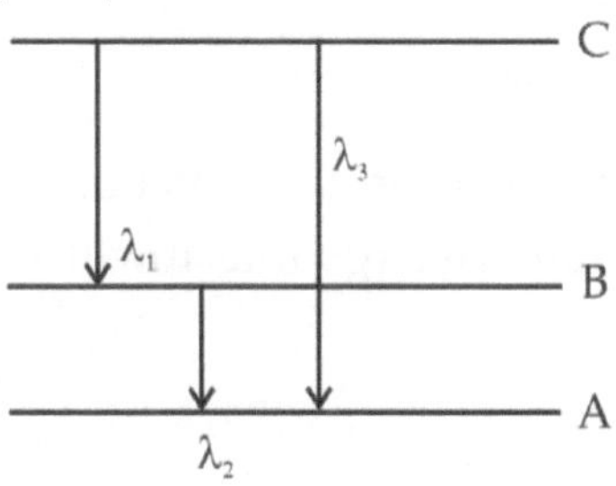

Ans. (i) According to Bohr's model of atom, only those orbits are permissible in which its angular momentum is integral multiple of $\frac{h}{2\pi}$.

(ii) From the figure given

$E_1 = hc/\lambda_1$ ....(1)

$E_2 = hc/\lambda_2$ ....(2)

$E_3 = hc/\lambda_3$ ......(3)

Now from the figure

$E_1 + E_2 = E_3$

or $\frac{hc}{\lambda_1} + \frac{hc}{\lambda_2} = \frac{hc}{\lambda_3}$

Or $\lambda_3 = \frac{\lambda_1\lambda_2}{\lambda_1+\lambda_2}$ .....(4)

129. What is the value of angular momentum of electron in the second orbit of Bohr's model of hydrogen atom? (2020)

Ans. The angular momentum of electron in nth orbit of hydrogen atom is given by

$mv_nr_n = \frac{nh}{2\pi}$

And angular momentum in 2nd orbit will be, $mv_2r_2 = \frac{2h}{2\pi} = \frac{h}{\pi}$

130. A hydrogen atom in the ground state is excited by an electron beam of 12.5 eV energy. Find out the maximum number of lines emitted by the atom from its excited state. (2019)

Ans. Ans. Energy in ground state of electron in hydrogen atom is $E_1 = -13.6$ eV.

After absorbing 12.5 eV, assume is jumps to nth orbit then energy will be

$E_n = -13.6 + 12.5 = -1.1$ eV and the value of 'n' will be -

$n^2 = -\frac{13.6}{-1.1} \approx 12.36$

Or $n \approx 3$ (Please note electron can not go to n = 4 with this energy thus it will jump to n = 3)

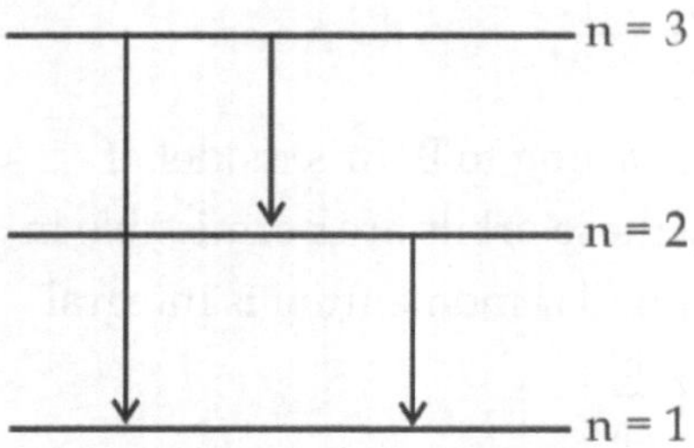

Now maximum number of lines emitted by this transition N = 3 as shown in the figure.

131. Write shortcomings of Rutherford atomic model. Explain, how these were overcome by the postulates of Bohr's atomic model. (2020)

Ans. Drawbacks of Rutherford's model.

(1) It could not explain stability of an atom. When an electron rotates in a given shell, it should radiate energy and fall into the nucleus.

(2) It was not able to explain the line spectrum obtained for various atoms.

Bohr's atomic model has three main postulates -

(1) When an electron revolves in a certain orbit it does not radiate energy. It's energy is fixed in a given shell. The centripetal force is provided by electrostatic force of the nucleus.

(2) Electron revolves only in those shell in which its angular momentum is integral multiple of $\frac{h}{2\pi}$.

(3) When electron jumps from one energy level to another energy level, it emits a photon which has energy equal to the energy difference between initial and final states. $E_f - E_i = h\upsilon$. This explains how line spectrums are obtained for various atoms.

132. The ground state energy of hydrogen atom is -13.6 eV. What are the kinetic and potential energies of electron in this state? (2010)

Ans. The ground state energy of hydrogen atom is given as-

$E = -13.6$eV

Now, Kinetic energy $= -$ Total Energy

$\therefore KE = -(-13.6eV) = 13.6eV$

And potential energy = 2 Total energy

$= 2 \times (-13.6) = -27.2eV$

133. Consider two different hydrogen atoms. The electron in each atom is in an excited state. Is it possible for the electrons to have different energies but same orbital angular momentum according to the Bohr model? Justify your answer. **(2021)**

Ans. If in two hydrogen atoms (i.e., same value of Z) the electrons have same orbital angular momentum that means-

$L_1 = L_2 = \frac{nh}{2\pi}$

Here $\frac{h}{2\pi}$ is a constant which means both the electrons are in same thell. (n is same).

Now the energy of an electron in n^{th} orbit is given by $E_n = -13.6 \frac{Z^2}{n^2}$ eV. Here 'Z' is same, n is also same as proven above, thus $E_1 = E_2$ both of them must have same energies.

XIII. Nuclei

134. The ratio of the nuclear densities of two nuclei having mass numbers 64 and 124 is **(2023)**

(a) $\frac{64}{125}$ (b) $\frac{4}{5}$

(c) $\frac{5}{4}$ (d) 1

Ans. (d) The formula for nuclear density is-

$\rho = \frac{m}{\frac{4}{3}\pi R_0^3}$ = A constant quantity

That means nuclear density of all elements is a constant quantity.

$\therefore \frac{\rho_{64}}{\rho_{125}} = 1$

135. The curve of binding energy per nucleon as a function of atomic mass number has a sharp peak for helium nucleus. This implies that helium nucleus is **(2023)**

(a) radioactive

(b) unstable

(c) easily fissionable

(d) more stable nucleus than its neighbours

Ans. (d) The sharp peak suggests that it has high BE/ nucleon than its neighbouring nuclei. Higher is the BE/nucleon, more stable is the nucleus. Thus, helium is more stable nucleus than its neighbouring atoms.

136. The radius of $^{27}_{13}X$ nucleus is R. The radius of $^{125}_{53}Y$ nucleus will be **(2023)**

(a) $\frac{5}{3}R$ (b) $\left(\frac{13}{53}\right)^{1/3} R$

(c) $\left(\frac{5}{3}R\right)^{1/3}$ (d) $\left(\frac{13}{53}R\right)^{1/3}$

Ans. (a) Assume the radius of $^{27}_{13}X$ is R

$\therefore R = R_0 A^{1/3} = R_0(27)^{1/3} = 3R_0$ ....(1)

And for the nucleus of $^{125}_{53}Y$ is R'

$\therefore R' = R_C(125)^{1/3} = 5R_0$ .....(2)

Thus from Eqs. (1) and (2), we get

$\frac{R'}{R} = \frac{5R_0}{3R_0} = \frac{5}{3}$

$R' = \frac{5}{3}R$

137. Binding energy per nucleon versus mass number curve is as shown. $^{A}_{Z}S$, $^{A_1}_{Z_1}W$, $^{A_2}_{Z_2}X$ and $^{A_3}_{Z_3}Y$ are four nuclei indicated on the curve. **(2019)**

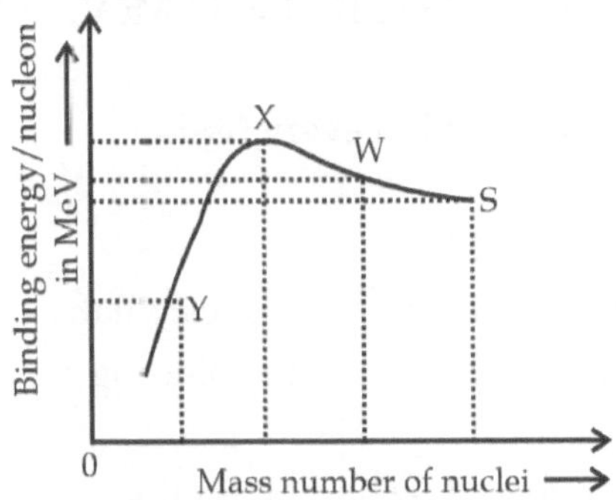

Based on the graph

(i) Arrange X, W and S in the increasing order of stability.

(ii) Write the relation between the relevant A and Z values for the following nuclear reaction.

S ⟶ X + W

(iii) Explain why binding energy for heavy nuclei is low.

Ans. (i) Since BE/nucleon for heavy nuclei decreases with mass number thus their stability decreases.
∴ The stability order of X, W and S is
$S < W < X$.
(ii) The nuclear reaction given here is-
$S \longrightarrow X + W$
Here the heavy nucleus S undergoes nuclear fission and forms X and W lighter nuclei which have more stability. Thus the relation between their respective A and Z values is given by
$A = A_1 + A_2$ and $Z = Z_1 + Z_2$
(iii) When number of nucleons increases the distance between nucleons decreases. The nuclear force remains attractive upto a distance of (0.8 fm) and becomes repulsive when the distance is lesser than 0.8 fm. Thus the binding energy for heavy nuclei becomes low.

138. A heavy nucleus X of mass number 240 and binding energy per nucleon 7.6 MeV is splitted into two fragments Y and Z of mass numbers 110 and 130 . The binding energy of nucleons in Y and Z is 8.5 MeV per nucleon. Calculate the energy released per fission in MeV. (2010)

Ans. Energy released in every fission
$= \Delta m \times c^2$
Here Δm = mass defect
Since energy is released, we can calculate final binding energy and initial binding energy.
The energy released = Gain in binding energy
$= (110 + 130) \times 8.5\text{MeV} - 240 \times 7.6\text{MeV}$
$= 240 \times (8.5 - 7.6)\text{MeV}$
$= 240 \times 0.9 = 216\text{MeV}$

139. (i) Differentiate between nuclear fission and fusion.
(ii) The fission properties of $_{94}Pu^{239}$ are very similar to those of $_{92}U^{235}$. How much energy (in MeV), is released, if all the atoms in 1 g of pure $_{94}Pu^{239}$ undergo fission? The average energy released per fission is 180 MeV. (2024)

Ans. (i) Differences between nuclear fission and nuclear fusion-
(a) Fission does not occur in nature normally but fusion is taking place in many stars.
(b) Nuclear fission is the splitting of a large nucleus into two or more smaller nuclei whereas fusion is combining of two or more lighter nuclei to form a heavier nucleus.
(c) In nuclear fusion the energy released is more than the energy released in nuclear fission.
(d) Fission requires critical mass of the substance and high speed neutrons but fusion requires high density and high temperature environment.
(ii) Number of atoms in one gram of pure Pu -
$= \frac{6.023\times10^{23}}{239} = 2.52 \times 10^{21}$
Now average energy released in one fission is 180 MeV thus the total energy released will be
$= 2.52 \times 10^{21} \times 180\text{MeV}$
$= 4.53 \times 10^{23}\text{MeV}$

140. Define the Q-value of a nuclear process. When can a nuclear process not proceed spontaneously? If both the number of protons and the number of neutrons are conserved in a nuclear reaction, in what way is mass converted into energy (or vice-versa) in the nuclear reaction? (2010)

Ans. The Q-value is used for a nuclear process in which the energy is released and its value is determined by using Einstein's mass-energy relation,
$E = mc^2$.
The Q-value is equal to the difference of

product mass and reactant mass times square of speed of light.

$Q = [m_p - m_r]c^2$

The nuclear process will not be spontaneous if Q-value of a process is negative.

When a nucleus is formed, though the number of protons and neutrons remain the same but the actual mass of the nucleus is always lesser than the sum of masses of its individual protons and neutrons. This difference is called mass defect and is converted into binding energy that hold the nucleus together.

141. Draw a graph showing the variation of potential energy of a pair of nucleons as a function of their separation. Indicate the region in which the nuclear force is (a) attractive and (b) repulsive. (2023)

Ans. The graph between potential energy of two nucleons and their distance is given here-

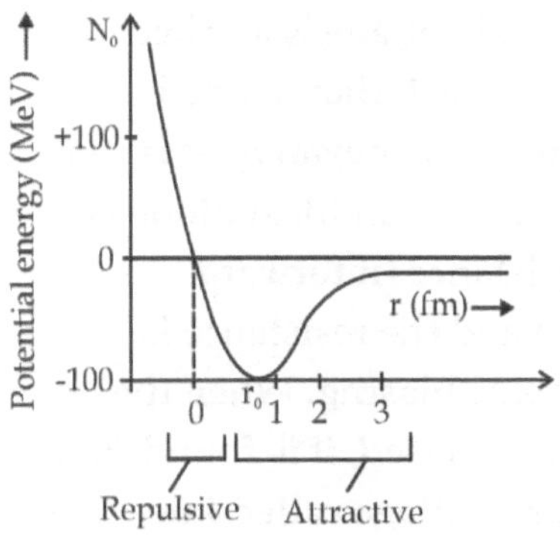

XIV. Semiconductor Electronics

142. The formation of depletion region in a p – n junction diode is due to

(a) movement of dopant atoms

(b) diffusion of both electrons and holes

(c) drift of electrons only

(d) drift of holes only (2023)

Ans. (b) Diffusion of both electrons and holes.

143. At a certain temperature in an intrinsic semiconductor, the electrons and holes concentration is $1.5 \times 10^{16}\ m^{-3}$. When it is doped with a trivalent dopant, hole concentration increases to $4.5 \times 10^{22}\ m^{-3}$. In the doped semiconductor, the concentration of electrons (n_e) will be (2023)

(a) $3 \times 10^6\ m^{-3}$ **(b) $5 \times 10^7\ m^{-3}$**

(c) $5 \times 10^9\ m^{-3}$ **(d) $6.75 \times 10^{38}\ m^{-3}$**

Ans. (c) Given

For intrinsic semiconductor -

$n_e = 1.5 \times 10^{16}\ m^{-3}$

$n_h = 1.5 \times 10^{16}\ m^{-3}$

Thus $n_i^2 = (1.5 \times 10^{16})^2$

And new hole concentration after doping

$n_h' = 4.5 \times 10^{22}\ m^{-3}$

$\because n_e n_h' = n_i^2$

$\therefore n_e = \frac{n_i^2}{n_h} = \frac{(1.5\times10^{16})^2}{4.5\times10^{22}}$

$= \frac{2.25\times10^{32}}{4.5\times10^{22}}$

$= 5 \times 10^9\ m^{-3}$

144. During the formation of a p – n junction

(a) diffusion current keeps increasing

(b) drift current remains constant

(c) both the diffusion current and drift current remain constant

(d) diffusion current remains almost constant but drift current increases till both currents become equal (2023)

Ans. (d) Diffusion current remains almost constant but drift current increases till both currents become equal.

145. In the following diagram, is the junction diode forward biased or reverse biased? (2017)

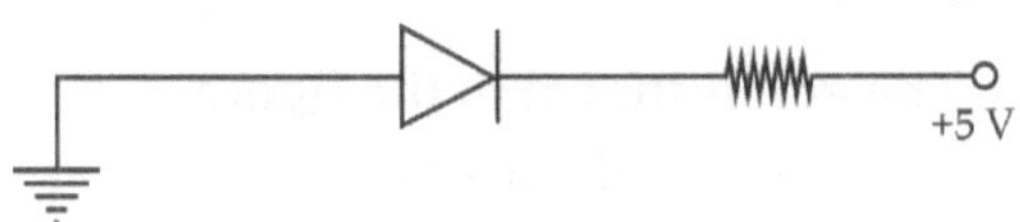

Ans. (i) Here the p-section is earthed i.e., connected to zero voltage and n-section is connected to +5 V end. The diode is reverse biased because n-side is connected to higher voltage end.

146. In a p – n junction diode, the forward bias resistance is low as compared to the reverse bias resistance. Give reason. (2020)

Ans. In forward biasing the width of depletion region reduces and a large current due to majority charge carriers flow due to which the resistance is reduced.

147. Assuming that the two diodes D_1 and D_2 used in the electric circuit shown in the figure are ideal, find out the value of the current flowing through 1Ω resistor. (2013)

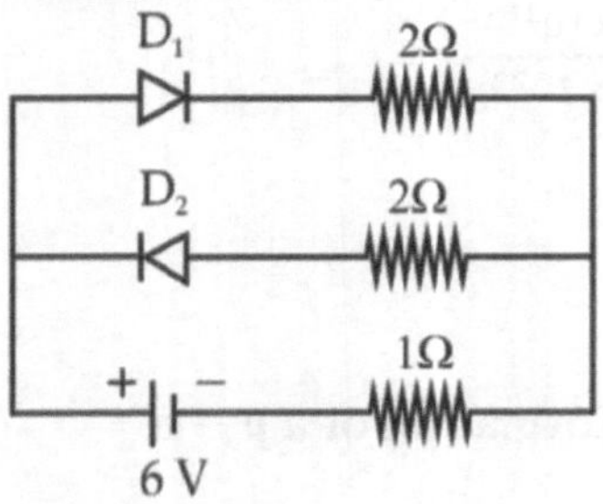

Ans. In the given circuit, diode D_2 is reverse biased thus current will not flow in that branch. Therefore the circuit diagram can be represented as -

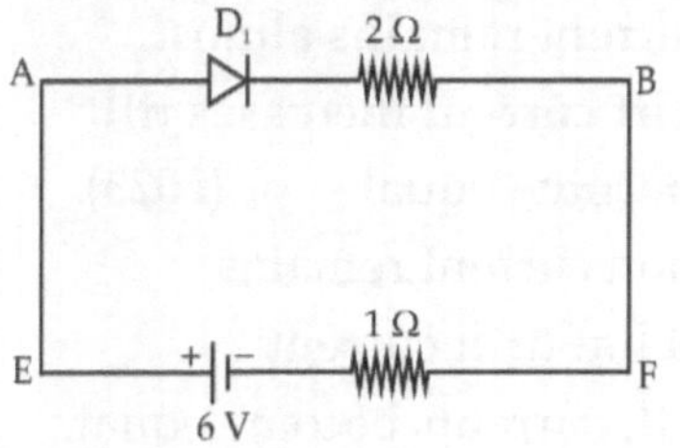

Now resistances 2Ω and 1Ω are in series,

$R = 2 + 1 = 3\Omega$

$\therefore$ Current, $I = \frac{V}{R} = \frac{6}{3} = 2$ A

148. The graph shown in the figure represents a plot of current versus voltage for a given semiconductor. Identify the region, if any over which the semiconductor has a negative resistance. (2013)

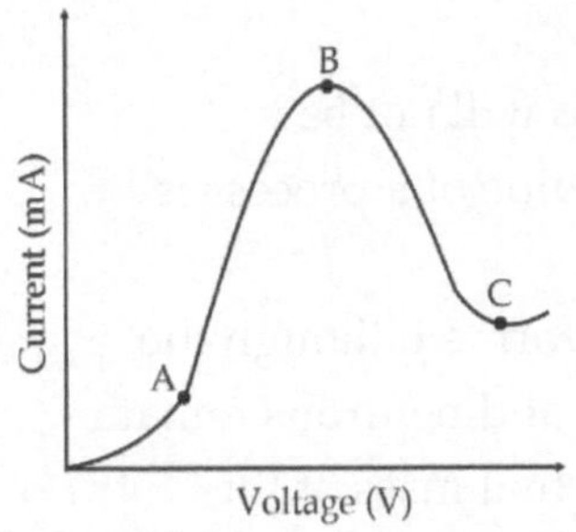

Ans. Part BC of the curve shows the negative resistance because in this region the current decreases by increasing the voltage.

149. A semiconductor diode is basically a p – n junction with metallic contacts provided at the ends for the application of an external voltage. It is a two terminal device. When an external voltage is applied across a semiconductor diode, such that p-side is connected to the positive terminal of the battery and n-side to the negative terminal, it is said to be forward biased. Whan an external voltage is applied across the diode, such that n-side is positive and p-side is negative, it is said to be reverse biased. An ideal diode is one whose resistance in forward biasing is zero and the resistance is infinite in reverse biasing. When the diode is forward biased, it is found that beyond forward voltage called knee voltage, the conductivity is very high. When the biasing voltage is more than the knee voltage, the potential barrier is overcome and the current increases rapidly with increase in forward voltage. When the diode is reverse biased, the reverse bias voltage produces a very small current about a few microamperes which almost remains constant with bias. This small current is reverse saturation current. (2023)

(i) **In the given figure, a diode D is connected to an external resistance R = 100Ω and an emf of 3.5 V . If the barrier potential developed across the diode is 0.5 V , the current in the circuit will be**

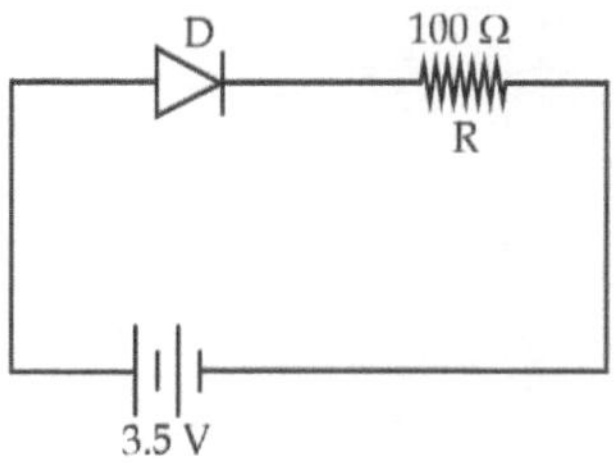

(a) 40 mA **(b) 20 mA**
(c) 35 mA **(d) 30 mA**

(ii) **In which of the following figures, the p – n diode is reverse biased?**

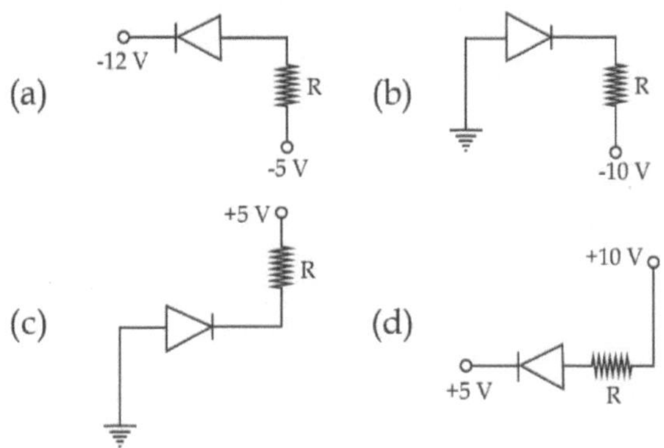

(iii) **Based on the V – I characteristics of the diode, we can classify diode as**
(a) bilateral device
(b) ohmic device
(c) non-ohmic device
(d) passive element
Or
Two identical p – n junctions can be connected in series by three different methods as shown in the figure. If the potential difference in the junctions is the same, then the correct connections will be

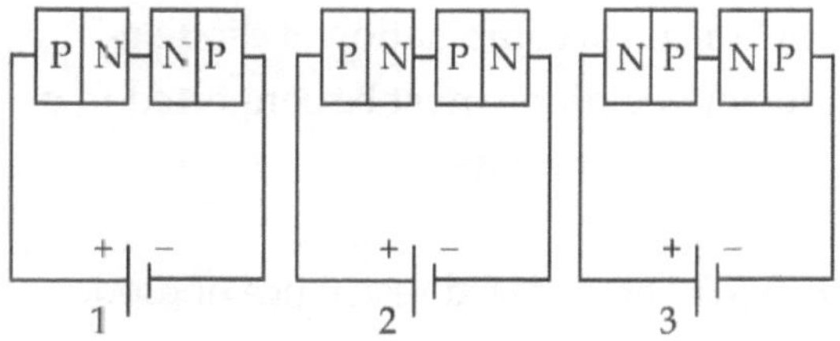

(a) in the circuits (1) and (2)
(b) in the circuits (2) and (3)
(c) in the circuits (1) and (3)
(d) only in the circuit (1)

(iv) **The V – I characteristic of a diode is shown in the figure. The ratio of the resistance of the diode at I = 15 mA to the resistance at V = −10 V is**

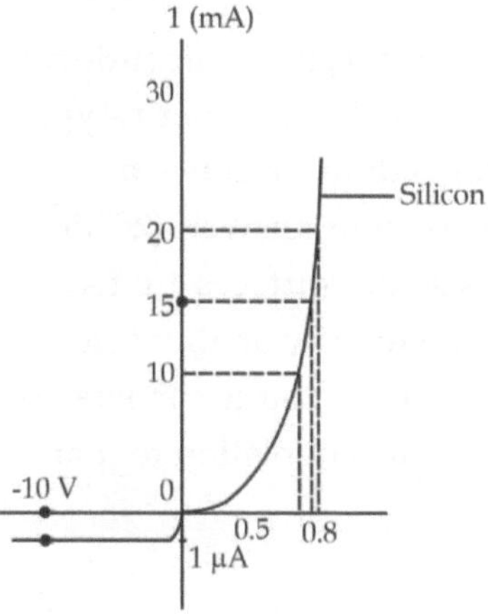

(a) 100 **(b) 10^6**
(c) 10 **(d) 10^{-6}**

Ans. (i) (d) Here the available voltage to the circuit is 3.5 – 0.5 = 3 V

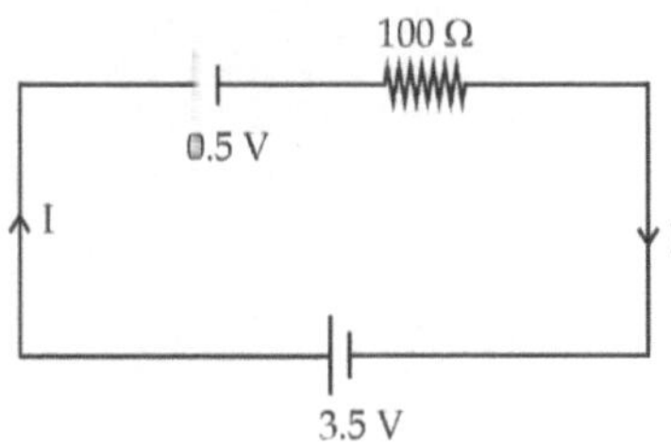

$\therefore I = \frac{V}{R} = \frac{3}{100} = 0.03$ A
Or I = 30 mA

(ii) (c) n-side of the diode is connected to higher potential. So, it is reversed biased.

(iii) (c) The diode is classified as non-ohmic device.
Or

(b) For series combination of diode n-side of one diode must be connected to p -side of other diode.

(iv) (d) The forward resistance of diode at current

$I = 15\ mA$

$$R_f = \frac{(0.8-0.7)}{(20-10)\times10^{-3}} = 10\ \Omega$$

And the backward resistance at

$V = -10\ V$

$$R_b = \frac{V}{I} = \frac{10}{10^{-6}} = 10^7\ \Omega$$

$$\therefore\ \frac{R_f}{R_b} = \frac{10}{10^7} = 10^{-6}$$

150. Can a slab of p-type semiconductor be physically joined to another n-type semiconductor slab to form p – n junction? Justify your answer. (2020)

Ans. No, because the surfaces are too rough to have continuity at atmoic level and thus the majority charge carriers do not flow and create a depletion region.